The Genus Bidens

Yamin Bibi • Kulsoom Zahara
Abdul Qayyum • Matthew A. Jenks
Editors

The Genus Bidens

Chemistry and Pharmacology

Springer

Editors
Yamin Bibi
Department of Botany
Rawalpindi Women University
Rawalpindi, Punjab, Pakistan

Kulsoom Zahara
Department of Botany
PMAS-Arid Agriculture University
Rawalpindi, Pakistan

Abdul Qayyum
Department of Agronomy
The University of Haripur
Haripur, Pakistan

Matthew A. Jenks
School of Plant Sciences, College of Agriculture and Life Sciences
University of Arizona
Tucson, AZ, USA

ISBN 978-981-96-4256-4 ISBN 978-981-96-4257-1 (eBook)
https://doi.org/10.1007/978-981-96-4257-1

This Springer imprint is published by the registered company Springer Nature Singapore Pte Ltd.
The registered company address is: 152 Beach Road, #21-01/04 Gateway East, Singapore 189721, Singapore

Contents

About the Editors

Yamin Bibi has made significant contributions in the field of botany. She is currently serving as a chairperson in the Department of Botany at Rawalpindi Women University. She has completed her PhD at Quaid-i-Azam University, Islamabad, with a focus on the evaluation of ethnobotanically important plants for anticancer and antibacterial activities, as well as phytochemical analysis. Her research is focused on plant-based natural products, particularly in the isolation and chemical characterization of active components and the evaluation of their biological activities. Her work aims to reveal novel bioactive compounds from plants, contributing to the understanding of their therapeutic potential and mechanisms of action.Throughout her career, Dr. Yamin Bibi has supervised numerous PhD and MPhil students, guiding them in their research endeavors. She has also completed several fully funded research projects and published extensively in national and international journals. Her expertise and dedication have earned her recognition and awards, including multiple paper incentive awards and a research award from PMAS-Arid Agriculture University Rawalpindi.Dr. Yamin Bibi has presented her research at various national and international conferences, sharing her findings with the global scientific community. Her contributions to the field of botany, combined with her commitment to teaching and mentoring, have made a significant impact on the academic and research landscape.

Kulsoom Zahara is an emerging scholar in the field of Botany. As an early-career researcher, she has already made notable contributions to the study of phytochemistry, with a particular focus on the genus *Bidens*. Her extensive work has significantly contributed to the understanding of the pharmacological properties and bioactive compounds of *Bidens* species, such as *Bidens biternata*, *Bidens pilosa*, and *Bidens bipinnata*.Dr. Kulsoom Zahara completed her academic journey at the Pir Mehr Ali Shah Arid Agriculture University in Rawalpindi, Pakistan. Her research primarily revolves around the isolation and identification of bioactive compounds from various plant species, utilizing advanced techniques like HPLC-DAD and GC-MS. One of her notable contributions includes evaluating pharmacological properties of genus *Bidens*, highlighting their potential as a nutraceutical. Additionally, her research has explored the anticancer potential of *Bidens* species, identifying key compounds responsible for their bioactivity.Beyond her work on *Bidens*, Dr. Kulsoom Zahara has also investigated other plant species, contributing to the broader field of plant biodiversity and phytochemistry. Her interdisciplinary approach and dedication to scientific excellence have earned her recognition in the academic community, with numerous publications and collaborations with researchers worldwide.Dr. Kulsoom Zahara's passion for phytochemistry and her commitment to advancing knowledge in this field continue to inspire her peers and students alike. Her work not only enhances our understanding of plant-based compounds but also paves the way for potential therapeutic applications.

Abdul Qayyum is currently serving as an Assistant Professor in the Department of Agronomy at the University of Haripur, Pakistan. Dr. Qayyum's research interests are centered on crop nutrition management, modern concepts of crop production, integrated agriculture, and stress agronomy. He has made significant contributions to understanding plant drought stress tolerance, particularly in wheat, through in-depth chemical analysis using advanced techniques like Gas Chromatography and Mass Spectrometry.His academic career includes a stint as a Visiting Scientist at Purdue University, USA, where he explored plant drought stress tolerance responses in diverse wheat genotypes. Dr. Qayyum has supervised numerous MS and PhD students, guiding them in their research projects and helping them develop their scientific careers. Dr. Qayyum has led several research projects funded by the Higher Education Commission (HEC) of Pakistan, focusing on the impact of nanoparticles on biochemical and agronomic attributes of maize and wheat, and the evaluation of fermenting microorganisms for oil production from rice bran. His work has been published in various peer-reviewed journals, contributing valuable knowledge to the field of agronomy.In addition to his research and teaching responsibilities, Dr. Qayyum is actively involved in the academic community, presenting his findings at national and international conferences. His dedication to advancing agronomic science and his commitment to education have earned him recognition and respect among his peers and students alike.

Matthew A. Jenks is a distinguished professor at the University of Arizona, holding the Bud Antle Endowed Chair for Excellence in the School of Plant Sciences. His research primarily focuses on understanding the cellular mechanisms that enable plants to adapt to both biotic and abiotic environmental stresses, with a particular emphasis on the plant cuticle.Dr. Jenks earned his bachelor's degree in Horticultural Science from Michigan State University, followed by a master's degree from the University of Florida, and a PhD from Purdue University, all in Horticultural Science. His work has significantly contributed to the field of plant industrial products, and he has an extensive publication record, reflecting his active engagement in research.Throughout his career, Dr. Jenks has been involved in various research projects that explore the adaptability of plants, particularly focusing on the cuticular waxes and their role in protecting plants from environmental stressors. His contributions have been recognized through numerous citations and collaborations with researchers worldwide.

Part I

Introduction

The Systematics and Biogeography of Genus *Bidens*

1

Sadia Abdullah, Mamoona Munir, Bisma Javed, Mushtaq Ahmad, Banzeer Ahsan Abbasi, Sumreen Dawood, and Lixin Zhang

Abstract

Genus *Bidens* belongs to the family Asteraceae and is a diverse genus with around 230 species throughout tropical and warm temperate zones. Despite its broad dissemination, the systematics and taxonomy of *Bidens* are often described as complex and disorganized. The taxonomic history of the genus is long and complex, with limited knowledge about its phylogenetic and biogeographic background. These plants produce fruits with a variety of fascinating names, including beggarticks, blackjack, burr marigolds, cobbler's pegs, Spanish needles, stickseeds, tickseeds, and tickseed sunflowers. These species have a unique feature reflected even in their scientific name, *Bidens*, which comes from the Latin language, meaning "two" and "tooth." The genus *Bidens*, which includes plants and shrubs, is associated with the American genus Coreopsis and has spread to numerous maritime islands. They have erect stems and cauline, opposite or whorled leaves with varying leaf blades. The Hawaiian Islands are home to several species, including those at risk of extinction due to limited ranges and

S. Abdullah · M. Munir (✉) · B. Javed · B. A. Abbasi
Department of Botany, Rawalpindi Women University, Rawalpindi, Pakistan
e-mail: mamoona.munir@f.rwu.edu.pk; banzeer.abbasi@f.rwu.edu.pk

M. Ahmad
Department of Plant Sciences, Quaid-e-Azam University, Islamabad, Rawalpindi, Pakistan
e-mail: mushtaq@qau.edu.pk

S. Dawood
College of Life Sciences, Northwest A&F University, Yangling, Shaanxi, People's Republic of China
e-mail: sdbot@f.rwu.edu.pk

L. Zhang
Department of Botany, University of Mianwali, Mianwali, Pakistan
e-mail: zhanglixin@nwsuaf.edu.cn

Y. Bibi et al. (eds.), *The Genus Bidens*,
https://doi.org/10.1007/978-981-96-4257-1_1

the absence of native mammals. *Bidens* seeds can disperse to new habitats by attaching to clothing, fur, or feathers.

Keywords

Genus *Bidens* · Beggarticks · Hawaiian Islands · Species · Zoochorous · Oceanic Islands

1.1 Introduction

Bidens belongs to the aster family, Asteraceae, and encompasses a variety of flowering plants (Reaume 2009). The genus *Bidens* (Asteraceae) consists of more than 230 distinct species found across the globe, thriving in tropical as well as warm temperate regions (Knope et al. 2020; Ganders et al. 2000). The fruits of the plants are called beggarticks, blackjack, burr marigolds, cobbler's pegs, Spanish needles, stickseeds, tickseeds, and tickseed sunflowers. The fruits are sometimes bristly and barbed. Their distinctive qualities are reflected in these names. The Latin words bis (meaning "two") and dens (meaning "tooth") are the sources of the generic name *Bidens* for these plants (Hall 1967).

Genus *Bidens* mostly includes herbaceous species that range in height from 10 to 150 cm. These plants can be annual or perennial. The Mexican-born fern-leaved beggar tick, or *Bidens ferulifolia*, is most popular as a summertime ornamental flower for pots on patios and balconies ranging from 15 to 40 cm. On upright, overhanging branches, the leaves are grouped oppositely and have single or divided leaf blades with a broad ovate to lanceolate form. A profusion of flower heads forms at the end of each stalk during its bloom, which occurs from May to October. *Bidens* blooms are star-shaped, with several tiny tubular flowers arranged in a center, which is typical of the daisy family's plants. It comprises a ring of five to seven broad ray florets that are colored yellow, white, orange, red, or pink. Nonetheless, the decorative corolla is absent in several species. The so-called achene fruits are lengthy, dark-shelled seeds that develop after pollination. Many common names for *Bidens* originate from their pappus, an appendage utilized for propagation. Typically consisting of two bristle-shaped teeth equipped with small hooks, this mechanism is used to transport seeds over long distances by attaching to the fur of passing animals that come into contact with the flowers (Knope et al. 2020).

The genus comprises approximately 230 species that are found all over the world in various warm-temperate and tropical regions. Asia, Europe, Polynesia, Africa, and the Americas are home to various species. Blackjack can be found growing wild from sea level to 2400 meters above sea level, or it can be found growing as a weed in crops like cotton, pulses, and cereals, in sunny or slightly shaded areas. For germination, a temperature of 25 to 30°C is ideal. Lower than 15°C and higher than 45°C are not ideal temperatures. It endures extremely elevated salinity levels, reaching as high as 100 mM NaCl, and grows best in soils with a pH range of 4–9. Flooding inhibits the sprouting of seeds.

Bidens exhibit zoochory, meaning their seeds disperse by attaching to animals. Because of this, they have settled in a large area, including numerous marine islands. Many of these species, particularly those found in the Hawaiian Islands, are currently in danger of going extinct, and some are only in tiny geographic areas. Since there are no native mammals on these islands, several species that are specific to oceanic islands have evolved adaptations that appear to facilitate wind-borne seed distribution, resulting in fewer burrs (Owino et al. 2014).

1.2 Taxonomy and Classification

Greene (1901) states that the name *Bidens* originated in 1583 when *Caesalpinus* divided Eupatorium from the common weed that is now known as *J. tripartita*. In 1753, Linnaeus reorganized his original classification of *Coreopsis*, shifting the majority of discoid species to *Bidens* (Table 1.1). Under his revised concept, which is members of the genus were likewise characterized by retrogradely barbed awns. Later researchers did not accept the findings of Necker in 1790 and Moench in 1794, which split the genus into two groups based on leaf morphology and awn count (Sherff 1920). De Candolle distinguished between *Bidens* and *Coreopsis* in his Prodromus (de Candolle 1836), assigning those species with retrorsely barbed awns to the former. Britton (Britton 1893) was the first to identify the direction of instability in barbs and used this information to distinguish the two genera. The division has been acknowledged by American botanists Wiegand (Wiegand 1899), Greene (Greene 1906a), and others who have examined the genus. Sherff characterized the antrorsely barbed species as either *Coreopsis* or *Bidens*, depending on other features to make the ultimate decision. Additionally, the Sheriff acknowledged that Greene (Greene 1906b) had classified *B. beckii*, the aquatic species, as a distinct genus (*Megladonta*). Weedon (1974), who investigated the genus in the north-central plains states, and Hall (1967), who evaluated the group's activities in North America, are two recent researchers in the genus (Gillett and Lim 1970).

Gillett and associates conducted experimental investigations on the Hawaiian Islands, and Carlquist (Carlquist 1966a) employed the group in his work on

Table 1.1 Classification of *Bidens* (Chiang et al. 2010)

Rank	Scientific name	Common name
Kingdom	*Plantae*	Plants
Subkingdom	*Tracheobionta*	Higher plants
Super division	*Spermatophyta*	Seed plants
Division	*Magnoliophyta*	Flowering plants
Class	*Magnoliopsida*	Dicotyledons
Subclass	*Asteridae*	
Order	*Asterales*	
Family	*Asteraceae* Bercht. & J. Presl	Aster family
Genus	*Bidens* L.	Beggarticks

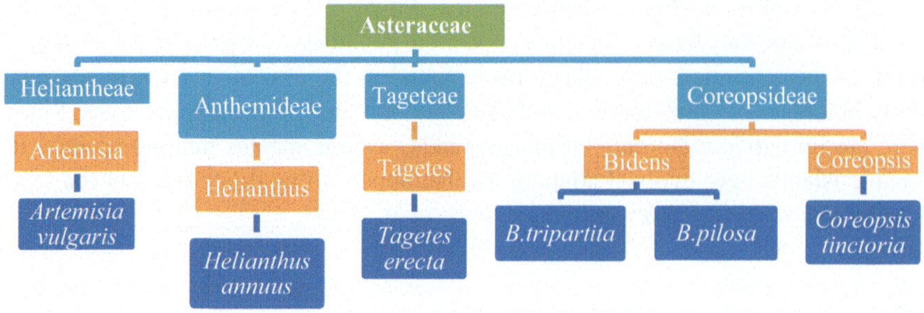

Fig. 1.1 Phylogenetic tree showing the evolutionary relationship between *Bidens* species and related genera within the Asteraceae family

dispersal. T.E. Melchert and his students are studying the *Coreopsidineae* in Mexico and the Southwest of the United States (Crawford 1970). N.C. Fassett, who as a student collaborated with Fernald in examining several members of the genus, determining their linkages and where specific intermediate forms are located, has long been a perplexing issue. Despite the genus's widespread distribution, its systematics and taxonomy have been characterized as convoluted and disorganized. Since the American genus *Coreopsis* and *Bidens* are closely related, it can occasionally be challenging to distinguish between the two; also, neither genus is monophyletic (Fig. 1.1) (Crawford and Mort 2005).

1.3 Key Species of Genus Bidens

1.3.1 *Bidens pilosa* L.

A flowering plant named *Bidens pilosa* L., with over 280 relatives in the Asteraceae family (also known as the daisy family), has a wide range of nicknames. These include chipaca, amor seco, masaquía sillcao, and more descriptive terms like Spanish needles, beggar's ticks, and cobbler's pegs *(*Arthur et al. 2012*)*. Though originally thought to be native only to South America, *Bidens pilosa* has spread throughout warm areas of the world, including Colombia, Brazil, Peru, Uganda, Kenya, China, Australia, and Hawaii. This wide distribution contrasts with its past assumption of a limited range. Interestingly, this plant has a long history of traditional use in the Americas, Africa, Asia, and Oceania, suggesting its spread may be linked to human activity *(*Xuan and Khanh 2016*)*.

1.3.2 *Bidens tripartita* L.

Bidens tripartita is an annual plant with yellow blooms that grows to a height of 30 to 100 cm (Pozharitskaya et al. 2010). Often referred to as trifid bur-marigold, leafy-bracted beggarticks, three-lobe beggarticks, or three-part beggarticks (Table 1.2)

Table 1.2 Some common species of *Bidens* (Ronzhina et al. 2021; Sharma et al. 2023; Khapugin et al. 2022a; Batool et al. 2020)

Sr no	Scientific name	Common name
1	*Bidens*	Beggarticks
2	*Bidens alba* (L.) DC.	Romerillo
3	*Bidens amplissima* Greene	Vancouver Island beggarticks
4	*Bidens aristosa* (Michx.) Britton	Bearded beggarticks
5	*Bidens aurea* (Aiton) Sherff	Arizona beggarticks
6	*Bidens bidentoides* (Nutt.) Britton	Delmarva beggarticks
7	*Bidens bipinnata* L.	Spanish needles
8	*Bidens cernua* L.	Nodding beggarticks
9	*Bidens cervicata* Sherff	Kauai beggarticks
10	*Bidens connata* Muhl. ex Wild.	Purples tem beggarticks
11	*Bidens cosmoides* (A. Gray) Sherff	Cosmos flower beggarticks
12	*Bidens cuneata* Sherff	Wedge beggarticks
13	*Bidens cynapiifolia* Kunth	West Indian beggarticks
14	*Bidens discoidea* (Torr. & A. Gray) Britton	Small beggarticks
15	*Bidens ferulifolia* (Jacq.) DC.	Apache beggarticks
16	*Bidens forbesii* Sherff	Coastal bluff beggarticks
17	*Bidens frondosa* L.	Devil's beggarticks
18	*Bidens gardneri* Baker	Ridge beggartick
19	*Bidens hawaiensis* A. Gray	Hawai'i beggarticks
20	*Bidens heterosperma* A. Gray	Rocky Mountain beggarticks
21	*Bidens hyperborea* Greene	Estuary beggarticks
22	*Bidens laevis* (L.).	Smooth beggarticks
23	*Bidens pilosa* L.	Hairy beggarticks
24	*Bidens populifolia* Sherff	Oahu beggarticks
25	*Bidens tripartita* L.	Three-lobe beggarticks

(Mtenga and Ripanda 2022). Globally, there are over 230 different species of *B. tripartita*, most of which are found in tropical and temperate regions (Lv and Zhang 2013).

It's a cultivable annual herbaceous plant. *B. tripartita* is a common plant that typically grows next to brooks, in vegetable gardens, and beside rivers and wetlands. The plant experiences a wide array of pharmacological and biological effects due to different BAS. Extracts of *B. tripartita* have antibacterial, antidiabetic, antiulcerogenic, antiradical, antioxidant, antimycotic, anti-inflammatory, and anticancer properties. The medicinal raw material from *B. tripartita* is essential for the creation of novel medications (Rodin et al. 2021).

1.3.3 *Bidens frondosa* L.

Bidens frondosa (L.) is found near river and lake coasts in areas with muddy sand soils or nutrient-rich mud soils. This plant goes by several popular names, such as pitchfork weed, bur marigold, sticktights, devil's beggarticks, devil's-pitchfork, tickseed sunflower (Benhura and Chitsiku 1997), leafy beggarticks (Sarg et al. 1991), and common beggarticks (Vranová and Coman 2013). It is currently found throughout China, but it originated in North America. Due to its many biological properties, including antibacterial (Rahman et al. 2011), antioxidant (Venkateswarlu et al. 2004), antimalarial (Brandão et al. 1997), and antidiarrheal properties, it has garnered a lot of interest (Zhao and Wang 1999).

1.3.4 *Bidens alba*

Bidens alba is generally recognized as shepherd's needles, Spanish needles, and butterfly needles (Table 1.2) (Tholl 2015). It thrives in tropical and subtropical areas across North America, Asia, South America, and Africa, often found in gardens, alongside roads, agricultural fields, and disturbed areas. In the United States (Sarg et al. 1991). *B. alba* is classified as a weed. "It occurs on an annual basis or as a perennial plant that typically lives for a brief period." The leaves of *B. alba*, however, are edible and have therapeutic uses (Luck et al. 2020).

B. alba is a nutrient-dense wild plant, comparable to kale, or leaf cabbage, which is cultivated for its edible leaves and is rich in fiber, protein, carotenes, folate, and magnesium. Indians, Zulus, and South Africans boil the fresh or dried leaves for consumption. It's also possible to eat *B. alba* young leaves in a salad. (Zeevaart and Creelman 1988).

1.3.5 *Bidens ferulifolia*

The most often used species of *Bidens* in horticulture are hybrids of *Bidens ferulifolia*, also known as burr marigold or beggar tick (Table 1.2) (Gershenzon and Dudareva 2007). *Bidens ferulifolia*, originating from Mexico and the Southern United States (García-Mendoza and Meave 2011), is a bushy plant known for its upright growth habit. It has been a popular choice in gardens, pots, and mixed ornamental hanging baskets for nearly three decades. Among the *Bidens* species commonly used in horticulture, *Bidens ferulifolia* and its hybrids are the most frequently cultivated (Pichersky and Gang 2000).

1.3.6 *Bidens aurea*

Bidens aurea is a highly varied perennial herbaceous plant that grows elongated rhizomes into one or more tall stems. It ranges from the Southwest region of the United States—Arizona to Guatemala via Mexico. It typically reaches a height of 50 to 100 cm; however, it has been known to exceed 250 cm. *B. aurea* is occasionally grown as a decorative and has striking blossoms. Seeds are easily dispersed because they stick to clothes and animal fur (Connelly et al. 2009). The plant may eventually become a weed because it has adapted in various Mediterranean nations

(Fern 1997). The leaves could be used in place of tea (G. K. 1984). The plant has anti-inflammatory and anti-ulcer properties. Research has demonstrated that the flavonoids present in this plant have a protective effect on the lining of the stomach (Subhuti Dharmananda 1996).

1.3.7 *Bidens bipinnata*

It is commonly known as Spanish needle (Korcelli-Olejniczak 2004). In the practice of traditional Chinese medicine, *Bidens bipinnata* has long been utilized for addressing various infectious disorders. Malaria, sore throat, severe nephritis, and diarrhea can all be treated with the entire plant. This plant's pharmacological properties, such as its anti-arthritis (Shen et al. 2015) and anti-hepatoprotective (Wang et al. 2014) effects, have been studied previously. However, little research has been done on its antimicrobial properties.

1.3.8 *Bidens parviflora* Wild

This species has previously been found to include flavones (Li et al. 2008), flavanones (Hoffmann and Hölzl 1988a), flavonoid glycosides (Li et al. 2008; Hoffmann and Hölzl 1988b), flavonol glycosides (De Tommasi et al. 1998; Sashida et al. 1991), chalcones (Redl et al. 1992; Bauer et al. 1992), aurones (Sashida et al. 1991; Redl et al. 1992), sterols (Chen et al. 1975), polyacetylene glucosides (Hoffmann and Hölzl 1988a; Bauer et al. 1992), and monoterpenes (Hoffmann and Hölzl 1988a; Chen et al. 1975). Previous research on this plant produces different phenolic compounds.

1.4 Phylogenetic Relationships

The publications of "Island Biology" by Carlquist (CARLQUIST 1951) and "The Theory of Island Biogeography" by MacArthur and Wilson (Macarthur and Wilson 1963) encouraged many scientists to conduct formal studies into the ecology and evolution of species on islands. In his research, Carlquist explored the distribution and diversity of the genus *Bidens* L., particularly within the isolated archipelagos of Polynesia. He noted that a significant proportion of *Bidens* species are restricted to individual islands within this region. By analyzing the fruit morphology, in particular the achenes, indehiscent fruits that are probably adapted for dispersal by attaching to bird feathers via barbed awns for long-distance transport, Carlquist concluded that this was a glaring example of diminished dispersal ability, a common phenomenon among island species (Arthur et al. 2012). The Polynesian *Bidens* represent a curious paradox: unlike their continental ancestors, they seem to have forfeited the capability to disperse via their achenes. Despite this, they have managed to spread extensively and diversify into one of Polynesia's most extensive plant radiations, comprising 41 documented species across the Hawaiian, Marquesas, Society, and Austral Islands. Compared to their mainland counterparts, Polynesian *Bidens* exhibit a diverse array of achene structures that hinder their dispersal potential, including reduced or absent barbs and shorter awns.

The evolution of Polynesian *Bidens* is a fascinating puzzle with pieces yet to be discovered. According to research by MacArthur & Wilson (1967), the isolation of far-off island groups like Polynesia would lead to high rates of speciation (the emergence of new species). This, along with immigration and extinction rates, would significantly influence the variety of life found on these islands. *Bidens* is a member of the Coreopsideae group, which is a subfamily of daisies with over 550 species, most of which are found in the Americas. Although Panero (2006) described these groups, a more thorough examination of *Bidens* and Coreopsis is still required. *Bidens* is the largest genus in its tribe, with an estimated 150–235 species; however, there are lingering doubts about whether it's a truly unified group (monophyletic) and the exact number of species it contains (Jagtap and Cholekar 2016). The poor resolution of morphological features used in attempts to construct a phylogenetic theory for the tribe has led several researchers to turn to molecular approaches to address their evolutionary history (Tadesse et al. 1996; Muth and Pigliucci 2006).

A study that used the nuclear ribosomal internal transcribed spacer (ITS) for phylogenetic analysis of *Bidens* and Coreopsis discovered a divergence between north temperate and subtropical *Bidens* inside Coreopsis. The study included 50 species in the group, including two from the Hawaiian Islands. This suggests that neither Coreopsis nor *Bidens* is monophyletic. Interestingly, the two Hawaiian species were found to be closely related to *Bidens pilosa* from Mexico in a particular clade (Kim et al. 1999). The information was obtained from the ITS region and consisted of three undisclosed *Bidens* species from the Marquesas and seven taxa from the Hawaiian Islands, two of which were suspected hybrids (Crawford et al. 2001). The study suggested that a group containing *B. pilosa* and *B. alba* from Mexico and the Caribbean was closely related to Polynesian species, except for the one found on Starbuck Island. However, the evidence that was provided did not show a clear correlation. The investigation also demonstrated that Coreopsis and *Bidens* are not monophyletic groups and that some parts of Coreopsis have stronger affinities with other Coreopsideae genera than with other species within *Bidens* or Coreopsis. Two Hawaiian species were found to be the sister group to an unidentified clade that includes *B. alba*, *B. pilosa*, and *B. hintonii* in their analysis of the ITS tree; however, the location of these two clades similarly lacked sufficient support. The study supported numerous clades while demonstrating a lack of organization among them to clarify the classifications in *Bidens* and Coreopsis (Kimball and Crawford 2004; Ganders et al. 2000).

The two chloroplast markers, matK and trnL-F, as well as the ITS region, did not support the monophyly of *Bidens* and Coreopsis. However, it should be emphasized that this analysis did not include any Pacific species (Mort et al. 2008). In addition to *Bidens* Henryi Sherff, the 19 endemic Hawaiian species included an unidentified Marquesan species. Previous studies have identified *B. pilosa* and *B. alba* as the sister clade, but recent data based on ITS suggests that the Hawaiian radiation and the two Marquesan species most likely shared an ancestor that migrated across the Pacific from Mexico or its adjacent areas (Knope et al. 2012). Tsai and colleagues carried out a pioneering study on molecularly identifying *Bidens* species. To distinguish between distinct species of *Bidens*, they used noncoding sections from the chloroplast genome (trnL intron and trnL-trnF) and nuclear ribosomal DNA segments (ITS1, 5.8S, and ITS2). According to their research, the only genes that could effectively discriminate between *B. biternata* and *B. pilosa* were ITS1, 5.8S, ITS2, and the trnL intron. Even after examining every sequence that could be accessed from the NCBI nucleotide database, it was discovered that neither ITS1 nor ITS2 had sufficient resolution for distinguishing between different *Bidens* species (Tsai et al. 2008).

1.5 Biogeography of Genus *Bidens*

The genus *Bidens* includes more than 230 species, usually inhabiting a wide range of tropical and warm temperate climates globally. Asia, Europe, Polynesia, Africa, and the Americas are home to various species (Fig. 1.2). Ten species are known to exist in China, according to records on flora. The fascinating thing about the distribution of *Bidens* species in China is that comparable physical and chemical traits were discovered, which hampered research on the species' evolutionary history and ability to discriminate one from the other (Knope et al. 2020; Funk and Wagner 1995; Carlquist 2001).

In contrast, a single progenitor that settled in Hawaii is believed to have given rise to several morphologically distinct species of Hawaiian plants, including *Schiedea* (Caryophyllaceae), *Mints* (Lamiaceae), *Lobeliads* (Campanulaceae), *Bidens* (Asteraceae), and *Cyrtandra* (Gesneriaceae). These lineages are currently found in a variety of habitats (Price and Wagner 2004). The limited geographic area

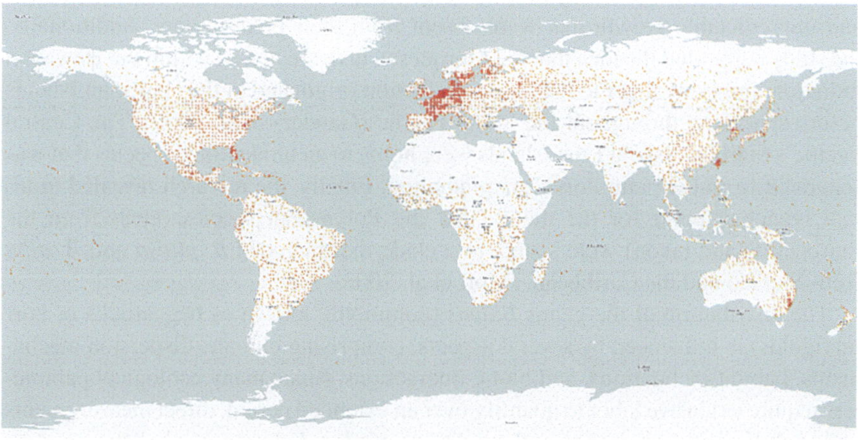

Fig. 1.2 Worldwide distribution map of *Bidens*. (GBIF 2024)

and brief time span over which speciation has occurred are two of these remarkable Hawaiian plant radiations' most remarkable characteristics (Baldwin and Wagner 2010). The species *B. pilosa* is widely naturalized throughout the world's warm temperate and tropical climates, having originated in tropical America (Chiang et al. 2010).

Comparing Polynesian *Bidens* to their mainland relatives, the former has a greater variety of achene morphologies, but the latter have characteristics like missing or vestigial barbs and shorter awn length that limit their dispersibility (Carlquist 1966b). Therefore, Polynesian *Bidens* present a clear paradox because, unlike their continental ancestors, they appear to have lost the capability to spread through their achenes. These plants were extremely successful throughout Polynesia, even though they had modest origins. With 41 recognized species found on islands including Hawaii, the Marquesas, the Societies, and the Australs, they currently comprise one of the region's most varied plant groups. The "radiation zone," which is believed to include these remote archipelagoes, among others, is where local extinction rates, immigration, and in situ speciation are anticipated to have a major impact on our knowledge of island biodiversity patterns. However, a thorough understanding of the Polynesian *Bidens* radiation's evolutionary history is still difficult. (MacArthur and Wilson 2001).

With a global distribution of about 550 species, the majority of this tribe's species are located in the Americas (Kadereit and Jeffrey 2007). Neither Bidens nor Coreopsis has received a thorough treatment. The tribe's largest genus, *Bidens*, is estimated to have 150–235 species (Strother and Weedon 2006); however, it is unclear how many species there are and whether the genus is monophyletic. Numerous researchers have turned to molecular methods to investigate the tribe's evolutionary history when their morphological characters failed to yield a phylogenetic hypothesis that could be reliably developed (Karis and Ryding 1994).

It seems that a collection of herbaceous species with determinate growth occurred in Central America, making them the continental sister group of Polynesian *Bidens*. This shows that Polynesian *Bidens* have apomorphic character states of woodiness and unpredictable growth that were absent from their predecessors. Additionally, the study supported the idea that the Marquesan taxa are nested within the Hawaiian radiation, implying that the Polynesian radiation originated in the Hawaiian Islands before spreading throughout the South Pacific (Ganders et al. 2000). The Central Pacific's Starbuck Island Line Islands were home to an unidentified species that was separated from the main Polynesian radiation. Finally, the research revealed minimal branch support for the theory that the Polynesian species (apart from the Starbuck Island taxon) were sisters to a clade that included *B. pilosa* and *B. alba* from Mexico and the Caribbean. (Mort et al. 2008).

The distribution of the genus *Bidens* (commonly known as beggarticks or burr marigolds) is influenced by several aspects, comprising climate, dispersion mechanisms, habitat inclinations, and biotic interactions. Since many ecological parameters require extensive labor to quantify over an extended period, direct measurements and analysis of the environmental conditions of plant habitats are challenging. As a substitute, techniques for indirect ordination predicated on vegetation type are extensively useful in ascertaining the parameters and fluctuations of the surroundings at the location in question. Therefore, there is a great deal of interest in and use for habitat evaluation utilizing environmental scales (Landolt et al. 2010). *Bidens* species can be found in a broad variety of climates, from tropical to temperate. For successful germination, growth, and reproduction, temperature is essential. Water availability is essential since many species of *Bidens* live in moist habitats like marshes and riverbanks. Their abundance and distribution are influenced by seasonal variations of rainfall. Since many *Bidens* species are found in humid locations, high humidity levels are beneficial to them (Khapugin et al. 2022b).

Dispersal mechanisms also affect the species of the *Bidens* genus. Some species are hydrochory and have seeds that can float and be dispersed by water, which is significant for those growing near water bodies (Leck et al. 1994). *Bidens* species have been spread to new regions by human activities such as agriculture, horticulture, and trade; occasionally, this has resulted in invasive behavior (Woo et al. 2021). *Bidens* seeds are frequently equipped with barbs or hooks that attach to animal fur or clothing, aiding in zoochory (Bullock and Primack 1977). Although many species of *Bidens* favor moist, well-drained, nutrient-rich soils, they can grow in a range of soil types (Cui and He 2009). Though they may tolerate some shade, these plants are typically found in bright, open spaces. Usually opportunistic, they settle in disturbed areas with less competition (Pan et al. 2017). Numerous *Bidens* species have adapted to live in wetland environments and depend on the recurring flooding and nutrient inflows that these habitats offer (Leck et al. 1994).

Bidens plants frequently face competition for resources from other herbaceous plants. Their advantage over rivals is their capacity to settle in disturbed regions. For many *Bidens* species, insects, especially bees, are essential pollinators that influence the species' spread and rate of reproduction. Herbivores can affect *Bidens* in many ways, although grazing pressure may occasionally restrict their range (Schmalhofer 1996; Gao et al. 2021). To quickly take over and colonize suitable places, several species of *Bidens* have short life cycles and rapid growth (Danuso et al. 2012). Biden seeds are long-stored, allowing them to withstand extreme environments. The seeds can then begin to sprout as the surrounding conditions become favorable (Helenurm and Ganders 1985). While certain Biden species can disrupt established ecosystems, they can also adapt to new environments generated by farming, urbanization, and development. In certain new locations, some *Bidens* species have even spread like invasive species, pushing away native plants and upsetting the delicate equilibrium (Mahmoud et al. 2015).

1.6 Ecology and Habitats

Numerous ecological niches are occupied by the species in the genus *Bidens*, which is widely dispersed. These niches are influenced by biotic interactions, the species' adaptations to specific environmental variables, and resource availability. Among the numerous species of *Bidens* that grow in wetlands, marshes, and along riverbanks are *Bidens cernua*, also known as nodding bur-marigold, and *Bidens laevis*, also known as smooth bur-marigold. High soil moisture content and periodic flooding are characteristics of these locations. Saturated soils and varying water levels are usually not a problem for the species that inhabit these niches. They might also contain air-filled tissues called aerenchyma, which aids in the transfer of oxygen to roots that are submerged (Gratani et al. 2009). Numerous *Bidens* species thrive in environments where human activity has caused disturbance, such as cropland, construction sites, ditches beside roads, and empty lots. These species can spread swiftly over bare or disturbed ground since they are usually opportunistic and fast-growing. They frequently provide vast quantities of seeds that can last in the soil seed bank (Asami et al. 1999). The common examples of species are *Bidens pilosa* and *Bidens frondosa* (Wang et al. 2019; Ronzhina 2017).

There is less competition from woody plants and full sunlight for *Bidens* species, which means they can grow well in open grasslands and meadows. Many soil types can be tolerated by these plants, which are suited for full sun. In these habitats, they frequently comprise the herbaceous layer. Two species that are examples are *Bidens connata* (Sherff 1962), which has a purple stem, and *Bidens tripartite* (Jan et al. 2022). Habitats with moderate light levels and comparatively high resource availability can be found in forest borders and clearings inside forests. In order to benefit from increased light after disturbances like tree falls or human clearance, species in these niches are frequently able to grow quickly and can withstand some shade (Sharma et al. 2023). Spanish needles, or *Bidens bipinnata*, and bearded beggartick, or *Bidens aristosa*, grow in such habitats and adapt themselves to shades and edges (Gehlhausen et al. 2000).

Certain species of *Bidens* can be found in coastal environments that are prone to tidal fluctuations and saltwater intrusion, such as salt marshes and brackish water habitats (León de la Luz and Medel-Narváez 2013). These species can survive on saline soils and show tolerance to salt. They frequently possess systems for compartmentalizing or excreting salt. *Bidens hyperborea*, or northern beggarticks, is a common species of the genus *Bidens* having a coastal and brackish habitat (Fassett 1925). A few species of *Bidens* are acclimated to higher altitudes when growing seasons are shorter and temperatures are lower (Huang et al. 2015). These plants have evolved to withstand freezing temperatures and can quickly finish their life cycles in the short alpine season like *Bidens beckii*, which is found in aquatic montane environments (Björk 2010). In the South West Region of Nigeria, *Bidens pilosa* is a common weed that spreads by sticking itself to the skin of animals and farmers' clothes. An analysis was performed on leaf meal from *Bidens pilosa* (BPLM) to determine its nutritional content, including proximate nutrients, amino acid composition, phytochemicals, and mineral elements (Alikwe et al. 2014).

The reproductive ecology of a few *Bidens* species in North and Central America has been examined. According to reports, hymenopteran and lepidopteran insects pollinate *Bidens pilosa* and *B. subalternans* in South America. These plants are protandrous and capable of self-pollination (Grombone-Guaratini et al. 2004). Fertile hermaphrodite disc florets and sterile ray florets are produced in capitula by *B. segetum* (Valentin-Silva et al. 2016). The Hawaiian Islands host a diverse array of *Bidens* species, each exhibiting unique morphological and ecological characteristics. Their capitula provide self-compatible, extremely protandrous sterile ray florets and fertile disc florets. Honey bees that have been imported are the main pollinators of these species (Budumajji and Solomon Raju 2018). These plants interact with other plants and their surroundings in a variety of ways, which are frequently impacted by their reproductive and ecological strategies. There may be competition between some *Bidens* species and other plants for light, water, and nutrients. Because of their ability to form dense stands and quick growth rates, they might be able to outcompete less vigorous species (Kato-Noguchi and Kurniadie 2024). Certain varieties of *Bidens* release chemicals known as allelochemicals, which inhibit neighboring plants from germinating and developing, hence reducing competition and promoting their own growth (Khanh et al. 2009).

Bidens plants have the ability to decompose rapidly, which helps increase soil fertility and benefits other plants by contributing to nutrient cycling in their habitats (Meisner et al. 2011). Their root systems can enhance the growing environment for other plants by reducing soil erosion and stabilizing the soil (Wei et al. 2017). While the *Bidens* species is not parasitic itself, it can host parasitic plants like the *Cuscuta australis* species, which could hinder the *Bidens* plant's ability to develop and survive (Yang et al. 2015). By increasing the complexity of their habitat and offering a home to a wide range of fauna, *Bidens* species can significantly contribute to the promotion of biodiversity. The distribution of plant groups and the properties of the soil are altered by the introduction of *Bidens pilosa* (Asteraceae) in the northwest Himalayan region (Sharma et al. 2023; Zongliang et al. 2023).

1.7 Conclusion

The genus *Bidens* is part of the Asteraceae family, which contains approximately 230 species that are found in warm temperate and tropical regions of the world. The fruits of these species are well-known for being bristly and thorny, which contributes to their zoochorous dissemination. There are many different types of *Bidens*, ranging from popular decorative flowers like *Bidens ferulifolia* to herbaceous plants. They do well in a range of habitats, such as marshes, disturbed places, grasslands, forest margins, and coastal regions. Their resilience to a range of environmental stressors, including flooding, shading, and high salinity, aids in their flexibility. *Biden's* taxonomic history is complicated by the substantial contributions made throughout decades by botanists such as Linnaeus, Greene, and De Candolle, who helped to enhance the categorization of the plant. Separating closely related genera like Coreopsis can be difficult due to the systematics and taxonomy of Bidens, which are nonetheless complicated and sometimes disorganized despite the species' widespread distribution.

Ecologically, *Bidens* species are remarkably adaptable, growing well on disturbed or nutrient-rich soils and adapting to a wide range of ecological circumstances. They spread widely because their seeds are skilled at traveling great distances, which is frequently made possible by water or animals. Though certain species can become invasive outside of their native areas, this flexibility also presents difficulties. Although neither *Bidens* nor Coreopsis is monophyletic, phylogenetic studies have shown this. Molecular methods are currently being used in research to shed light on their evolutionary ties. The Polynesian *Bidens* are notable for their distinct evolutionary characteristics. They have undergone extensive radiation inside solitary archipelagos, even though they have lost some of the dispersion capacities present in their mainland relatives. All things considered, the genus *Bidens* is a fascinating and diverse group of plants with important ecological, taxonomic, and biogeographic significance. It will take more investigation to properly comprehend the evolutionary history and environmental adaptations of these widespread and adapted species, namely in the areas of molecular phylogenetics and ecological interactions.

References

Alikwe PCN, Ohimain EI, Omotosho SM (2014) Evaluation of the proximate, mineral, phytochemical, and amino acid composition of Bidens pilosa as potential feed/feed additive for non-ruminant livestock. Anim Vet Sci 2(2):18–21

Arthur G, Naidoo K, Coopoosamy R (2012) Bidens pilosa L.: agricultural and pharmaceutical importance. J Med Plants Res 6(17):3282–3287

Asami K, Yamato M, Hattori T, Akamatsu H, Takeda Y (1999) Floristic composition and process of establishment of the Bidens pilosa var. minor-Imperata cylindrica community maintained by cutting on non-arable land in Okinawa Prefecture. Veg Sci 16(1):1–11

Baldwin BG, Wagner WL (2010) Hawaiian angiosperm radiations of North American origin. Ann Bot 105(6):849–879

Batool A, Batool Z, Qureshi R, Raja NIJPS (2020) Phytochemicals, pharmacological properties and biotechnological aspects of a highly medicinal plant: Datura stramonium. J Plant Sci 8(2):29–40

Bauer R, Redl K, Davis B (1992) Four polyacetylene glucosides from Bidens campylotheca. Phytochemistry 31(6):2035–2037

Benhura MA, Chitsiku IC (1997) The extractable β-carotene content of Guku (Bidens pilosa) leaves after cooking, drying and storage. Int J Food Sci Tech 32(6):495–500

Björk CR (2010) Distribution patterns of disjunct and endemic vascular plants in the interior wetbelt of northwest North America. Botany 88(4):409–428

Brandão M, Krettli A, Soares L, Nery C, Marinuzzi H (1997) Antimalarial activity of extracts and fractions from Bidens pilosa and other Bidens species (Asteraceae) correlated with the presence of acetylene and flavonoid compounds. J Ethnopharmacol 57(2):131–138

Britton EG (1893) Contributions to American bryology. III. Notes on the North American species of *Orthotrichum*. Bull Torrey Bot Club 20(10):393–405

Budumajji U, Solomon Raju AJ (2018) Pollination ecology of Bidens pilosa L.(Asteraceae). Taiwania 63(2)

Bullock SH, Primack RB (1977) Comparative experimental study of seed dispersal on animals. Ecology 58(3):681–686

CARLQUIST E (1951) Modern sociology, its structure and significance. III. Soc Med Tidskr 28(12):245–249. contd

Carlquist S (1966a) Wood anatomy of compositae: a summary, with comments on factors controlling wood evolution. Aliso 6(2):25–44

Carlquist S (1966b) The biota of long-distance dispersal. I. Principles of dispersal and evolution. Q Rev Biol 41(3):247–270

Carlquist S (2001) Wood anatomy of the endemic woody Asteraceae of St Helena I: phyletic and ecological aspects. Bot J Linn Soc 137(2):197–210

Chen AH, Lin SR, Hong CH (1975) Hua Hsueh 9:42–48

Chiang G, Insel P, Tosun D, Schuff N, Truran-Sacrey D, Raptentsetsang S et al (2010) Hippocampal atrophy rates and CSF biomarkers in elderly APOE 2 normal subjects. Neurology 75(22):1976–1981

Connelly W, Shenton F, Lethbridge N, Leurs R, Waldvogel H, Faull R et al (2009) The histamine H4 receptor is functionally expressed on neurons in the mammalian CNS. Br J Pharmacol 157(1):55–63

Crawford AR (1970) Continental drift and un-continental thinking. Econ Geol 65:11–16

Crawford DJ, Mort ME (2005) Phylogeny of Eastern North American Coreopsis (Asteraceae-Coreopsideae): insights from nuclear and plastid sequences, and comments on character evolution. Am J Bot 92(2):330–336

Crawford D, Kimball R, Tadesse M (2001) The generic placement of a morphologically enigmatic species in Asteraceae: evidence from ITS sequences. Plant Syst Evol 228:63–69

Cui QG, He WM (2009) Soil biota, but not soil nutrients, facilitate the invasion of Bidens pilosa relative to a native species Saussurea deltoidea. Weed Res 49(2):201–206

Danuso F, Zanin G, Sartorato I (2012) A modelling approach for evaluating phenology and adaptation of two congeneric weeds (Bidens frondosa and Bidens tripartita). Ecol Model 243:33–41

de Candolle AP (1836) Prodromus Systematis Naturalis Regni Vegetabilis [...] Pars quinta

De Tommasi N, Piacente S, Pizza C (1998) Flavonol and chalcone ester glycosides from Bidens andicola. J Nat Prod 61(8):973–977

Fassett NC (1925) Bidens hyperborea and its varieties. Rhodora 27(321):166–171

Fern K Notes from observations, tasting etc at Plants For A Future and on field trips. permanent publication. 1997

Funk VA, Wagner WL (1995) Biogeographic patterns in the Hawaiian Islands. In: Hawaiian biogeography: evolution on a hot spot archipelago

G. K (1984) An excellent book for the dedicated. A comprehensive listing of Latin names with a brief list of edible parts

Ganders FR, Berbee M, Pirseyedi M (2000) It's base sequence phylogeny in Bidens (Asteraceae): evidence for the continental relatives of Hawaiian and Marquesan Bidens. Syst Bot 25(1):122–133

Gao F-L, He Q-S, Xie R-Q, Hou J-H, Shi C-L, Li J-M et al (2021) Interactive effects of nutrient availability, fluctuating supply, and plant parasitism on the post-invasion success of Bidens pilosa. Biol Invasions 23(10):3035–3046

García-Mendoza A, Meave J (2011) Diversidad Florística de Oaxaca: de Musgos a Angispermas 1–351. Universidad Nacional Autónoma de México, Ciudad Universitaria

GBIF (2024) Global Biodiversity Information Facility. Available from: https://www.gbif.org/what-is-gbif

Gehlhausen SM, Schwartz MW, Augspurger CK (2000) Vegetation and microclimatic edge effects in two mixed-mesophytic forest fragments. Plant Ecol 147:21–35

Gershenzon J, Dudareva N (2007) The function of terpene natural products in the natural world. Nat Chem Biol 3(7):408–414

Gillett GW, Lim EK 1970 An experimental study of the genus Bidens (Asteraceae) in the Hawaiian Islands. (No Title)

Gratani L, Crescente MF, Fabrini G, Bonito A, Varone L (2009) Growth analysis of a reestablished population versus a natural population of Bidens cernua L. Plant Biosyst 143(sup1):S63–SS8

Greene EL (1906a) A study of Rhus glabra: the academy. Proc. Washington Acad Sci 8:167–196

Greene R. Growing conifers from the seed. 1906b

Grombone-Guaratini MT, Solferini VN, Semir J (2004) Reproductive biology in species of Bidens L.(Asteraceae). Sci Agric 61:185–189

Hall GW (1967) A biosystematic study of the North American complex of the genus Bidens (Compositae). Indiana University

Helenurm K, Ganders FR (1985) Adaptive radiation and genetic differentiation in Hawaiian Bidens. Evolution 39(4):753–765

Hoffmann B, Hölzl J (1988a) Weitere acylierte chalkone aus Bidens pilosa. Planta Med 54(05):450–451
Hoffmann B, Hölzl J (1988b) A methylated chalcone glucoside from Bidens pilosa. Phytochemistry 27(11):3700–3701
Huang H-L, Huang Y-L, Wu T-C, Kao W-Y (2015) Phenotypic variation and germination behavior between two altitudinal populations of two varieties of Bidens pilosa in Taiwan. Taiwania 60(4):194–203
Jagtap D, Cholekar M (2016) Bidens bachulkarii (Asteraceae-Heliantheae): a new species from Western Ghats, India. Trop Plant Res 3(3):681–685
Jan I, Yaqoob S, Reshi ZA, Rashid I, Shah MA (2022) Risk assessment and management framework for rapidly spreading species in a Kashmir Himalayan Ramsar site. Environ Monit Assess 194(3):175
Kadereit JW, Jeffrey C (2007) The families and genera of vascular plants. Springer
Karis P, Ryding O (1994) Asteraceae Cladistics And Classification. K Bremer, Ed, pp 559–569
Kato-Noguchi H, Kurniadie D (2024) The invasive mechanisms of the noxious alien plant species Bidens pilosa. Plan Theory 13(3):356
Khanh T, Cong L, Xuan T, Uezato Y, Deba F, Toyama T et al (2009) Allelopathic plants: 20. Hairy beggarticks (Bidens pilosa L.). Allelopath J 24(1):243–259
Khapugin AA, Kuzmin IV, Ivanov LA, Ronzhina DA, Ivanova LA (2022a) Environmental preferences of an invasive plant species, Bidens frondosa (Asteraceae), European Russia and Western Siberia. Diversity 14(8):598
Khapugin AA, Kuzmin IV, Ivanov LA, Ronzhina DA, Ivanova LA (2022b) Environmental preferences of an invasive plant species, Bidens frondosa (Asteraceae), in European Russia and Western Siberia. Diversity 14(8):598
Kim S-C, Crawford DJ, Tadesse M, Berbee M, Ganders FR, Pirseyedi M et al (1999) ITS sequences and phylogenetic relationships in Bidens and Coreopsis (Asteraceae). Syst Bot 24:480–493
Kimball RT, Crawford DJ (2004) Phylogeny of Coreopsideae (Asteraceae) using ITS sequences suggests lability in reproductive characters. Mol Phylogenet Evol 33(1):127–139
Knope ML, Morden CW, Funk VA, Fukami T (2012) Area and the rapid radiation of Hawaiian Bidens (Asteraceae). J Biogeogr 39(7):1206–1216
Knope ML, Funk VA, Johnson MA, Wagner WL, Datlof EM, Johnson G et al (2020) Dispersal and adaptive radiation of Bidens (Compositae) across the remote archipelagoes of Polynesia. J Syst Evol 58(6):805–822
Korcelli-Olejniczak E (2004) Funkcje metropolitalne Berlina i Warszawy w latach 1990–2002: współzależność pozycji w systemie miast Europy. IGiPZ PAN, Środkowej
Landolt E, Bäumler B, Erhardt A, Hegg O, Klötzli F, Lämmler W et al (2010) Flora indicativa= Ecological inicator values and biological attributes of the flora of Switzerland and the Alps: ökologische Zeigerwerte und biologische Kennzeichen zur Flora der Schweiz und der. Haupt Verlag, Alpen
Leck MA, Baskin CC, Baskin JM (1994) Germination ecology of Bidens laevis (Asteraceae) from a tidal freshwater wetland. Bull Torrey Bot Club 230-9:230
León de la Luz JL, Medel-Narváez A (2013) A new species of Bidens (Asteraceae: Coreopsidae) from Baja California Sur, Mexico. Acta botánica mexicana 103:119–126
Li Y-L, Li J, Wang N-L, Yao X-S (2008) Flavonoids and a new polyacetylene from Bidens parviflora Willd. Molecules 13(8):1931–1941
Luck K, Kim D-K, Lambourne L, Spirohn K, Begg BE, Bian W et al (2020) A reference map of the human binary protein interactome. Nature 580(7803):402–408
Lv J-L, Zhang L-B (2013) Flavonoids and polyacetylenes from the aerial parts of Bidens tripartita. Biochem Syst Ecol 48:42–44
Macarthur RH, Wilson EO (1963) An equilibrium theory of insular zoogeography. Evolution **17**:373–387. 2 RH Mac Arthur and EO Wilson. 1967
MacArthur RH, Wilson EO (2001) The theory of Island biogeography. Princeton University Press
Mahmoud T, Gairola S, El-Keblawy A (2015) Parthenium hysterophorus and Bidens pilosa, two new records to the invasive weed flora of The United Arab Emirates. J New Biol Rep 4(1):26–32

Meisner A, de Boer W, Verhoeven KJ, Boschker HT, van der Putten WH (2011) Comparison of nutrient acquisition in exotic plant species and congeneric natives. J Ecol 99(6):1308–1315
Mort ME, Randle CP, Kimball RT, Tadesse M, Crawford DJ (2008) Phylogeny of Coreopsideae (Asteraceae) inferred from nuclear and plastid DNA sequences. Taxon 57(1):109–120
Mtenga DV, Ripanda ASJH (2022) A review on the potential of underutilized Blackjack (Biden Pilosa) naturally occurring in sub-Saharan. Africa 8(6):e09586
Muth NZ, Pigliucci M (2006) Traits of invasives reconsidered: phenotypic comparisons of introduced invasive and introduced noninvasive plant species within two closely related clades. Am J Bot 93(2):188–196
Owino J, Habimana D, Gakuru E, Ongol M (2014) Phytochemical screening and antibacterial activity of Bidens pilosa. Asian J Pharm Biol Res 4(1):34
Pan Y, Tang S, Wei C, Li X (2017) Comparison of growth, photosynthesis and phenotypic plasticity between invasive and native Bidens species under different light and water conditions. Biodivers Sci 25(12):1257
Pichersky E, Gang DR (2000) Genetics and biochemistry of secondary metabolites in plants: an evolutionary perspective. Trends Plant Sci 5(10):439–445
Pozharitskaya O, Shikov A, Makarova M, Kosman V, Faustova N, Tesakova S et al (2010) Anti-inflammatory activity of a HPLC-fingerprinted aqueous infusion of aerial part of Bidens tripartita L. Phytomedicine 17(6):463–468
Price JP, Wagner WL (2004) Speciation in Hawaiian angiosperm lineages: cause, consequence, and mode. Evolution 58(10):2185–2200
Rahman A, Bajpai VK, Dung NT, Kang SC (2011) Antibacterial and antioxidant activities of the essential oil and methanol extracts of Bidens frondosa Linn. Int J Food Sci Technol 46(6):1238–1244
Reaume T (2009) 620 wild plants of North America: fully illustrated. University of Regina Press
Redl K, Davis B, Bauer R (1992) Chalcone glycosides from Bidens campylotheca. Phytochemistry 32(1):218–220
Rodin M, Bokov D, Kovaleva TY, Bobkova N, Sergunova E, Strelyaeva A et al (2021) Composition of biologically active compounds, biological and pharmacological activity of the three-part beggarticks (Bidens tripartita L.). Nat Vol Essent Oil J 8(5):11039–11053
Ronzhina D (2017) Distribution, competitive ability, and seed production of Bidens frondosa L. in the Middle Urals. Russ J Biol Invasions 8:351–359
Ronzhina DA, Ivanova LA, Ivanov LA, Khapugin AA (2021) Bidens frondosa (Asteraceae), a new alien invasive plant species in the Kurgan Region (Russia). Contribuţii Botanice 56:53
Sarg T, Salama O, El-Domiaty M, Bishr M, Mansour SES, Weight E (1991) Iridoid glucosides from Gentiana kurroo Royle. Alex J Pharm Sci 5:84–86
Sashida Y, Ogawa K, Kitada M, Karikome H, Mimaki Y, Shimomura H (1991) New aurone glucosides and new phenylpropanoid glucosides from Bidens pilosa. Chem Pharm Bull 39(3):709–711
Schmalhofer VR (1996) The effects of biotic and abiotic factors on predator-prey interactions in old-field flower-head communities: Rutgers The State University of New Jersey. School of Graduate Studies
Sharma A, Kaur A, Kohli RK, Singh HP, Batish DR (2023) Bidens pilosa (Asteraceae) invasion reshapes the pattern of plant communities and edaphic properties across the north-western Himalayan landscape. Eco Inform 77:102281
Shen A-Z, Li X, Hu W, Chen F-H (2015) Total flavonoids of Bidens bipinnata L. ameliorate experimental adjuvant-induced arthritis through induction of synovial apoptosis. BMC Complement Altern Med 15:1–9
Sherff EE (1920) Studies in the Genus Bidens. V Bot Gaz 70(2):89–109
Sherff EE (1962) Further notes on the distribution of Bidens connata vars. pinnata and gracilipes. Rhodora 64(757):23–28
Strother JL, Weedon RR (2006) Bidens. In: Flora of North America, vol 21, pp 205–218
Subhuti Dharmananda PD. A leaflet produced by the Institute for Traditional Medicine, Portland, Oregon. Institute for Traditional Medicine, Portland, Oregon. 1996

Tadesse M, Crawford D, Smith E (eds) (1996) Generic concepts in Bidens and Coreopsis (Compositae): an overview. The biodiversity of African plants: proceedings XIVth AETFAT Congress 22–27 August 1994. Springer, Wageningen

Tholl D (2015) Biosynthesis and biological functions of terpenoids in plants, pp 63–106

Tsai L-C, Wang J-C, Hsieh H-M, Liu K-L, Linacre A, Lee JC-I (2008) Bidens identification using the noncoding regions of chloroplast genome and nuclear ribosomal DNA. Forensic Sci Int Genet 2(1):35–40

Valentin-Silva A, Godinho M, Cruz K, Lelis S, Vieira M (2016) Three psychophilous Asteraceae species with distinct reproductive mechanisms in southeastern Brazil. N Z J Bot 54(4):498–510

Venkateswarlu S, Panchagnula GK, Subbaraju GV (2004) Synthesis and antioxidative activity of 3′, 4′, 6, 7-tetrahydroxyaurone, a metabolite of Bidens frondosa. Biosci Biotechnol Biochem 68(10):2183–2185

Vranová E, Coman D, Gruissem W (2013) Network analysis of the MVA and MEP pathways for isoprenoid synthesis. Annu Rev Plant Biol 64(1):665–700

Wang X-Y, Chen G-R, Pan C-X, Deng Z-Y, Ge J-F, Li N et al (2014) Polyacetylenes from Bidens bipinnata L. and their biological activities. Phytochem Lett 7:198–201

Wang J, Chen W, Zhu H (2019) Ecological stoichiometry and invasive strategies of two alien species (Bidens pilosa and Mikania micrantha) in subtropical China. Ecol Res 34(5):612–623

Wei H, Yan W, Quan G, Zhang J, Liang K (2017) Soil microbial carbon utilization, enzyme activities and nutrient availability responses to Bidens pilosa and a non-invasive congener under different irradiances. Sci Rep 7(1):11309

Wiegand KM (1899) A revision of the genus Listera. Bull Torrey Bot Club 26(4):157–171

Woo S, Lee D, Cho YC, Lee S, Kim E (2021) Differential responses to fertilization and competition among invasive, noninvasive alien, and native Bidens species. Ecol Evol 11(1):516–525

Xuan TD, Khanh TD (2016) Chemistry and pharmacology of Bidens pilosa: an overview. J Pharm Investig 46:91–132

Yang B, Li J, Zhang J, Yan M (2015) Effects of nutrients on interaction between the invasive Bidens pilosa and the parasitic Cuscuta australis. Pak J Bot 47(5):1693–1699

Zeevaart JAD, Creelman RA (1988) Metabolism and physiology of abscisic acid. Annu Rev Plant Physiol Plant Mol Biol 39(1):439–473

Zhao J, Wang C (1999) Antidiarrheal activity of Bidens frondosa. J Nanjing Univ TCM 15:299–300

Zongliang Q, Hua X, Yue W, Tao P (2023) Effect of bidens alba on interspecific association and the stability of companion species. 四川林业科技 44(2):31–37

Morphological Characteristics of *Bidens* Species: Vegetative Morphology

2

Neelum Naheed, Aqsa Jabeen, and Hira Aslam

Abstract

Bidens cosmopolitan genus belongs to the family Asteraceae and has great medicinal and ornamental significance, e.g., *Bidens ferulifolia* grown in gardens for beautiful flowers, approximately 230 to 240 species. *Bidens* plants are sticktights, tickseed sunflowers, beggarticks, and bur marigolds. *Bidens* species are widely distributed in tropical, subtropical regions of the world, especially in temperate South America and North America: mostly annual, shrubs, climber, and perennial woody herbs. The average height of *Bidens* species is 60 cm and may exceed 150 cm when environmental conditions favor the growth. These species can grow in dry, arid, and barren soil at different elevations. Most species of this genus have branched, erect, and hairy stems. Leaves simple or lobed to pinnately compound (*Bidens mitis*), mostly petiolate, occasionally sessile (*Bidens laevis*), curved or dissected cauline usually opposite, rarely alternate, sometimes whorled. *Bidens* genus has capitulum inflorescence, heads terminal, radiate, discoid on peduncle having white or yellow, ligulate flowers, and hermaphrodite. Most of the species involved have more or less herbaceous and foliaceous bracts, these bracts constitute calyculi, and usually, they suppress the phyllaries. The torus is flat to slightly convex, yellow to orange in color. Flowers have a both disc and ray florets. Ray florets are commonly 1–21+ in number, and the most frequent number is 3, 5, 8, or 13, with a diameter of 2–3.5 mm. White, yellow, and pinkish-colored lobed corolla is often present. Disc florets are 5–20 in number, but in different species, the number varies from 20 to 60. It is bisexual, fertile, tubular,

N. Naheed (✉) · A. Jabeen
Department of Botany, PMAS Arid Agriculture University, Rawalpindi, Pakistan

H. Aslam
Department of Plant Breeding and Genetics, PMAS Arid Agriculture University, Rawalpindi, Pakistan
e-mail: 13-arid-327@student.uaar.edu.pk

Y. Bibi et al. (eds.), *The Genus Bidens*,
https://doi.org/10.1007/978-981-96-4257-1_2

and mostly orange-yellow, and rarely white-purplish corolla is present. Stamens are usually five or numerous, exerted, anther has various lengths containing lobes, triangular-ovate connectives, and anther base may vary from long to short truncate. The style-arm apices are truncated to widely conic, with a few sweeping hairs at the tip. Genus *Bidens* has tricolporate or tetracolporate pollen grains with sizes ranging from 17 to 33 μm. Cypsela is the characteristic fruit of *Bidens* species, mostly flat, angled (3, 4), oblanceolate or obovate, and sometimes linear fusiform (*Bidens cernua*, *Bidens aristosa*, and *Bidens polylepis*), smooth, and hairy. Pappus with 2 to 4 bristles (barbed) stick with human clothes and animal fur.

Keywords

Bidens · Cypsela · Involucre · Hermaphrodite · Pappus

2.1 Introduction

German botanist Paul Dietrich Giseke originally named the family Asteraceae (Compositae) in 1792. Asteraceae is one of the biggest plant families, sometimes referred to as the daisy or sunflower family, having more than 1600 genera and 25,000 species. In 1976, it was further divided into two large subfamilies based on morphology, Asteroideae and Cichorioideae (Bremer and Jansen 1992). Its species are found throughout the world, except Antarctica and the deep Arctic, particularly in Central America, Eastern Brazil, the Mediterranean, the Levant, Southwest China, South Africa, and Central Asia. In addition to more tropical and subtropical areas, it has a range of medicinal plant species, including wormwood, dandelion, chamomile, sunflower, lettuce, chicory, coreopsis, dahlias, and daisy (Nikolić and Stevović 2015). A large number of Asteraceae members are a good source of food as well as medicine (Rahman et al. 2008). Asteraceae includes a small number of shrubs, vines, and trees in addition to certain annual and perennial herbs (Easmin et al. 2021). The family's worldwide distribution allows its species to live anywhere from the subpolar to tropical regions. In the dry and semiarid areas of subtropical and lower temperate latitudes, the species is most prevalent. In open, arid settings, asterace are very common (Judd et al. 1999). Asteraceae has great economic importance due to its valuable products, such as cooking oils, lettuce, artichokes, sweeteners, coffee alternatives, sunflower seeds, and herbal teas. *Bidens* is a widespread genus of family Asteraceae, with 230 species, and in North America, it is commonly known as beggar ticks or Spanish needles. In Africa and New, with annual rainfall falling of less than 0.3 mm and annual precipitation rate exceeding world, there is a great diversity of genus *Bidens* (Knope et al. 2020). Due to an annual precipitation rate of more than 7 mm and an annual rainfall falling of less than 0.3 mm, its habitat is semiarid to arid. *Bidens* species may grow well up to 2200 m above the sea level. Their species are inhabitants of a diversity of habitats; they have diverse morphology including leaf shape, texture, inflorescence, flower, and fruit type, and they are well adapted to various environmental challenges (Zahara et al. 2022).

2.2 Morphological Characteristics of Some Key Species of Genus *Bidens*

2.2.1 *Bidens aristosa*

Habitat: Grows in pine forests and in marshes.
Habit: Annual herb.
Stem: Herbaceous, 30–150 cm tall.
Leaves: Petiolate, glabrous, pinnately divided, 3–7 lobs, lamina 10–30 mm long, linear to lanceolate, margins laciniate to serrate, acuminate tip, cuneate base (Fig. 2.1) (Flora of North America Editorial Committee 1993).

2.2.2 *Bidens biternata*

Habitat: Gardens, villages, along roadsides, on banks of small streams, and cultivated areas.
Habit: Weed.
Vernacular: Phutium or Dipmal, Yinpan and Phutjom, Kerrai, Kathori, Kuro (Zahara et al. 2015).
Stem: Hairy, grooved, and quadrangular. The leaves have dentate edges and are acuminated, trifoliate, imparipinnate, and opposite (Bhatt et al. 2011; Sharma et al. 2009).

2.2.3 *Bidens bipinnata*

Habitat: Distributed in subtropical and temperate areas, endemic to China. It mostly grows in fields and forests and is well distributed on roadsides.
Habit: Annual, herbaceous weed.
Stem: Herbaceous, broad, flat, membranaceous, acuminate, hairy, and joined at the base, 40–85 cm tall (Fig. 2.2) (Hu et al. 2018; Wang et al. 2017).
Leaves: Petiolate (20–50 mm), leaf lamina is rounded, dentate, ovate or lanceolate, cuneate to truncate leaf base, entire or ciliolate (sometimes) leaf margin, rounded, acute or attenuate leaf tip, mostly glabrous or hirtellous (sometimes) faces (Wiegand 1899).

2.2.4 *Bidens cernua*

Habitat: Flood plains, sedge bogs, peat, and marshes.
Stem: Erect, glabrous, stramineous or reddish, and occasionally sparsely hispid, 20–70 cm tall, rarely a meter, short branches, seldom grow more than the subtending leaves and become shorter than the stem.

Fig. 2.1 *Bidens aristosa*, leaves, inflorescence, fruit. (Flora of North America Editorial Committee 1993)

Leaves: Narrow, long, sessile, lanceolate or linear-lanceolate, undivided, acuminate, margins are serrated, coarse, sharp, distant teeth and more or less connate by a broad base, slightly confined below the center (Fig. 2.3) (6–16 cm long) (Wiegand 1899).

2.2.5 *Bidens comosa*

Habitat: Marshes and other wet regions.

Stem: Stout, glabrous, and erect stem of 30–80 cm in height; internodes are stramineous or sometimes reddish; short spreading branches.

Fig. 2.2 *Bidens bipinnata*, branches, inflorescence, and achenes. (Flora of North America Editorial Committee 1993)

Leaves: Pale yellow (8 cm long), undivided, narrowly elliptic or elliptic-lanceolate, apex is acute, attenuate, margined petiole, serrate or toothed regular, and sometimes entire margin (Wiegand 1899).

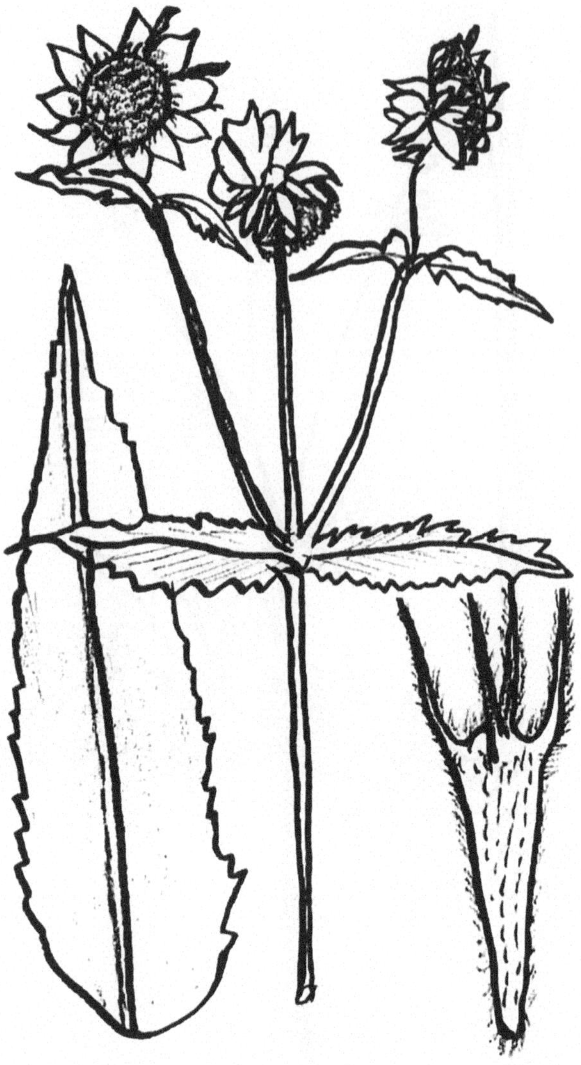

Fig. 2.3 *Bidens cernua*, leaves, inflorescence, fruit. (Flora of North America Editorial Committee 1993)

2.2.6 *Bidens decipiens*

Habitat: Distributed in temperate biome.

Stem: Tall, glabrous, and purplish and moderately branched stem with 5–14 mm high, log internodes, spreading branches.

Leaves: Bright-green, petiolate, undivided, elliptical, elliptic-lanceolate, or lanceolate, large (12–35 mm × 70–130 mm) at base deeply parted or tapering base, acuminate, margins are coarsely and serrated, sharp teeth (Wiegand 1899).

2.2.7 *Bidens dentate*

Habitat: Grows in marshy places.

Stem: Glabrous, sparsely hairy, stout (40–100 cm), and short ascending branches. **Leaves**: Pale, large, broad, oval, lanceolate, oblanceolate leaves with 8 cm length, leaf tip is acute or short (Flora of North America Editorial Committee 1993), and upper leaves are larger than lower (Wiegand 1899).

2.2.8 *Bidens discoidea*

Habitat: Widely distributed in wet sites like ponds, swamps, and marshy places.

Stem: Annual herb, soft, branched, 10–180 cm tall.

Leaves: Petiolate, 10–60 mm tall, lamina delicate to lance-ovate, mostly 3-foliate, petiolate leaflets, serrate margin, glabrous, attenuate to acuminate, cuneate base (Flora of North America Editorial Committee 1993).

2.2.9 *Bidens eatonii*

Habitat: Grows in tidal flats and marshes.

Habit: Annual herb.

Stem: Soft, herbaceous, 25–150 cm tall.

Leaves: Petiolate, narrow, 10–35 mm tall, lobed near base, serrate margin, cuneate base, glabrous, leaf tip attenuate (Flora of North America Editorial Committee 1993).

2.2.10 *Bidens forbesii*

Habitat: Woody, perennial.

Stem: Horizontal, ascending lateral branches, inconspicuously quadrangular, 0.75–1.5 m tall.

Leaves: 3–7 foliolate or pinnately compound, lanceolate to ovate-lanceolate, dark green, coriaceous, serrate, glabrous, mostly 50–120 mm long, and 20–45 mm broad (Fig. 2.4) (Ganders and Nagata 1983).

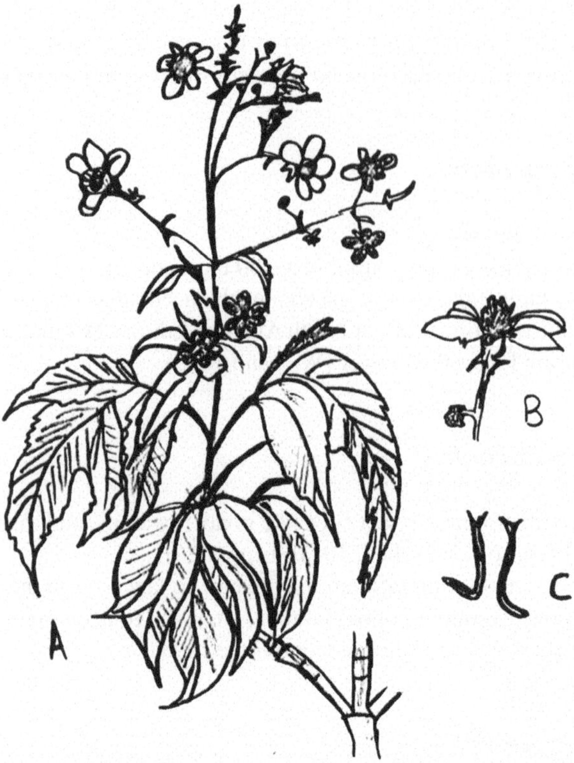

Fig. 2.4 *Bidens forbesii*: (*A*) inflorescence on a lateral branch, (*B*) flowering head showing outer involucral bracts, and (*C*) achenes. (Ganders and Nagata 1983)

2.2.11 *Bidens frondosa*

Habitat: Distributed in ponds, sloughs, ditches, swamps, meadows, roadsides, and railroads.

Habit: Annual, herb.

Stem: Herbaceous, glabrous, erect, branched, 0.5–1.5 or (2.0) m tall.

Leaves: Petiolate, opposite, petiolate, pinnate, one or two pairs of leaflets, acute or acuminate, lanceolate or oblong-lanceolate, coarsely serrate, light-green below and deep dull-green above, 100 mm long, and 40 mm wide (Fig. 2.5) (Šumberová et al. 2004).

2.2.12 *Bidens hirsute*

Habitat: Mesic and wet forests.

Stem: Herbaceous, branched, 2–3 feet tall, sometimes hirsutely pubescent, quadrangular, and grooved.

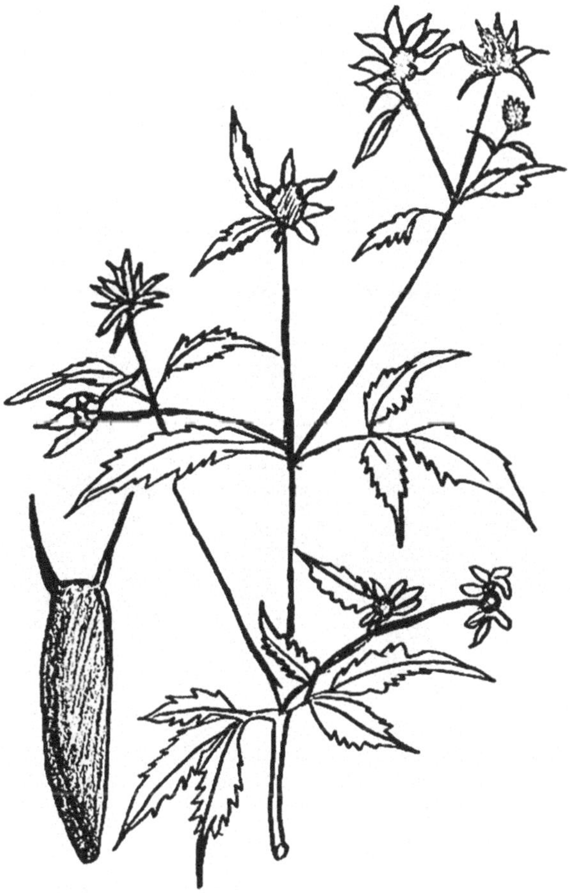

Fig. 2.5 *Bidens frondosa*: Inflorescence on the lateral branch and Achenes. (Flora of North America Editorial Committee 1993)

Leaves: Opposite, pseudopinnate leaves, occasionally ternate, lanceolate, acuminate, sharply serrated leaflets, obliquely attenuated base (Nuttall 1841).

2.2.13 *Bidens hyperborea*

Habitat: Distributed at the border of estuaries and marshes.
Stem: Soft, herbaceous, broad, flat, membranaceous, 5–70 cm tall.
Leaves: Sessile or petiolate, lamina linear to lanceolate, entire or serrate margin, rounded, attenuate or obtuse tip, glabrous face, base cuneate (Flora of North America Editorial Committee 1993).

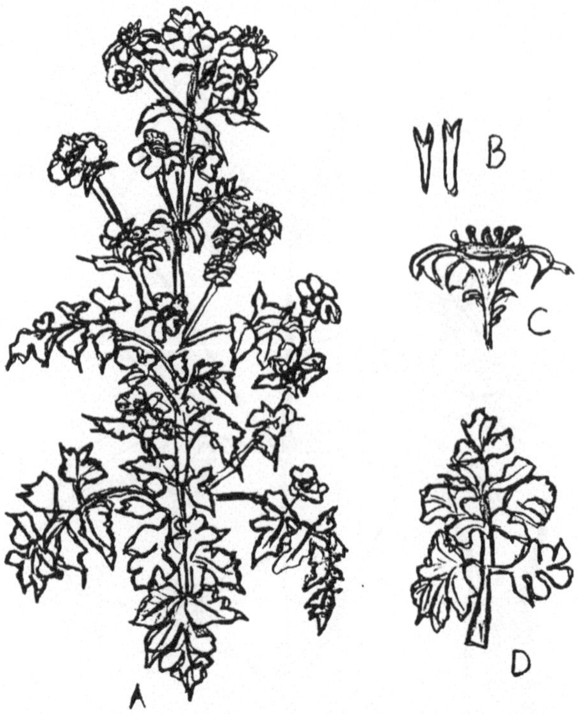

Fig. 2.6 *Bidens hillebrandiana*: (*A*) terminal inflorescence, (*B*) achenes, (*C*) flowering head showing outer involucral bracts, and (*D*) stem leaf. (Ganders and Nagata 1983)

2.2.14 *Bidens hillebrandiana*

Habitat: Widespread in the island of Hawaii, windward eastern coast of Kohala near the northern tip of the island.

Stem: Perennial, decumbent, base woody, branched, less than 1 m tall (Fig. 2.6).

Leaves: Simple or compound, fleshy, pinnately or bipinnately divided, leaf length 80–110 mm long (petiole and lamina both), leaflets 2–15 mm wide, 5–25 mm long, margins crenate, lobed or toothed, leaf lamina glabrous, or meagerly pubescent (Ganders and Nagata 1983).

2.2.15 *Bidens laevis*

Habitat: Marshes, estuaries, and pool margins.

Stem: Erect, tinged with purple, glabrous, half to one-meter length and branched.

Leaves: Simple, glabrous, scarcely paler beneath, lanceolate or more commonly elliptic-lanceolate, contracted sessile base, sometimes clasping, acute or short acuminate, margins being serrated, inconspicuous teeth, rarely connate, small or medium size (7–13 cm) (Wiegand 1899).

2.2.16 *Bidens mutica*

Habitat: Grow on Hawaiian Islands.
Stem: Smooth, herbaceous.
Leaves: Complex or simple, pinnate or ternately divided, ovate-lanceolate, acuminate, serrated (Nuttall 1841).

2.2.17 *Bidens melanocarpa*

Habitat: Wildly distributed across much of the United States, Mexico, and Canada.
Stem: Calendrical, bushy-branched, spread branches, leaves: petiolate, tapering long stalk, lanceolate to oblong-lanceolate, acuminate glabrous.
Leaves: 3–5 in number, pinnately divided, scabrous, and scarce margins (3–4 cm long), 4–8 cm terminal leaflets. Long, tapering into a long stalk, sometimes trilobed at the base, serrate, or coarse teeth margin; lateral leaflets are short and more truncate, less acuminate, and very oblique at the base (Wiegand 1899).

2.2.18 *Bidens micrantha*

Habitat: Distributed in dry, sheer rocks, subalpine and alpine shrubland, mesic and wet forest.
Stem: Erect, perennial herb, ascending branches that flower both terminally and on lateral branches; 0.5–1.5 m tall, stems are not distinctly quadrangular (Fig. 2.7).
Leaves: Compound, pinnately divided, rarely simple, 3–7 leaflets, lanceolate, serrate, ciliate margin, 25–100 mm tall, and 5–30 mm broad (Nuttall 1841).

2.2.19 *Bidens nashii*

Habitat: Distributed in pools and stream margins.
Stem: succulent, 30–80 cm tall, glabrous, erect, branched, long intermodal distance.
Leaves: 5–12 cm thick leaves, oblong-lanceolate, acute to acuminate leaf tip, contracted, sessile, base scarcely connate, lamina broad, serrate, toothed, or entire margin (Wiegand 1899).

2.2.20 *Bidens roraimensis*

Habitat: Distributed in fields, and on roadsides.
Stem: Erect, herbaceous, tiny size, slender to filiform, small, ascending, vittate, scaly, caducous, thicker.
Leaves: Short-incised or variably lobed at the apex (Wiegand 1899).

Fig. 2.7 *Bidens micrantha*: Portion of flowering plant, stem leaf, flowering head showing outer involucral bracts, and achene. (Ganders and Nagata 1983)

2.2.21 *Bidens phyllobola* Spruce

Habitat: Well distributed on roadsides.

Stem: Erect, brownish, 6 cm long, branched.

Leaves: Subimbricate, bifid, ascendant, 0.5 mm broad at the base, 0.75–1.5 mm long, oval, vittate, rounded-entire or acute tip, lunulate sinus, small teeth, margins crenulate, serrate (Wiegand 1899).

2.2.22 *Bidens pilosa*

Habitat: Distributed in temperate, tropical regions.

Vernacular: Broomsticks, pitchforks, Spanish needles, beggar's ticks, devil's needles, and cobbler's pegs (Bartolome et al. 2013).

Stem: Perennial weed, herb, erect, 150 cm tall.

Leaves: Broad, petiolate, alternate or opposite, green, trifoliate, dissected, serrate, glabrous or hairy, exstipulate, lanceolate, or ovate (Food and Nutrition Division 1997; Rokaya et al. 2012).

Fig. 2.8 *Bidens sandvicensis*: (*A*) terminal inflorescence, (*B*) flowering head showing outer involucral bracts, and (*C*) achene (Ganders and Nagata 1983)

2.2.23 *Bidens sandvicensis*

Habitat: Distributed in Hawaiian Islands, shrubland, and mesic forest.

Stem: Erect, perennial, branched herb, 0.5–1.0 m height, lateral branches horizontally ascending to near ground (Fig. 2.8).

Leaves: Petiolate, glabrous or globate, compound, pinnately or occasionally bipinnately divided, 50–160 mm leave length, 5–9 leaflets <8 mm width, and 15–80 mm height (Rodin et al. 2024).

2.2.24 *Bidens tenuisecta*

Habitat: Grow along streams and meadows.

Habit: Annual herb.

Stem: Herbaceous, 20–40 cm tall.

Leaves: Petiolate, lamina 5–40 mm long, ovate to rounded lamina, pinnatisect, linear, glabrous, leaf tip acute to attenuate, entire margin (Flora of North America Editorial Committee 1993).

References

Bartolome AP, Villaseñor IM, Yang W-C (2013) Bidens pilosa L. (Asteraceae): botanical properties, traditional uses, phytochemistry, and pharmacology. Evid Based Complement Alternat Med 2013(1):340215

Bhatt J, Singh JS, Singh S, Tripathi R, Kohli R (2011) Invasive alien plants an ecological appraisal for the Indian subcontinent. CABI, Wallingford

Bremer K, Jansen RK (1992) A new subfamily of the Asteraceae. Ann Mo Bot Gard 79(2):414–415

Easmin F, Al Faria L, Rani R, Rahman AM (2021) Asteraceae: a taxonomically and medicinally important sunflower family. Am Int J Biol Life Sci 3(1):1–17

Flora of North America Editorial Committee (1993) Flora of North America: volume 2: pteridophytes and gymnosperms. Oxford University Press, Oxford

Food and Nutrition Division (1997) Agriculture, food and nutrition for Africa: a resource book for teachers of agriculture. FAO Food and Nutrition Division, Rome

Ganders FR, Nagata KM (1983) New taxa and new combinations in Hawaiian Bidens (Asteraceae). Ganders 2(1):1–16

Judd WS, Campbell CS, Kellogg EA, Stevens PF, Donoghue MJ (1999) Plant systematics: a phylogenetic approach. Ecología mediterránea 25(2):215

Hu HM, Bai SM, Chen LJ, Hu WY, Chen G (2018) Chemical constituents from Bidens bipinnata Linn. Biochemical Systematics and Ecology, 79, 44–49

Knope ML, Funk VA, Johnson MA, Wagner WL, Datlof EM, Johnson G et al (2020) Dispersal and adaptive radiation of Bidens (Compositae) across the remote archipelagoes of Polynesia. J Syst Evol 58(6):805–822

Nikolić M, Stevović S (2015) Family Asteraceae as a sustainable planning tool in phytoremediation and its relevance in urban areas. Urban For Urban Green 14(4):782–789

Nuttall T (1841) Descriptions of new species and genera of plants in the natural order of the Compositae, collected in a tour across the continent to the Pacific, a residence in Oregon, and a visit to the Sandwich Islands and Upper California, during the years 1834 and 1835. Trans Am Philos Soc VII:283–453

Rahman A, Alam M, Khan S, Ahmed F, Islam A, Rahman MM (2008) Taxonomic studies on the family Asteraceae (Compositae) of the Rajshahi division. Res J Agric Biol Sci 4(2):134–140

Rodin MN, Baeva VM, Bokov DO, Voronov VM, Samylina IA (2024) Comparative study of morphological and anatomical features of Bidens L. species. Farmaciya (Pharmacy) 73(2):25–33

Rokaya MB, Münzbergová Z, Timsina B, Bhattarai KR, Rheum australe D. (2012) Don: a review of its botany, ethnobotany, phytochemistry, and pharmacology. J Ethnopharmacol 141(3):761–774

Sharma A, Bargali K, Pande N (2009) The allelopathic potential of bryophyte extract on seed germination and seedling growth of Bidens biternata. Nat Sci 7:30–38

Singh, R., Singh, G., Tiwari, A. K., Patel, S., Agrawal, R., Sharma, A., & Singh, B. B. Diversity of Host Plants of Aphids (Homoptera: Aphididae) Infesting Asteraceae in India. International Journal of Zoological Investigations. s Vol. 1, No. 2, 137-167 (2015)

Šumberová K, Tzonev R, Vladimirov V (2004) Bidens frondosa (Asteraceae) – a new alien species for the Bulgarian flora. Phytol Balcan 10(2–3):179–181

Tereza G-GM, Mansanares ME, Semir J, Solferini VN. Chromosomal studies of three species of Bidens (L.) (Asteraceae). Caryologia 2006;59(1):14–18

Wang R, Yan W, Quan G, Liu S, Zhang J (2017) Effects of light intensity on morphology and physiology of exotic invasive Bidens pilosa L. and non-invasive congener Bidens bipinnata L. Allelopath J 42:157–168

Wiegand K (1899) Some species of Bidens found in the United States and Canada. Bull Torrey Bot Club 26(8):399–422

Zahara K, Bibi Y, Tabassum S, Bashir T, Haider S, Araa A et al (2015) A review on pharmacological properties of Bidens biternata: a potential nutraceutical. Asian Pac J Trop Dis 5(8):595–599

Zahara K, Bibi Y, Arshad M, Kaukab G, Al Ayoubi S, Qayyum A (2022) In-vitro examination and isolation of antidiarrheal compounds using five bacterial strains from invasive species Bidens bipinnata L. Saudi J Biol Sci 29(1):472–479

Morphological Characteristics of *Bidens* Species: Generative Morphology

3

Huma Qureshi and Tauseef Anwar

Abstract

Bidens, a diverse genus within the Asteraceae family, exhibits significant variation in floral morphology and reproductive strategies. Through systematic examination and scientific analysis, the chapter aims to elucidate *Bidens* reproductive organs' intricate features, including flowers, inflorescences, and fruits. *Bidens* flowers are characterized by their daisy-like appearance, with composite heads comprised of disc and ray florets arranged in distinct patterns. The chapter examines the morphology and anatomy of these florets, emphasizing their adaptations for efficient pollination and seed production. It discusses the diverse pollination mechanisms in *Bidens*, including self-pollination and various forms of insect-mediated pollination, and their implications for reproductive success and genetic diversity. Special attention is given to the structural features of achenes, such as bristles and hooks, facilitating dispersal over long distances. Furthermore, the role of fruit morphology in seed dormancy and germination is examined, providing insights into *Bidens* species' reproductive ecology and evolutionary dynamics. By synthesizing current research findings and rigorous morphological observations, this chapter advances our understanding of *Bidens* reproductive biology and evolutionary adaptations, highlighting the importance of morphological studies in elucidating the ecological and evolutionary significance of plant reproductive structures.

H. Qureshi (✉)
Department of Botany, University of Chakwal, Chakwal, Pakistan
e-mail: huma.qureshi@uoc.edu.pk

T. Anwar
Department of Botany, The Islamia University of Bahawalpur, Bahawalpur, Pakistan

Y. Bibi et al. (eds.), *The Genus Bidens*,
https://doi.org/10.1007/978-981-96-4257-1_3

Keywords

Bidens species · Floral morphology · Reproductive strategies · Pollination mechanisms · Fruit development

3.1 Introduction

Bidens is a genus within the Asteraceae family, consisting of approximately 230–240 species. This genus is widely distributed across tropical and temperate regions, notable for its diverse morphology and ecological adaptability (Table 3.1) (Bartolome et al. 2013). The morphological study of *Bidens* species, particularly their generative (reproductive) structures, is essential for taxonomic classification, ecological research, and understanding of evolutionary relationships within the genus. Generative morphology encompasses the reproductive organs and structures, such as flowers, inflorescences, seeds, and fruits. In *Bidens* species, these characteristics exhibit considerable variation, contributing to their adaptability and success in various habitats. This chapter provides an in-depth analysis of the generative morphological characteristics of *Bidens* species, highlighting the distinct features and variations that aid in their identification and classification. A detailed examination of these reproductive structures can reveal significant insights into the evolutionary pathways and ecological strategies employed by different *Bidens* species. By

Table 3.1 Diversity of *Bidens* species

Species name	Common name	Geographic distribution	Habitat types	Notable morphological traits
Bidens pilosa	Black jack	Tropical and subtropical regions	Roadsides, fields	Small, white ray florets, barbed achenes
Bidens alba	Shepherd's needles	Southeastern USA, Central America	Grasslands, gardens	White ray florets, yellow disc florets
Bidens tripartita	Trifid bur-marigold	Europe, Asia, North America	Wetlands, marshes	Deeply divided leaves, barbed achenes
Bidens cernua	Nodding beggarticks	North America, Europe	Wetlands, marshes	Yellow ray florets, nodding heads
Bidens aristosa	Bearded beggarticks	Eastern North America	Wet meadows, ditches	Long, narrow leaves, bristly achenes
Bidens connata	Purple-stem beggarticks	North America, Central America	Wetlands, ponds	Purple stems, serrated leaves
Bidens frondosa	Devil's beggarticks	North America, Europe	Wetlands, fields	Opposite leaves, bristly achenes
Bidens laevis	Smooth beggarticks	North America, South America	Wetlands, marshes	Smooth stems, bright yellow flowers
Bidens beckii	Water marigold	North America	Freshwater habitats	Submerged leaves, yellow flowers
Bidens bipinnata	Spanish needles	Worldwide	Disturbed areas, gardens	Bipinnate leaves, white or yellow flowers

understanding the generative morphology, researchers can better grasp how these species adapt to diverse environmental conditions and how they interact with their ecosystems. This knowledge is crucial for conservation efforts, particularly in maintaining biodiversity and supporting sustainable ecosystems. Additionally, the study of generative morphology in *Bidens* species can inform breeding programs and the development of species with desirable traits for horticultural and agricultural purposes. By exploring the intricate details of flowers, seeds, and fruits, we can uncover the mechanisms behind their reproductive success and resilience, paving the way for future research and practical applications.

The inflorescence in *Bidens* is typically a capitulum (head), a hallmark of the Asteraceae family. The capitulum consists of a central disc with numerous small flowers (florets) surrounded by involucral bracts. The bracts are arranged in one or more series and can vary significantly in shape, size, and color among different species. They often serve to protect the developing florets and aid in attracting pollinators. The peduncle (stalk of the inflorescence) can be short or long, sometimes contributing to the identification of species within the genus. *Bidens* species exhibit dimorphic florets: disc florets and ray florets. Disc florets are tubular, actinomorphic (radially symmetrical), and usually hermaphroditic, containing both male and female reproductive organs. The corolla is typically five-lobed, with stamens united by their anthers (syngenesious) and a bifurcated style. Ray florets are zygomorphic (bilaterally symmetrical) and pistillate or neuter, often functioning primarily to attract pollinators. The corolla of ray florets is ligulate, resembling a petal, and can be white, yellow, or orange, depending on the species. Table 3.2 shows a morphological comparison of some *Bidens* species (Grombone-Guaratini et al. 2004), (Budumajji and Solomon Raju 2018).

3.2 Floral Morphogenesis and Ontogeny

Floral morphogenesis and ontogeny encompass the developmental processes that lead to the formation of flowers in *Bidens* species. Understanding these processes provides critical insights into the genetic, hormonal, and environmental factors that shape floral architecture. This section explores the stages of flower development, from floral induction to mature flower formation, highlighting the key morphogenetic and ontogenetic events in *Bidens* species.

3.2.1 Environmental and Genetic Triggers

Floral induction is the initial phase of flower development, where vegetative meristems are converted into reproductive meristems. In *Bidens* species, this process is influenced by a combination of environmental factors such as photoperiod, temperature, and hormonal signals, as well as genetic regulation. Photoperiod plays a crucial role as many *Bidens* species are photoperiodic, meaning that the length of day and night influences flowering, with short-day or long-day conditions triggering

Table 3.2 Comparison of floral morphology in *Bidens* species

Species name	Disc florets characteristics	Ray florets characteristics	Floret color	Floret shape	Floret size (disc)
Bidens pilosa	Tubular, perfect	Ligulate, sterile	White, yellow	Cylindrical	3–5 mm
Bidens alba	Tubular, perfect	Ligulate, sterile	White, yellow	Ligulate	5–8 mm
Bidens tripartita	Tubular, perfect	Absent	Yellow	Cylindrical	4–6 mm
Bidens cernua	Tubular, perfect	Ligulate, fertile	Yellow	Ligulate	5–7 mm
Bidens aristosa	Tubular, perfect	Ligulate, sterile	Yellow	Ligulate	4–6 mm
Bidens connata	Tubular, perfect	Ligulate, sterile	Yellow	Ligulate	4–6 mm
Bidens frondosa	Tubular, perfect	Ligulate, sterile	Yellow	Ligulate	3–5 mm
Bidens laevis	Tubular, perfect	Ligulate, fertile	Yellow	Ligulate	6–8 mm
Bidens beckii	Tubular, perfect	Ligulate, sterile	Yellow	Ligulate	4–6 mm
Bidens bipinnata	Tubular, perfect	Ligulate, sterile	White, yellow	Ligulate	5–7 mm

the transition from vegetative to reproductive growth. Temperature also affects floral induction, with vernalization, the requirement for a period of cold temperatures, being necessary for some species. Additionally, hormonal signals are integral to the regulation of floral induction. Hormones such as gibberellins, cytokinins, and auxins play key roles, with gibberellins particularly promoting flowering in some *Bidens* species (Chailakhyan and Khrianin 2012; Gaspar et al. 2002; Yamasaki et al. 2005) (Table 3.3).

3.2.2 Genetic Regulation

The genetic regulation of floral induction involves a network of floral identity genes. Key genes such as *LEAFY* (*LFY*), *APETALA1* (*AP1*), and *FLOWERING LOCUS T* (*FT*) are central to this process. These genes interact to initiate and control the development of floral meristems.

Following floral induction, the vegetative meristem transitions into a floral meristem, which gives rise to all the floral organs (sepals, petals, stamens, and carpels) through a series of well-coordinated developmental stages. Key meristem identity genes play a crucial role in this process. *LEAFY* (*LFY*) is a key regulator that promotes the transition from vegetative to floral meristem. *APETALA1* (*AP1*) and *CAULIFLOWER* (*CAL*) act downstream of *LFY* to establish floral meristem identity.

Table 3.3 Genetic and environmental factors influencing floral induction

Factor	Description	Key species
Photoperiod	Length of day/night affecting flowering	*Bidens pilosa*
Temperature	Optimal temperatures for floral induction	*Bidens alba*
Vernalization	Cold treatment is required for flowering	*Bidens tripartita*
LFY (*LEAFY*) gene	Floral meristem identity gene	*Bidens pilosa*
AP1 (*APETALA1*) gene	Controls floral organ development	*Bidens alba*
FT (*FLOWERING LOCUS T*) gene	Integrator of flowering signals	*Bidens tripartita*
GA (gibberellic acid)	Hormones affecting flowering time	*Bidens cernua*
CO (*CONSTANS*) gene	Regulates photoperiodic flowering	*Bidens aristos*
SOC1 (*SUPPRESSOR OF OVEREXPRESSION OF CONSTANS 1*) gene	Integrates signals for floral transition	*Bidens frondosa*
AGL24 (AGAMOUS-LIKE 24) gene	Regulates flowering time and development	*Bidens laevis*

Floral organogenesis, the process by which floral organs are formed from the floral meristem, involves the differentiation and growth of sepals, petals, stamens, and carpels. The formation of these floral organs is governed by the ABC model of floral development, which categorizes genes into three classes: A, B, and C. Class A genes, such as *APETALA1*, determine the identity of sepals and petals. Class B genes, including *APETALA3* and *PISTILLATA*, work in combination with A and C genes to specify petal and stamen identity. Class C genes, such as *AGAMOUS*, control the development of stamens and carpels.

3.2.3 Stages of Floral Development

Floral development in *Bidens* can be divided into distinct stages:

Floral induction: Transition from vegetative to reproductive meristem
Floral meristem formation: Establishment of the floral meristem
Floral organ initiation: Differentiation of sepals, petals, stamens, and carpels
Floral organ development: Growth and maturation of floral organs

Sepals are typically the first floral organs to form, arising from the outermost whorl of the floral meristem. They serve to protect the developing inner floral organs. Petals form next from the second whorl, guided by the interaction of Class A and B genes. These petals often exhibit bright colors and patterns to attract pollinators. Stamens, the male reproductive organs, develop from the third whorl under the control of Class B and C genes. Each stamen consists of a filament and an anther, where pollen is produced. Carpels, the female reproductive organs, arise from the innermost whorl and are regulated by Class C genes. The carpels fuse to form the ovary, style, and stigma, which are critical for seed development.

3.2.4 Ontogeny of Bidens Flowers

In the early stages of *Bidens* flower development, the floral meristem undergoes rapid cell division and differentiation. The floral organs begin to emerge sequentially from the periphery towards the center of the meristem. During mid-development, these floral organs elongate and assume their characteristic shapes and structures. This process involves tightly regulated cell proliferation, expansion, and differentiation, guided by genetic and hormonal signals. In the final stages, the floral organs mature, achieving their full size and functionality. This maturation includes the development of reproductive structures such as pollen grains in anthers and ovules in carpels.

3.2.4.1 Developmental Plasticity and Environmental Influence

Bidens species exhibit developmental plasticity, allowing them to adapt to varying environmental conditions. Factors such as nutrient availability, water stress, and photoperiod can influence the rate and success of floral development.

3.2.4.2 Evolutionary Perspectives

The diverse floral morphologies observed in *Bidens* species reflect their evolutionary adaptations to different pollination syndromes and ecological niches. Variations in floral ontogeny contribute to reproductive isolation and speciation within the genus (Helenurm and Ganders 1985; Ritland and Ganders 1985; Sun and Ganders 1988).

3.3 Reproductive Strategies in Asteraceae: Insights from *Bidens*

Within Asteraceae, the genus *Bidens* is notable for its widespread distribution and diverse reproductive strategies. Understanding these strategies provides insights into the ecological success and adaptability of *Bidens* species. This section explores the reproductive strategies in Asteraceae, with a focus on *Bidens*, highlighting their floral structures, pollination mechanisms, seed dispersal methods, and ecological adaptations.

3.3.1 Floral Structures and Reproductive Units

The capitulum, or flower head, is the characteristic inflorescence of the Asteraceae family, including *Bidens*. This compact structure consists of numerous small flowers (florets) arranged on a common receptacle.

3.3.1.1 Ray Florets

Often sterile or female, these zygomorphic (bilaterally symmetrical) florets are located on the periphery and function primarily to attract pollinators.

3.3.1.2 Disc Florets

Centrally located, these actinomorphic (radially symmetrical) florets are usually hermaphroditic, containing both male and female reproductive organs.

The specialization of florets within the capitulum enables efficient pollination and reproductive success. The diversity in floral morphology, color, and nectar production among *Bidens* species enhances their attractiveness to a variety of pollinators.

3.3.2 Pollination Mechanisms

Bidens species exhibit a range of pollination strategies that reflect their ecological versatility and adaptability (Table 3.4).

3.3.2.1 Entomophily (Insect Pollination)

Insect pollination is the predominant strategy in *Bidens* species, facilitated by various floral traits. Brightly colored ray florets serve as visual attractants for pollinators such as bees, butterflies, and beetles. Nectar production acts as a reward, enticing

Table 3.4 Summary of pollination mechanisms in *Bidens*

Pollination mechanism	Associated floral traits	Examples of *Bidens* species
Entomophily (insect)	Bright colors, nectar, scent	*Bidens pilosa*
Autogamy (self)	Small, inconspicuous flowers	*Bidens alba*
Anemophily (wind)	Reduced petals, large quantities of pollen	*Bidens tripartita*
Hydrophily (water)	Floating flowers, waterborne pollen	*Bidens cernua*
Ornithophily (birds)	Tubular flowers, bright colors, no scent	*Bidens aristosa*
Chiropterophily (bats)	Nocturnal opening, strong scents	*Bidens connata*
Cantharophily (beetles)	Robust flowers, strong scents	*Bidens frondosa*
Myophily (flies)	Small, unscented, exposed nectar	*Bidens laevis*
Psychophily (butterflies)	Brightly colored, fragrant flowers	*Bidens beckii*
Melittophily (bees)	Yellow/blue flowers, nectar guides	*Bidens bipinnata*

pollinators to visit the flowers and ensure pollen transfer. Additionally, the structure of disc florets and the timing of pollen release are adapted to maximize pollen transfer by insect visitors (Kevan and Baker 1983).

3.3.2.2 Self-Pollination (Autogamy)

While cross-pollination is common, some *Bidens* species exhibit self-pollination as a reproductive assurance mechanism, especially in environments where pollinators are scarce. Cleistogamy, the production of closed, self-pollinating flowers that do not open, ensures seed production under adverse conditions.

3.3.2.3 Wind Pollination (Anemophily)

Although less common in *Bidens*, wind pollination can occur, particularly in species with reduced floral structures that do not rely on visual or nectar rewards to attract pollinators.

3.3.3 Seed and Fruit Dispersal Mechanisms

The dispersal of seeds and fruits is a critical component of the reproductive strategy in *Bidens*, contributing to their wide distribution and colonization of diverse habitats (Spjut 1994) (Table 3.5).

Table 3.5 Achene dispersal mechanisms in *Bidens*

Dispersal mechanism	Morphological adaptations of achenes	Examples of species	References
Epizoochory	Barbed achenes that attach to animals	*Bidens pilosa*	Sánchez et al. (2023)
Hydrochory	Buoyant achenes with air pockets	*Bidens tripartita*	Wagatsuma et al. (2021)
Anemochory	Lightweight achenes with pappus for wind dispersal	*Bidens alba*	Kompała-Bąba et al. (2023)
Autochory	Explosive mechanisms to release seeds	*Bidens cernua*	Hall and Nguyen (2021)
Endozoochory	Achenes eaten by animals and excreted elsewhere	*Bidens connata*	Smith et al. (2020)
Myrmecochory	Achenes with elaiosomes attracting ants	*Bidens aristosa*	Wagatsuma et al. (2021)
Synzoochory	Achenes collected by animals and stored	*Bidens frondosa*	Sánchez et al. (2023)
Barochory	Gravity dispersal, heavy achenes	*Bidens laevis*	Smith et al. (2020)
Ornithochory	Achenes dispersed by birds	*Bidens beckii*	Bargielski (2020)
Ballochory	Achenes mechanically ejected	*Bidens bipinnata*	Douglas et al. (2021)

3.3.3.1 Epizoochory (Animal Dispersal)

Many *Bidens* species have developed specialized structures for epizoochorous dispersal.

Barbed Achenes

The pappus bristles or awns are often barbed, enabling the achenes to attach to the fur or feathers of animals. This facilitates long-distance dispersal as animals move through the environment.

3.3.3.2 Hydrochory (Water Dispersal)

Adaptation to wetland habitats has led to the development of hydrochorous dispersal mechanisms in some *Bidens* species.

Buoyant Achenes

These achenes can float on water, allowing them to be dispersed by rivers, streams, and other bodies of water. This adaptation is particularly beneficial for colonizing riparian and aquatic environments.

3.3.3.3 Autocory (Self-Dispersal)

In some cases, *Bidens* species exhibit self-dispersal mechanisms.

Explosive Dehiscence

Certain species have fruits that burst open, propelling seeds away from the parent plant to reduce competition and promote colonization of new areas.

3.3.4 Ecological Adaptations

The reproductive strategies of *Bidens* species are closely linked to their ecological adaptations, enabling them to thrive in a wide range of environments.

3.3.4.1 Habitat Specificity

Bidens species occupy diverse habitats, from wetlands and riparian zones to grasslands and disturbed sites. Their reproductive strategies are often fine-tuned to these specific environments.

Wetland Adaptations

Species in aquatic and semi-aquatic habitats tend to have buoyant seeds for water dispersal and mechanisms to tolerate flooding.

Disturbed Sites

Species in disturbed habitats may exhibit rapid growth and prolific seed production, enabling them to quickly colonize and dominate these areas.

3.3.5 Genetic Diversity and Adaptation

The genetic diversity within *Bidens* species is enhanced by their varied reproductive strategies, promoting adaptability and resilience.

3.3.5.1 Hybridization

Hybridization between species can occur, leading to the emergence of new genetic combinations and potentially new species. This contributes to the evolutionary flexibility of the genus.

3.3.5.2 Polyploidy

Some *Bidens* species exhibit polyploidy, a condition where the organism has more than two sets of chromosomes. This can result in increased genetic variation and adaptability.

3.3.5.3 Reproductive Isolation and Speciation

The reproductive strategies of *Bidens* species play a crucial role in reproductive isolation and speciation.

Prezygotic Isolation

Temporal Isolation

Differences in flowering times can prevent cross-pollination between species.

Spatial Isolation

Habitat preferences and geographic barriers can limit gene flow between populations.

Postzygotic Isolation

Hybrid Viability

Hybrids may have reduced fitness, limiting the success of interspecific crosses and promoting the maintenance of distinct species.

3.4 Pollination Ecology of *Bidens*: Mechanisms and Adaptations

The pollination ecology of *Bidens* species is a critical aspect of their reproductive success and ecological adaptation. Understanding the mechanisms and adaptations involved in pollination provides insights into the evolutionary strategies that have enabled these species to thrive in diverse environments. This section explores the various pollination mechanisms in *Bidens*, the floral adaptations that facilitate pollination, and the ecological interactions between *Bidens* species and their pollinators.

3.4.1 Pollination Mechanisms in Bidens

3.4.1.1 Entomophily (Insect Pollination)

Insect pollination is the primary pollination mechanism in *Bidens* species. The structure and function of the capitulum are highly adapted to attract and utilize insects for pollen transfer.

Visual Attraction

Brightly colored ray florets act as visual signals to attract insect pollinators such as bees, butterflies, and beetles.

Nectar Rewards

Many *Bidens* species produce nectar, which serves as a reward for pollinators. The presence of nectar guides and markings that direct pollinators to the nectar source enhances the efficiency of pollination.

3.4.1.2 Autogamy (Self-Pollination)

Self-pollination is a supplementary mechanism in *Bidens*, ensuring reproductive success when pollinators are scarce.

Cleistogamous Flowers

Some *Bidens* species produce cleistogamous flowers that remain closed, facilitating self-pollination without the need for external pollinators. This mechanism is particularly advantageous in stable or adverse environments where pollinator visits are unpredictable.

3.4.1.3 Anemophily (Wind Pollination)

Although less common, some *Bidens* species may utilize wind pollination, particularly in open or disturbed habitats where insect activity is low.

Reduced Floral Structures

Species that rely on wind pollination often exhibit reduced floral structures, with exposed stamens and pistils to facilitate pollen release and capture by the wind.

3.4.2 Floral Adaptations for Pollination

The floral morphology of *Bidens* species is highly specialized to enhance pollination efficiency. These adaptations can be broadly categorized into structural and functional adaptations.

3.4.2.1 Structural Adaptations

Capitulum Structure

The compact arrangement of florets in the capitulum maximizes the surface area available for pollinator visits. The differentiation between ray and disc florets allows for specialized functions within the inflorescence.

Ray Florets

Typically, sterile or pistillate, ray florets are brightly colored and larger than disc florets, serving primarily to attract pollinators.

Disc Florets

These hermaphroditic florets are the primary reproductive units, containing both male (stamens) and female (pistils) organs. The positioning of disc florets ensures that pollinators come into contact with both pollen and stigma during visits.

3.4.2.2 Functional Adaptations

Nectar Guides

These are visual patterns on the ray florets that guide pollinators to the nectar source, ensuring effective pollen transfer.

Temporal Separation

Temporal separation of male and female phases within the same flower (protandry or protogyny) reduces self-pollination and promotes cross-pollination.

Scent Production

Some *Bidens* species produce floral scents that attract specific pollinators, enhancing pollination efficiency.

3.4.2.3 Ecological Interactions with Pollinators

The ecological interactions between *Bidens* species and their pollinators are diverse and complex, involving mutualistic relationships that benefit both the plants and their pollinators.

Bees

Bees are the most common pollinators of *Bidens* species. They are attracted by the bright colors and nectar rewards of the flowers. Bees exhibit specific foraging patterns, visiting multiple flowers in a single foraging trip. This behavior increases the likelihood of cross-pollination. Bees collect pollen as a food source, and in the process, they inadvertently transfer pollen between flowers, facilitating fertilization.

Butterflies and Moths

Butterflies and moths are also important pollinators for many *Bidens* species. They are particularly attracted to flowers with accessible nectar sources. Butterflies and moths have long proboscises that allow them to reach deep nectar sources, ensuring they come into contact with reproductive structures during feeding. While butterflies are active during the day, moths can pollinate *Bidens* flowers at night, extending the period of pollinator activity.

Beetles

Beetles are less specialized pollinators but can be effective in transferring pollen, especially in species with open, easily accessible flowers. Beetles often visit flowers for food or shelter, and their robust structure can dislodge pollen as they move within the capitulum. Beetles are generalist foragers, which can increase the diversity of pollen they carry, promoting cross-pollination.

3.4.2.4 Adaptations to Pollinator Behavior

Bidens species exhibit various adaptations to align their reproductive strategies with the behavior of their pollinators.

Timing of Flowering

The timing of flowering in *Bidens* species is often synchronized with the activity patterns of their primary pollinators.

Diurnal Pollinators

Species pollinated by diurnal insects, such as bees and butterflies, often have flowers that open during the day.

Nocturnal Pollinators

Species that rely on nocturnal pollinators, such as moths, may have flowers that open in the evening or at night.

Flower Longevity

The longevity of individual flowers in *Bidens* species can be adapted to maximize pollination opportunities.

Extended Bloom Period

Longer-lived flowers provide extended opportunities for pollinator visits, increasing the chances of successful pollination.

Synchronous Blooming

Some species exhibit synchronous blooming, where many flowers open simultaneously, creating a large, attractive display for pollinators.

3.4.3 Ecological and Evolutionary Implications

The pollination ecology of *Bidens* species has significant ecological and evolutionary implications, contributing to their adaptability and diversification.

3.4.3.1 Genetic Diversity

Cross-pollination facilitated by diverse pollinators increases genetic diversity within *Bidens* populations. This genetic diversity enhances the adaptability and resilience of species to changing environmental conditions.

3.4.3.2 Reproductive Isolation

Pollination mechanisms and adaptations can contribute to reproductive isolation between *Bidens* species, promoting speciation.

Floral Divergence

Differences in floral morphology and phenology can reduce hybridization and maintain species boundaries.

Pollinator Specialization

Specialized relationships with specific pollinators can lead to reproductive isolation and divergence between species.

3.5 Achene Development and Dispersal Mechanisms in *Bidens* Species

Achenes are the primary reproductive units in *Bidens* species, serving as the means for seed development and dispersal. The development and dispersal mechanisms of achenes are critical for the survival and spread of *Bidens* species across diverse habitats. This section examines the processes involved in achene development, the morphological characteristics of achenes, and the various dispersal mechanisms that contribute to the ecological success of *Bidens* species (Yeo et al., 2022).

Achene development in *Bidens* follows a sequence of stages from pollination to mature fruit formation (Table 3.6). This process involves complex interactions between genetic, hormonal, and environmental factors.

3.5.1 Pollination and Fertilization

The development of achenes begins with successful pollination and fertilization (Table 3.7). Pollinators transfer pollen from the anthers of one flower to the stigma of another, facilitating fertilization. Once pollen grains germinate on the stigma, pollen tubes grow down the style to the ovary, where sperm cells fertilize the ovules, initiating seed development.

Table 3.6 Key stages of achene development

Development stage	Key events and characteristics	Timeframe (days postpollination)
Pollination	Transfer of pollen to the stigma	0
Fertilization	Fusion of gametes	1–3
Embryo formation	Development of zygote into embryo	4–10
Seed maturation	Accumulation of nutrients, hardening of seed coat	11–30
Dispersal readiness	Formation of dispersal structures, drying	31–45
Germination	Seedling emergence from achene	46–60

Table 3.7 Morphological features of achenes in *Bidens* species

Species name	Achene characteristics	Achene size (mm)	Surface texture	Pappus presence	Seed color
Bidens pilosa	Barbed achenes, hooked awns	3–4	Rough, barbed	Present	Brown
Bidens alba	Smooth achenes, lacking awns	2–3	Smooth	Absent	White
Bidens tripartita	Achenes with 4–8 ribs, notched edges	3–5	Ribbed, notched	Absent	Brown
Bidens cernua	Longitudinally ribbed achenes	2–4	Ribbed	Absent	Black
Bidens aristos	Achenes with 2–4 awns, rough surface	4–5	Rough	Present	Brown
Bidens connata	Smooth achenes, lacking awns	2–3	Smooth	Absent	Black
Bidens frondosa	Achenes with 5–7 ribs, triangular shape	3–4	Ribbed, triangular	Absent	Brown
Bidens laevis	Smooth achenes, no visible ribs	2–3	Smooth	Absent	Black
Bidens polylepis	Achenes with numerous small scales	1–2	Scaly	Absent	Brown
Bidens vulgata	Achenes with 3–5 ribs, elliptical shape	2–3	Ribbed, elliptical	Absent	Black

3.5.2 Ovule and Seed Development

After fertilization, the ovules develop into seeds through a series of stages.

3.5.2.1 Embryo Formation

The zygote undergoes mitotic divisions to form the embryo, which will develop into the future plant.

3.5.2.2 Endosperm Development

Concurrently, the endosperm develops, providing nutritional support for the growing embryo.

3.5.2.3 Seed Coat Formation

The outer layers of the ovule harden to form the seed coat, protecting the developing seed.

3.5.3 Achene Maturation

The ovary wall develops into the achene pericarp, encasing the seed. This process involves:

3.5.3.1 Cell Differentiation

Cells in the ovary wall differentiate to form the distinct layers of the achene.

3.5.3.2 Nutrient Accumulation

Nutrients are transported to the developing achene, supporting seed maturation.

3.5.3.3 Desiccation

As the achene matures, it undergoes desiccation, reducing its water content and preparing it for dispersal.

3.5.4 Morphological Characteristics of Achenes

Achenes in *Bidens* species exhibit a range of morphological features that are adapted to their dispersal mechanisms.

3.5.4.1 Shape and Size

Achenes can be flattened, cylindrical, or angular, depending on the species. The shape influences how achenes interact with their dispersal agents. Achene size varies among species, typically ranging from 3 to 10 mm. Smaller achenes may be more easily dispersed by wind, while larger achenes may be better suited for animal dispersal.

3.5.4.2 Surface Texture

Smooth or Ridged

The surface of achenes can be smooth or ridged. Ridged surfaces can enhance attachment to animal fur or feathers.

Hairy or Glabrous

Some achenes have hair-like structures that aid in wind dispersal or attachment to animals.

3.5.4.3 Pappus Structures

The pappus is a modified calyx that aids in dispersal.

Awns

Many *Bidens* species have barbed awns on their achenes, which can hook onto animal fur or feathers.

Bristles

Some species have pappus bristles that catch the wind, facilitating wind dispersal.

3.5.5 Dispersal Mechanisms

Bidens species employ a variety of dispersal mechanisms to ensure the spread of their achenes across different environments.

3.5.5.1 Epizoochory (Animal Dispersal)

Epizoochory is a primary dispersal mechanism in many *Bidens* species, involving the attachment of achenes to animals. The barbed awns of *Bidens* achenes effectively cling to the fur, feathers, or even clothing of animals. This mechanism allows achenes to be transported over long distances. As animals move through different habitats, they inadvertently disperse achenes, promoting the colonization of new areas.

3.5.5.2 Hydrochory (Water Dispersal)

Hydrochory involves the dispersal of achenes by water, which is particularly important for species in wetland habitats. Achenes adapted for water dispersal often have air-filled tissues that enable them to float. This buoyancy allows them to be carried by currents to new locations. Hydrochorous dispersal is especially effective in riparian zones, where flowing water can transport achenes downstream.

3.5.5.3 Anemochory (Wind Dispersal)

Anemochory, or wind dispersal, is another strategy employed by some *Bidens* species. Achenes with pappus bristles can be caught by the wind and carried away from the parent plant. This type of dispersal is effective in open or disturbed habitats where wind can travel unobstructed. Smaller, lighter achenes are more easily lifted by the wind, increasing the range of dispersal.

3.5.5.4 Autochory (Self-Dispersal)

Autochory involves self-dispersal mechanisms that do not rely on external agents. Some *Bidens* species have achenes that are released through explosive dehiscence, where the fruit splits open forcefully to scatter the seeds. This mechanism ensures that seeds are dispersed away from the parent plant.

3.5.6 Ecological Implications of Dispersal Mechanisms

The dispersal mechanisms of *Bidens* species have significant ecological implications, influencing their distribution, colonization, and adaptability.

3.5.6.1 Habitat Colonization

Effective dispersal mechanisms allow *Bidens* species to colonize a wide range of habitats, from wetlands and riparian zones to grasslands and disturbed areas.

Gene Flow

Dispersal promotes gene flow between populations, enhancing genetic diversity and reducing the risks of inbreeding.

Population Expansion

The ability to disperse over long distances facilitates population expansion and the colonization of new environments.

3.5.6.2 Adaptability and Resilience

The diverse dispersal strategies of *Bidens* species contribute to their ecological adaptability and resilience.

Environmental Variability

Dispersal mechanisms enable *Bidens* species to respond to environmental variability by moving to more favorable conditions.

Disturbance Recovery

In disturbed habitats, effective dispersal allows *Bidens* species to quickly recolonize and establish new populations, contributing to their success as pioneer species.

3.5.7 Evolutionary Perspectives

The evolution of dispersal mechanisms in *Bidens* species reflects their adaptive strategies and evolutionary history.

3.5.7.1 Adaptive Radiation

The diversification of dispersal mechanisms among *Bidens* species is an example of adaptive radiation, where different species evolve unique adaptations to exploit various ecological niches.

Speciation
Dispersal mechanisms can contribute to speciation by promoting reproductive isolation and allowing populations to adapt to different environments.

3.6 Niche Differentiation

The specialization of dispersal strategies enables *Bidens* species to occupy distinct ecological niches, reducing competition and promoting coexistence.

The reproductive biology of *Bidens* species exemplifies the intricate interplay between floral morphogenesis, pollination ecology, and seed dispersal mechanisms. Through a detailed examination of their diverse floral structures, sophisticated pollination strategies, and adaptive achene development, this chapter illuminates the evolutionary innovations that have enabled *Bidens* species to thrive across varied ecological niches. The adaptability of *Bidens*, facilitated by their versatile reproductive strategies, highlights the importance of morphological and ecological studies in understanding plant success and diversification. By synthesizing current research, this chapter not only advances our comprehension of *Bidens*' reproductive ecology but also underscores the broader implications for biodiversity conservation and ecological resilience in changing environments.

3.7 Conclusion

The generative morphology of *Bidens* species is a rich field of study, offering insights into their reproductive strategies, ecological adaptations, and evolutionary dynamics. Understanding these characteristics is essential for taxonomy, conservation, and ecological research. The diversity in floral structures, fruit morphology, and dispersal mechanisms underscores the complexity and adaptability of this widespread genus. Through continued study, we can further disentangle the relationships and evolutionary pathways that define *Bidens* species.

References

Bargielski RC (2020) Fields Brook superfund site: race, class, and environmental justice in a blasted landscape. University of South Florida

Bartolome AP, Villaseñor IM, Yang W (2013) Bidens pilosa L. (Asteraceae): botanical properties, traditional uses, phytochemistry, and pharmacology. Evid Based Complement Alternat Med 2013(1):340215

Budumajji U, Solomon Raju AJ (2018) Pollination ecology of Bidens pilosa L. (Asteraceae). Taiwania 63(2)

Chailakhyan MK, Khrianin VN (2012) Sexuality in plants and its hormonal regulation. Springer

Douglas MR, Chafin TK, Claussen JE, Philipp DP, Douglas MEJME (2021) Are populations of economically important bonefish and queen conch 'open' or 'closed' in the northern caribbean basin? Mar Ecol 42(2):e12639

Gaspar T, Hausman J, Faivre-Rampant O, Kevers C, Dommes J (2002) Auxins in the biology of roots. In: Waisel Y, Eshel A, Kafkafi U (eds) Plant roots, 3rd edn, pp 587–616
Grombone-Guaratini MT, Solferini VN, Semir J (2004) Reproductive biology in species of Bidens L. (Asteraceae). Sci Agric (Piracicaba, Brazil) 61:185–189
Hall D, Nguyen TA. Radical narratives in the American Jobs Plan (AJP) and the global potential of 'Biden plans'. 2021.
Helenurm K, Ganders FR (1985) Adaptive radiation and genetic differentiation in Hawaiian Bidens. Evolution 39(4):753–765
Kevan P, Baker HG (1983) Insects as flower visitors and pollinators. Annu Rev Entomol 28(1):407–453
Kompała-Bąba A, Bąba W, Ryś K, Hanczaruk R, Radosz Ł, Prostański D et al (2023) Taxonomic diversity and selection of functional traits in novel ecosystems developing on coal-mine sedimentation pools. Sustainability 15(3):2094
Ritland K, Ganders FR (1985) Variation in the mating system of Bidens menziesii (Asteraceae) in relation to population substructure. Heredity 55(2):235–244
Sánchez J, Estrada Castillón E, García Aranda MA, Duarte Hérnandez MF, García González F, Valenzuela Nuñez LM et al (2023) Arid and semi-arid environments: their relationship with the dispersion and germination of species. Rev mex de cienc forestales 14(75):35–67
Smith P, Doyle G, Lemly J (2020) Revision of Colorado's floristic quality assessment indices. Colorado State University
Spjut RW (1994) A systematic treatment of fruit types. Mem New York Bot Gard 70:1–181
Sun M, Ganders FR (1988) Mixed mating systems in Hawaiian Bidens (Asteraceae). Evolution 42(3):516–527
Wagatsuma T, Tanaka K, Iino Y, Satou T, Khan MSH, Tawaraya K et al (2021) Burr marigold (Bidens tripartita L.) roots directly and immediately scavenge rhizosphere methane with highly exuded hydrogen peroxide via a rhizosphere Fenton reaction. Plant Soil 459:289–313
Yamasaki K, Kigawa T, Inoue M, Tateno M, Yamasaki T, Yabuki T et al (2005) Solution structure of an Arabidopsis WRKY DNA binding domain. Plant Cell 17(3):944–956

Part II

Ethnobotany and Traditional Medicinal Practices

The Genus *Bidens* L. in Chinese Traditional System of Medicine

4

Kulsoom Zahara, Yamin Bibi, Azeesha Faisal, and Hira Rashid

Abstract

Traditional Chinese medicine (TCM) is a centuries-old system of medicine that was established on the origin of reinstating yin-yang balance. The *Bidens* species has been used as a curative herb identified as guizhencao. This genus is documented in ancient Chinese books, e.g., *The Bencao Gangmu*, *Oriental Materia Medica*, *Handbook of Chinese Medicinal Herbs*, and *Prescriptions Worth a Thousand Gold* as a therapeutic herb. According to traditional Chinese medicine, its species has properties of curing inflammation, common cold, fever, bronchitis, sore throat, appendicitis, sputum removal, and treating cough and asthma. They are used in different medicnal preparations, i.e., Xian Feng Cao Tang, ITM Formulary, guizhencao, Gan Mao Qing, and etc. This chapter aims to assemble the data on utilizing the genus *Bidens* in the traditional Chinese medicine system.

Keywords

Bidens · TCM · Chinese · Medicine · *pilosa*

4.1 Introduction

Traditional Chinese medicine (TCM) is a 23 centuries-old system of medicine that was developed on the basis of restoring yin-yang balance. According to their belief, yin and yang are the two complementary forces that pervade the human body and

K. Zahara (✉)
Department of Botany, PMAS Arid Agriculture University Rawalpindi, Rawalpindi, Punjab, Pakistan

Y. Bibi · A. Faisal · H. Rashid
Department of Botany, Rawalpindi Women University, Rawalpindi, Punjab, Pakistan

Y. Bibi et al. (eds.), *The Genus Bidens*,
https://doi.org/10.1007/978-981-96-4257-1_4

Table 4.1 *Bidens* species used in Chinese herbal medicine (Wu et al. 2022)

Plant name	Chinese name and translation	Comments
Bidens tripartite	(三裂双齿) longbacao	Chronic dysentery, anti-inflammatory and analgesic effects
Bidens pilosa	(双齿) sanye guizhencao	Enteritis, bacterial dysentery, and pharyngitis
Bidens parviflora	(小花双齿): Xiaohua guizhencao Lujiaocao, cizhencao	Substitute for *Bidens bipinnata*
Bidens biternata = *Bidens chinensis*	Jinzhan yinpan	Substitute for *Bidens bipinnata* and *Bidens pilosa*
Bidens bipinnata	(双齿毛茼蒿) nianshencao	Inflammation and pain
Bidens frondosa	(羽裂双齿)	Anti-inflammatory and pain-relieving effects
Bidens alba	(白花双齿)	Anti-inflammatory properties

the universe. TCM believes in the concept of vital energy, i.e., QI. In Traditional Chinese Medicine (TCM), Qi (pronounced "chee") is considered the vital life force or energy that flows through every living being and sustains all life. It is believed to circulate throughout the body via pathways called meridians. When the balance and flow of Qi is maintained, health is achieved, and when there is disruption in this balance, illness is seen. Some of the key components of the Chinese traditional system of medicine are acupuncture, Tai Chi and Qigong, cupping and moxibustion, and herbal medicine (Jin et al. 2022). Herbal medicine uses plants, minerals, and animal products to formulate remedies according to individual needs. For preparing therapeutic formulations, the natural products are compounds combined into formulas, i.e., few to several herbs. These formulations are usually designed to work synergistically (Jin et al. 2022).

The *Bidens* species "双齿" (Shuāngchǐ) in Chinese have been utilized from ancient times in China. They are used as a medicinal herb known as guizhencao. This name is given due to the characteristic barbs on the fruits (Pennacchio et al. 2010). One of the most commonly used species in TCM is *Bidens pilosa*, sometimes named sanye guizhencao, as its leaves are in the form of three leaflets. Other species of *Bidens* are also utilized in Chinese herbal medicine (Table 4.1).

In case of unavailability of *Bidens* species, *Desmodium* species are utilized as their substitutes as they have the same applications.

4.2 *Bidens* Genus in Ancient Traditional Chinese Medicine Texts

Genus *Bidens* has a rich history in traditional Chinese medicine and has been documented in ancient texts as a valuable medicinal resource.

4.2.1 The Bencao Gangmu (1596 A.D.)

This book, also identified as Great Pharmacopoeia, is a notable work gathered and edited by Li Shizhen during the late sixteenth century. This book has listed two types of *Bidens*: *Bidens pilosa* (guizhencao) for treating insect and snake bites and *Bidens tripartita* (longbacao), mostly used for chronic diarrhea (Smith and Stuart 1973).

4.2.2 Oriental Materia Medica

This book is an inclusive compendium that investigates the rich history of Oriental healing and the sources and formulations of various natural remedies. This book has listed *Bidens pilosa* among the herbs that clear heat and eliminate stagnancy. The uses of *Bidens* species recorded include enteritis, dysentery, jaundice, influenza, intestinal carbuncle, epilepsy in children, malnutrition in infants, and hemorrhoids (Hsu HongYen et al. 1996).

Bidens bipinnata is listed as herbs that have a cooling effect and resolve the surface. It is stated to disperse stagnant blood, remove wind-dampness, and invigorate blood. It is reported to be used in treating stomach ache, rheumatoid arthritis, diarrhea, dysentery, acute nephritis, sore throat, hepatitis, sprain, insect, and scorpion sting (Hsu HongYen et al. 1996).

4.2.3 Handbook of Chinese Medicinal Herbs

It is documented that the decoction prepared from *Bidens* species is used to cure dysentery, nausea, cardiac spasm laryngalgia, dysphagia, and esophageal dilatation (Chang and Bai 1992).

4.2.4 Prescriptions Worth a Thousand Gold

This book holds a significant place in the history of traditional Chinese medicine. This monumental work was edited by Sun Simiao in the year 652 AD. To cure finger cut, *Bidens* is blended with pig fat and applied externally (Chang and Bai 1992).

4.3 Preparations of Genus *Bidens* in Chinese Traditional System of Medicine

Genus *Bidens* has been used in a large number of formulations. According to traditional Chinese medicine, *Bidens* species are good for inflammation, common cold, fever, bronchitis, sore throat, appendicitis, sputum removal, and treating cough and asthma.

The three species of this genus, i.e., *Bidens bipinnata*, *B. pilosa*, and *B. biternata*, are named guizhencao, and all these species are documented to have the following properties:

- Reduce superficial heat (common cold, influenza).
- Reduce toxic materials (appendicitis, sore throat, snake bite, and centipede bite).
- Reduce dampness and heat from the gastrointestinal tract (diarrhea, dysentery, and stomach ache of heat type).

Bidens species are utilized in various medicinal preparations and some of them are described in Table 4.2.

In Chinese Medicinal Herbs of Hong Kong, *Bidens pilosa* is listed to cure influenza, gastroenteritis, dyspepsia, acute appendicitis, colds, fever, sore throat and hemorrhoids, and pruritus.

4.4 Usage of Genus *Bidens* in Different Communities in China

Genus *Bidens* is widely used in Chinese culture as a vegetable and for treating ailments. Communities that are using genus *Bidens* are:

- **Guangxi Province:** *Bidens pilosa* is listed to be used in the treatment of inflammation and infections (Addi et al. 2024).
- **Hainan Island (Li ethnic group):** *Bidens bipinnata* is documented for treating skin diseases and gastrointestinal issues (Pennacchio et al. 2010).
- **Sichuan Province (Tibetan communities):** *Bidens tripartita* is used for its anti-inflammatory and wound-healing properties (Hu et al. 1992).
- **Xishuangbanna:** The edible flowers of *Bidens* are used among different linguistic groups as an anti-inflammatory agent (Mon et al. 2024).
- **Guizhou Province:** *Bidens pilosa* is used for treating fever and infections, and as an anti-inflammatory agent (Yang et al. 2022).
- **Hunan Province:** *Bidens bipinnata* is as anti-inflammatory and antimicrobial drug, particularly in treating skin conditions and gastrointestinal issues (Lin et al. 2020).
- **Jiangxi Province (Hakka people):** *Bidens bipinnata* is documented for its application in treating digestive disorders and skin conditions and is used as a general anti-inflammatory agent (Xie et al. 2022).
- **Fujian Province (She ethnic group):** *Bidens pilosa* is used for its anti-inflammatory and antipyretic properties, particularly in treating fever and infections (Huang et al. 2011).
- **Inner Mongolia (Mongolian communities):** *Bidens tripartita* is used for its wound-healing properties and for treating respiratory and gastrointestinal ailments (Wang et al. 2013).

Table 4.2 Different formulations prepared in TCM using *Bidens* species

	Ingredients	Indications	Dosage	
(Xian Feng Cao tang) (咸豐草湯)	*Bidens pilosa* (hairy beggarticks) whole plan Licorice root (*Glycyrrhiza uralensis*) *Scutellaria baicalensis* root *Forsythia suspensa* fruit	Used for heat-clearing and detoxification Effective for acute infections, especially those affecting the throat and intestines	Typically decocted and taken orally	Affolter (1987)
(Xian Feng Cao Tu Fu Ling tang)	*Bidens pilosa* (hairy beggarticks) whole plant *Smilax glabra* rhizome *Sophora flavescens* root *Scutellaria baicalensis* root	Used for damp-heat conditions with skin eruptions, such as eczema and pruritus Supports blood circulation and clears heat	Decocted and taken orally	Smith (1871)
(Xian Feng Cao Jin yin Hua tang)	*Bidens pilosa* (hairy beggarticks) whole plant Honeysuckle flowers (*Lonicera japonica*) *Forsythia suspensa* fruit *Scutellaria baicalensis* root	Used for febrile diseases, especially those with toxic heat Supports the immune system and clears heat	Decocted and taken orally	Smith (1871)
(Xian Feng Cao Huang Qin tang)	*Bidens pilosa* (hairy beggarticks) whole plant *Scutellaria baicalensis* root (Huang Qin) *Gardenia jasminoides* fruit (Zhi Zi)	Used for heat-clearing, especially in cases of damp-heat affecting the skin Beneficial for skin conditions like acne and eczema	Decocted and taken orally	Chang and But (1987)
(Xian Feng Cao Lian Qiao tang)	*Bidens pilosa* (hairy beggarticks) whole plant *Forsythia suspensa* fruit (Lian Qiao) *Mentha haplocalyx* herb (Bo he)	Used for febrile diseases with symptoms like sore throat and headache Supports the immune system and clears heat	Decocted and taken orally	Chang and But (1987)

(continued)

Table 4.2 (continued)

	Ingredients	Indications	Dosage	
(Xian Feng Cao Jin yin tang)	*Bidens pilosa* (hairy beggarticks) whole plant Honeysuckle flowers (*Lonicera japonica*) (Jin yin Hua) *Scutellaria baicalensis* root	Used for acute febrile diseases with toxic heat Supports the Respiratory system and clears heat	Decocted and taken orally	SA (1992)
ITM formulary	*Bidens* 25%, Hu zhang 20%, Zi hua di ding 15%, ban lan gen 15%, Ju hua 15%, Gan cao 10%. All the herbs in this formula have anti-infection activities A dried hot water extract of *Bidens* with other herbs has been prepared in tablet form under the name *Bidens* (seven forests)	9–60 g (60–120 g for acute appendicitis)		Bo et al. (2012)
Guizhencao	*Bidens bipinnata*, *B. Pilosa*, and *B. Biternata*			Zahara et al. (2021)
Gan Mao Qing (感冒清)	*Bidens pilosa* (Spanish needles) whole plant *Andrographis paniculata* (Chuanxinlian) *Baphicacanthus cusia* (Banlangen) *Helicteres angustifolia* (Shanzhima) whole plant *Ilex asprella* (Gangmeigen) root *Isatis indigotica* (Daqingye) leaf	Symptoms associated with the common cold	3–4 tablets 3 times daily	Zahara et al. (2021) and Tomczykowa et al. (2011)

- **Guangdong Province (Cantonese people).** *Bidens pilosa* is used for its anti-inflammatory and antimicrobial properties, particularly in treating skin infections and digestive issues (Wang and Ladegaard 2008).
- **Ningxia Hui Autonomous Region (Hui communities):** *Bidens tripartita* is used for its wound-healing properties and for treating gastrointestinal and respiratory ailments (Ye et al. 2019).
- **Hubei Province (Tujia ethnic group).** *Bidens pilosa* is used for its anti-inflammatory and antimicrobial properties, particularly in treating skin infections and digestive issues (Brown 2001).

- **Anhui Province (Hui ethnic group).** *Bidens bipinnata* is noted for its application in treating respiratory infections and fever and is used as a general anti-inflammatory agent (Yu 2020).
- **Xinjiang Uygur Autonomous Region (Uygur communities).** *Bidens tripartita* is used for its wound-healing properties and for treating gastrointestinal and respiratory ailments (Cappelletti 2020).

Generally, the population in China have good knowledge of herbal medicine. This knowledge has been transferred from elders to young people for generations. However, the survival of medicinal species and their traditional knowledge are in danger. Presently, most of the traditional medicines are prepared from wild plants. The vast need for herbal medicines and the resulting swift demand from the herbal medicine market, along with over-harvesting, habitat damage, and climate change, have threatened the viable use of wild plants over the last few years.

4.5 Conclusion

Bidens is an important medicinal herb in China, chiefly used as a folk medicine, meaning that it is utilized by locals for treating common illnesses, i.e., dysentery, common cold, poisonous bites, skin infections, and digestive issues. In Chinese tradition, its two types were included in the authoritative compendium of the sixteenth century, the Bencao Gangmu. But these species are not stated in the traditional preparations documented in modern textbooks.

References

Addi Y-W, Ren Z-X, Rutherford S, Ding X-Y, Guo C-A, Zhang X et al (2024) Ethnobotanical study on medicinal plants used by the Yi people in Xiaoliangshan, Yunnan Province, SW China. J Ethnopharmacol 323:117683

Affolter J. Chinese medicinal herb garden dedicated. 1987

Bo Y, Yuan L-p, Zhang J-j, Meng D-d, Jing H, H-jJJop D et al (2012) Total flavonoids of *Bidens bipinnata* L. a traditional Chinese medicine inhibits the production of inflammatory cytokines of vessel endothelial cells stimulated by sera from Henoch–Schönlein purpura patients. J Pharm Pharmacol 64(6):882–887

Brown MJ (2001) Ethnic classification and culture: the case of the Tujia in Hubei, China. Asian Ethnicity 2(1):55–72

Cappelletti A (2020) Socio-economic development in Xinjiang uyghur autonomous region. Springer

Chang H-M, But PP-H (1987) Chouwutong. Pharmacology and applications of Chinese Materia Medica: (Volume II). World Scientific, pp 999–1004

Chang M, Bai Y (1992) Anticancer medicinal herbs. Human Science and Technology Publishing House

Hsu HongYen HH, Chen YuhPan CY, Sheu ShuennJyi SS, Hsu ChauShin HC, Chen ChienChih CC, Chang HsienChang CH. Oriental materia medica: a concise guide. 1996

Hu ST, Hannaway DB, Youngberg HW (1992) Forage resources of. Pudoc Wageningen, China

Huang J, Zhang W, Li X, Zhou J, Gao Y, Cai Y et al (2011) Analysis of the prevalence and risk factors of hypertension in the she population in Fujian. China. 34(2):69–74

Jin S, Zhang S, Shad N, Naeem A, Yang Y, Wu SK (2022) Ethnobotanical investigation of medicinal plants used in Lingchuan county, Shanxi, China. Braz J Biol 82:e260774

Lin Y-x, Duan R-y, Tan Z-x, Ma Y-h, Wu HJMDPB (2020) The complete chloroplast genome of the invasive and cd-hyperaccumulator herb *Bidens pilosa* L. Asteraceae 5(1):436–438

Mon AM, Hein PP, Zaw M, Kyaw MT, Yang Y, Yang X et al (2024) Ethnobotanical surveys reveal the crucial role of medicinal plants in the primary healthcare system of the Shan people in Myanmar 327:117875

Pennacchio M, Jefferson L, Havens K (2010) Uses and abuses of plant-derived smoke: its ethnobotany as hallucinogen, perfume, incense, and medicine. Oxford University Press

SA MC (1992) Ethnobotany and medicinal plants: July 1991–July 1992. National Agricultural Library

Smith FP (1871) Contributions towards the materia medica & natural history of China: for the use of medical missionaries & native medical students. American Presbyterian Mission Press

Smith FP, Stuart GA (1973) Chinese medicinal herbs. Georgetown Press

Tomczykowa M, Leszczyńska K, Tomczyk M, Tryniszewska E, Kalemba DJJomf. (2011) Composition of the essential oil of *Bidens tripartita* L roots and its antibacterial and antifungal activities. J Med Food 14(4):428–433

Wang L, Ladegaard HJ (2008) Language attitudes and gender in China: Perceptions and reported use of Putonghua and Cantonese in the southern province of Guangdong. Lang Aware 17(1):57–77

Wang J, Brown DG, Agrawal AJ (2013) Climate adaptation, local institutions, and rural livelihoods: A comparative study of herder communities in Mongolia and Inner Mongolia, China. Glob Environ Chang 23(6):1673–1683

Wu L, Nie L, Guo S, Wang Q, Wu Z, Lin Y et al (2022) Identification of medicinal Bidens plants for quality control based on organelle genomes. Front Pharmaco 13:842131

Xie G, Zhou Y, Liu CJS (2022) Spatial distribution characteristics and influencing factors of Hakka traditional villages in Fujian, Guangdong, and Jiangxi. China 14(19):12068

Yang Q, Jin B, Zhao X, Chen C, Cheng H, Wang H et al (2022) Composition, distribution, and factors affecting invasive plants in grasslands of Guizhou Province of Southwest China. Diversity 14(3):167

Ye C, Chen C, Butler OM, Rashti MR, Esfandbod M, Du M et al (2019) Spatial and temporal dynamics of nutrients in riparian soils after nine years of operation of the three gorges reservoir, China. Sci Total Environ 664:841–850

Yu F (2020) Muslim diversity and localization: the hui ethnic group in Yunnan. In: Ethnicity and religion in southwest. Routledge, China, pp 119–142

Zahara K, Bibi Y, Masood S, Nisa S, Qayyum A, Ishaque M et al (2021) Using HPLC–DAD and GC–MS analysis isolation and identification of anticandida compounds from Gui Zhen Cao Herbs (Genus Bidens): an important Chinese medicinal formulation. Molecules 26(19):5820

The Genus *Bidens* L. in African Traditional Medicine

5

Tomi Lois Adetunji, Olalekan Bukunmi Ogunro, Olamide Omolafe Ogunremi, and Olumayowa Rebecca Andrew

Abstract

For centuries, the genus *Bidens* L. (Asteraceae), comprising 246 species, has been used as herbal medicine in many parts of the world, including Africa, to treat several ailments. In this chapter, the role of *Bidens* species in African traditional medicine, phytochemistry, and their potential therapeutic applications are discussed. A comprehensive and systematic literature review was conducted on *Bidens* species in Africa using various scientific databases and resources such as Google Scholar, Scopus, Web of Science, PubMed, and books. With *Bidens pilosa* (blackjack) being the most used species, the literature indicates that leaves of some *Bidens* species are consumed as vegetables, and different plant parts of *Bidens* species are utilized in African folk medicine for treating various conditions such as diabetes, inflammation, diarrhea, malaria, leprosy, infections, and

---

Tomi Lois Adetunji and Olalekan Bukunmi Ogunro have shared the authorship of this chapter.

T. L. Adetunji (✉)
Unit for Environmental Sciences and Management, Faculty of Natural and Agricultural Sciences, North-West University, Potchefstroom, South Africa

O. B. Ogunro
Drug Discovery and Pharmacology Laboratory, Department of Biological Sciences, KolaDaisi University, Ibadan, Nigeria

O. O. Ogunremi
Institute for Sustainable Development (I4SD), Abiola Ajimobi Technical University, Ibadan, Oyo, Nigeria

O. R. Andrew
Research and Development Department, Kellogg Tolaram Nigeria Limited, Surulere, Lagos, Nigeria

Y. Bibi et al. (eds.), *The Genus Bidens*,
https://doi.org/10.1007/978-981-96-4257-1_5

digestive disorders, among others. They are often prepared as decoctions, infusions, poultices, and as paste, depending on the type of ailment. Overall, current evidence from the literature suggests the potential use of *Bidens* species in modern medicine and their continuous use in African traditional medicine. However, more research on toxicology is warranted to establish the safety of *Biden* species.

Keywords

Anti-hyperglycemic · *Bidens* · Pharmacology · Phytochemicals · Toxicity · Traditional medicine

5.1 Introduction

Traditional medicine, according to the World Health Organization (WHO) in 2023, encompasses all knowledge, skills, and practices that different cultures have developed to keep people healthy, prevent illnesses, diagnose health problems, and treat physical and mental ailments (Fokunang et al. 2020). African traditional medicine (ATM) is, therefore, the traditional medicine native to the African cultures and communities. This includes herbal products, herbal medications, herbal ingredients, and preparations.

Plants as natural sources of herbal medicine have been widely explored in folk medicine in most African countries (Kissanga et al. 2021), where they serve as an entity for innovative bioactive compounds. This is because the African continent boasts an abundant and diverse indigenous flora, which serves as a source of herbal remedies for many illnesses and contributes to strengthening the healthcare system of the developing regions (Gurib-Fakim 2006).

Within African traditional medicine (ATM) systems, specific plant species often attract greater ethnopharmacological significance due to their versatility and ease of cultivation. These plants exhibit a broader spectrum of therapeutic applications, treating a wider range of both internal and external ailments (Zahara et al. 2022a). Among these plant species is the genus *Bidens* L., a flowering plant in the Asteraceae family. *Bidens* species have been used in ATMs to treat a variety of illnesses across different African countries. The genus *Bidens* is widely distributed around the world, with about 240 known species (Lima Silva et al. 2011).

The genus is characterized by its thorny bristles, which readily attach to animal fur or human clothing. Some species have long ray flowers with sharp segments, while others have simpler leaves with short or nonexistent ray flowers, and most have yellow disk flowers. These unique features set aside numerous species in the genus from others in the Asteraceae family (Mtenga and Ripanda 2022).

Several *Bidens* species, including *B. pilosa, B. augustata, B. tripartita, B. bipinnata, B. biternata, B. macroptera, B. magnifolia, B. engleri, and B. schimperi*, have been applied in ATM practices and beyond. However, *B. pilosa, B. bipinnata*, and *B. biternata* are regarded as herbal *Bidens* due to their multifunctional potentials, with *B. pilosa* being the most utilized species in ATM. For instance, Bartolome et al.

(2013) documented the use of *B. pilosa* as both a stand-alone herb and an ingredient in herbal remedies and beverages. The shoots and leaves, whether fresh or dried, are used in sauces and teas. Hence, in 1970, the Food and Agriculture Organization (FAO) advocated for the widespread cultivation of *B. pilosa* in Africa due to its ease of growth, nutritional value, and safety for human consumption.

Bidens pilosa has been traditionally employed in various African nations to address a range of health conditions, including cough, diabetes, hypertension, obesity, diarrhea, dysentery, wound healing, respiratory infections, indigestion, immunological disorders, and metabolic syndrome (Mtenga and Ripanda 2022). Similarly, *B. biternata* has been applied in the cure of several diseases, which are not limited to asthma, fever, cough, liver infection, diabetes, toothache, skin infections, and leprosy. In addition to its anti-inflammatory, febrifuge, diuretic, anthelmintic, and wound-healing efficacy, the species has been utilized for stimulating purposes (Zahara et al. 2015).

Many of the *Bidens* species contain various active chemical compounds like flavonoids, sesquiterpenes, acetylenes, and lactones, which have notably sparked significant scientific and medical interest (Kuo et al. 2021). Furthermore, the extensive presence of bioactive compounds with anti-parasitic, antiulcer, antidiabetic, antifungal, and antioxidant properties in *Bidens* species has propelled a surge in their application in ATMs (Zahara et al. 2015).

Despite the potential and application of *Bidens* species in traditional and modern medicine, many of the species are still endangered in Africa because they are classified as weeds and a threat to other important plants due to their invasive and adaptive nature (Arthur et al. 2012). Thus, they are destroyed to conserve resources for cultivated florals. In light of this, this chapter aims to document the ethnomedicinal uses, phytochemistry, and pharmacological studies of *Bidens* species used in ATM. This can help to promote sustainable utilization and conservation practices of *Bidens* species.

5.2 Ethnomedicinal Uses of *Bidens* Species in ATM

The notable species of *Bidens* utilized in ATM for medicinal purposes include *B. pilosa, B. augustata, B. tripartita, B. bipinnata, B. biternata, B. macroptera, B. magnifolia, B. engleri,* and *B. schimperi*. Among them, *B. pilosa* is particularly well recognized and explored due to its widespread availability, efficacy, cultural significance, multifaceted use, and affordability (Lima Silva et al. 2011). There is little or no information on the medicinal applications of other species due to their low distribution in the African regions. Over forty different categories of ailments have been reported to be treated by various formulations of the whole plant or any plant part of species of the genus *Bidens* in ATM. The wide range of ethnomedicinal applications of available *Bidens* species in Africa includes treatment of stomach ache, colics, catarrh, diarrhea, constipation, dysentery/bacillary dysentery, choleretic, inflammation, asthma, acute appendicitis, enteritis, pruritus, conjunctivitis, otitis, pharyngitis, gastritis, diabetes, headache, diuretic, rheumatism, uterine contraction, intestinal injury, high blood pressure, conjunctivitis, and smallpox, colds, yellow fever, influenza, acute infectious hepatitis, intestinal worms, malaria, eye infection, pulmonary tuberculosis, bacterial infections in gastrointestinal tracts, renal infection, sore throat, cough, the coolness of the uterus, menstrual irregularities, dysmenorrhea, hyperemesis gravidarum (morning sickness), hemorrhoids, nose bleeds, stomach ulcers cuts, burns, and skin problems, wounds, and snake bites, among others (Bartolome et al. 2013).

The leaves of *Bidens* species are the most commonly used plant part in ATM. Herbal remedies from these species are mostly prepared by decoction and mostly administered orally. The leaves and young branches of *B. pilosa* are also cooked and consumed as vegetables in some parts of Africa, for example, in Angola (Kissanga et al. 2021), Benin (Shayanowako et al. 2021; Klotoé et al. 2012), Liberia (Arthur et al. 2012), Sierra Leone (Arthur et al. 2012), and Zimbabwe (Shayanowako et al. 2021; Otang-Mbeng and Mashabela 2020), for their nutraceutical benefits and to combat malnutrition. Table 5.1 shows some *Bidens* species used in ATM, the region used, and their traditional uses.

5.3 Pharmacological Potential

Different studies have reported the pharmacological activities of extracts of the whole plant or any plant part of species in the genus *Bidens*. For instance, studies on *B. pilosa,* which is the most extensively studied of the genus *Bidens* have demonstrated antitumor (Sundararajan et al. 2006; Wu et al. 2022), antioxidant (Chang et al. 2001; Deba et al. 2008), anti-inflammatory (Horiuchi and Seyama 2006), antihypertensive, vasodilatory (Tan et al. 2000; Nguelefack et al. 2006), immunomodulatory (Chiang et al. 2004), antimalarial (Kumari and Kakkar 2008), antidiabetic, antihyperglycemic (Chang et al. 2007), antifungal (Deba et al. 2008), and antiulcerative (Tan et al. 2000) properties. Additionally, *B. frondosa* has been reported for its antibacterial (Rahman et al. 2011) and anti-inflammatory properties (Antoniak et al. 2023).

5.4 Conclusion and Future Perspectives

Although often regarded as weeds in Africa, evidence from the current study revealed that species in the genus *Bidens* are valuable resources in traditional and modern medicine for the treatment of various ailments. The ethnomedicinal survey revealed that various species of the genus have been used as traditional medicine in ATM to treat a wide array of diseases including diabetes, asthma, pain, malaria, skin diseases, and diarrhea, to mention a few. The genus is also rich in a variety of bioactive phytochemicals, including flavonoids, terpenoids, phenolic acids, alkaloids, polyacetylenes, etc. To date, at least 180 chemical compounds have been isolated in the genus, mostly from *B. pilosa*. Further, different pharmacological activities have been reported mainly through in vitro and in vivo assessments. Summarily, species

Table 5.1 Ethnomedicinal uses of *Bidens* species used in ATM

Bidens species	Country/local name	Ailment	Part	Preparation	Mode	
B. Angustata (Sherff)	Kenya/Tanzania	Sunstroke	L	Pulverized	Topical	Lye et al. (2008)
B. Biternata (Lour.) Merr. And Sherff	Tanzania (Kagera region)/ rwongwe/ongwe kakurura/	Malaria and yellow fever	L	Decoction	Oral	Moshi et al. (2010) and Sharma et al. (2009)
		Skin disease	WP	Paste	Topical	
B. Bipinnata L.	NA	Muscle relaxation, asthma, respiratory conditions, cold, and eye problems	L	Juicing Decoction	Oral	Orwa et al. (2008), Nole et al. (2016), and Zahara et al. (2022b)
B. Engleri O. E. Schulz	Burkina Faso (Kinkirs-sabtulga)	Diarrhea, skin rash, bronchitis, dysentery, prostate tumor, stomach ache, epilepsy, diabetes, cardiovascular diseases, inflammatory and respiratory disease, smallpox, and liver diseases. Tonsillitis, tooth decay, cramps, bronchitis, malaria	L/WP	Decoction	Topical for skin rash, anal for diarrhea	Lye et al. (2008), Bonkian et al. (2017) and Bayala et al. (2019)
B. Macroptera (Sch.Bip. Ex Chiov.) Mesfin	Ethiopia (Adeyabeba/ Keelloo)	Abscess, athlete's foot	F/L	Pulverized	Topical	Mahalakshmi et al. (2019), Chekole (2017), Beyi (2018) and Kasali et al. (2021)
		Brain cancer			Oral	
B. Magnifolia Sherff	Tanzania	Cough, pneumonia	R	Chewing	Oral	Lye et al. (2008)
B. pilosa L.	Angola/Jisongo-jia-ngoma, olokoso, osungua.	Malnutrition	L/B	Cooked as vegetables	Oral	Kissanga et al. (2021)
	Benin/Vodunga	Malnutrition External bleeding	L	Cooked/ poultice/ decoction	Oral Topical	Shayanowako et al. (2021), Klotoé et al. (2012) and Otang-Mbeng and Mashabela (2020)

(continued)

Table 5.1 (continued)

Bidens species	Country/local name	Ailment	Part	Preparation	Mode	
	Burundi/Icanda	Microbial diseases	L/AP/ WP	Decoction	Oral	Ngezahayo et al. (2015)
				Paste	Topical	
	Cameroun (Bamboutos)/"Lietmik/ LipiliepMetsemi/Atchiti" in Bulu, "Njim njim" in Bassa, And "Fouwan" in Bamoun	Typhoid, fever, jaundice, threatened abortion, conjunctivitis, toothache, cough, intestinal helminthiasis, leg ulcer, injuries, skin problems, headaches, hypotensive, dysmenorrhea, placenta retention	L/WP/L	Succus Decoction Infusion	Oral/ topical	Njamen et al. (2013), Adjolohoun et al. (2008) and Tsobou et al. (2013)
	Congo/Kashisha (Ma)	Diarrhea, poisoning, malaria	L/F	Decoction with a few drops of *Aloe vera* juice	Oral	Kimpouni et al. (2013), Bi et al. (2018), Kasika et al. (2014) and Syamasamba et al. (2022)
	Cote d'Ivoire Abissawa, Anasipagne	Jaundice/dysentery, burns, anesthetic, swollen spleens, diarrhea Hypertension	L/SA/ SE/WP	Crushed leaves: Pulverized; decoction of the whole plant of *B. pilosa*, fresh leaves of *Moringa oleifera,* and salt	Oral Topical	Mtenga and Ripanda (2022), Bi et al. (2018), Kasika et al. (2014) and Abrogoua et al. (2016)
	Eswatini	Malnutrition	L	Cooked as vegetables	Oral	Van Damme (1996)
	Ethiopia (Addis Ababa)/ Yesetan Merfe/Chogogitii	Malaria and fungal diseases. Psychosis	WP R	Decoction Root with root and leaves of *Zehneria scabra* are boiled	Oral Inhalation	Bartolome et al. (2013), Kefalew et al. (2015), Wubetu et al. (2018) and Woldeamanuel et al. (2022)

	Ghana/Kurofidie, Aseduro, Dzani pipi	Tuberculosis	F	NA	Oral	Tabuti et al. (2010)
	Kenya (Giriama)/ Chepkotiwot	Infected skin injuries, upper respiratory tract infections, hemorrhage	L	Boiled and water is discarded; pulverized	Oral	Arthur et al. (2012), Shayanowako et al. (2021) and Otang-Mbeng and Mashabela (2020)
	Liberia/Niani, Zikilliwissi	Malnutrition	L	Cooked as vegetables	Oral	Arthur et al. (2012)
	Madagascar/Anatraka/ Tsipolitra	Hypertension, diabetes, wound	L/AP	Decoction/soup	Oral	Novy (1997) and Rakotondrafara et al. (2018)
	Malawi	AIDS management	L	Boiled and water-discarded	Oral	Lamorde et al. (2010)
	Mozambique	Malnutrition	L	Cooked as vegetables	Oral	Shayanowako et al. (2021) and Otang-Mbeng and Mashabela (2020)
	Nigeria/Abere, Langanran	Anesthetic, easy labor, abdominal pain, stomach disorders/ jaundice/dysentery, burns, and swollen spleens	L/SA/ SE/WP	Succus Pulverized	NA	Mtenga and Ripanda (2022)
	Rwanda	Poisoning, furuncle, hepatitis, otitis, wound	L/F	Decoction of the leaves and young flowers and the addition of a few drops of *Aloe vera* juice	NA	El-Kamali (2009)

(continued)

Table 5.1 (continued)

Bidens species	Country/local name	Ailment	Part	Preparation	Mode	
	Sierra Leone/dada, Kadane, Agberi-oku, Nana	Malnutrition	L	Cooked as vegetable	Oral	Arthur et al. (2012)
	South Africa (KwaZulu-Natal)/blackjack	Fever; umbilical cord; stomach aches, vomiting, raising, sunken frontal fontanelles, cold, influenza, wounds, ear infections, antirheumatic, colics, dysentery, tuberculosis	F/R/L/WP	Succus Paste	Oral and topical	Bartolome et al. (2013), Nguta et al. (2015) and Goudoum et al. (2016)
	Tanzania/Nyaweza, Akakurura/kakurura/obukurura	AIDS management	L	Boiled and water-discarded	Oral and topical	Moshi et al. (2010), Amri and Kisangau (2012) and Nyamukuru et al. (2017)
		Jaundice/dysentery, anesthetic, swollen spleens	SA	Decoction of the bark		
		Burns, wounds, and warts	SE/B	A decoction from boiled leaves with *Solanum nigrum*		
	Uganda (West Nile and Madi District)	Diarrhea, nose bleeds, sore throat, eye infection, influenza, yellow fever, colds, stomach ulcer, AIDS management, malaria, and tuberculosis	F/R/ L/WP	Fresh juice or decoction	Oral	Tabuti et al. (2010) and Borokini and Lawal (2014)
	Zambia/Bumpububu	Malnutrition	L/R	Boiling	Oral	Otang-Mbeng and Mashabela (2020)
	Zimbabwe	Malnutrition	L	Cooked as vegetables	Oral	Shayanowako et al. (2021)
B. Schimperi Sch.Bip. Ex Walp.	Tanzania/orwangwa)	Yellow fever treatment, relief of chest pain	L/R	Leaves/roots are boiled in water	Oral	Moshi et al. (2010)
B. Tripartite L.	North Africa regions (Algeria, Libya, Morocco, Tunisia, and Egypt)/tripartite bur-marigold	Bladder and kidney diseases, uterine conditions, hematuria, hemorrhage	L/WP	NA	NA	Lye et al. (2008) and Orhan (2016)

L leaves; *WP* whole plant; *R* root; *F* flowers; *SA* sap; *SE* seed; *BA* bark; *AE* aerial parts; *B* branches of the genus *Bidens* have ethnomedicinal applications and a wide range of pharmacological activities resulting from the presence of various bioactive constituents present in different parts of the plant.

The World Flora Online (WFO) has listed 246 species of the genus, yet only about 20 species, notably *B. pilosa*, have been extensively studied for their phytochemical and pharmacological properties, leaving a significant knowledge gap for the remaining species. Furthermore, a few plant parts, mostly leaves, have been investigated for their biological activities. The genus *Bidens* harbors numerous unexplored species, holding vast potential for the discovery of novel biologically active compounds, particularly those with a history of traditional medicinal use in ATMs. Notably, many traditional herbal remedies and plant extracts contain a diverse array of active constituents, which may interact synergistically to produce enhanced therapeutic effects, warranting further investigation into their untapped potential.

Future research on species in the genus *Bidens* demands a significant investment of effort to expand the understanding of their vast potential. Firstly, in-depth phytochemical and pharmacological exploration of the species diversity as well as different plant parts should be prioritized to unlock the genus' full resources and potential. Secondly, studies should assess other ethnomedicinal uses of the genus using modern pharmacological assays since only a few reported traditional medicinal uses have been investigated via in vivo and in vivo studies. Thirdly, human clinical trials should be done to further discover the mechanism of action of extracts and isolated compounds in the genus. Finally, the genus has a notable knowledge gap in toxicological research, with only a limited number of studies addressing safety aspects. Future investigations should prioritize evaluating the toxicity of the plants in question, aiming to establish a scientific foundation for the safe traditional and clinical use of various species within the genus, thereby ensuring the well-being of users and the responsible development of botanical therapies.

References

Abrogoua DP, Kamenan BAT, Ahui BJM, Doffou EJT (2016) Pharmaceutical interventions in the management of tuberculosis in a pneumophtisiology department, Ivory Coast. Ther Clin Risk Manag 12:1749–1756

Adjolohoun S, Bindelle J, Adandédjan C, Buldgen A (2008) Agronomie, Société et Environnement Some suitable grasses and legumes for ley pastures in Sudanian Africa: the case of the Borgou region in Benin. Agronomie, Société Environnement. 12(4)

Amri E, Kisangau DP (2012) Ethnomedicinal study of plants used in villages around Kimboza forest reserve in Morogoro, Tanzania. J Ethnobiol Ethnomed 8:1–9

Antoniak K, Studzińska-Sroka E, Szymański M, Dudek-Makuch M, Cielecka-Piontek J, Korybalska K (2023) Antiangiogenic, anti-inflammatory and antioxidant properties of Bidens tripartite herb, Galium verum Herb and Rumex hydrolapathum Root. Molecules 28(13):4966

Arthur G, Naidoo K, Coopoosamy RJ (2012) Bidens pilosa L: agricultural and pharmaceutical importance. J Med Plant Res 6(17):3282–3287

Bartolome AP, Villaseñor IM, Yang W-CJEBC, Medicine A (2013) Bidens pilosa L.(Asteraceae): botanical properties, traditional uses, phytochemistry, and pharmacology. eCAM 2013(1):340215

Bayala B, Sow B, Millogo V, Ouattara Y, Tamboura HH (2019) Toxicity, cytotoxicity and biological activities of seeds of Carapa procera (DC), a native oil tree. Int J Biol Chem Sci 13(1):49–62

Beyi MW (2018) Traditional medicinal plants in Ethiopia. International Journal of Biology, Physics & Mathematics. 1(1):80–87

Bi N, Jamwal M, Devi S, Sharma N (2018) Documentation, Ethno-botanical and Ethno-medicinal Survey of Wild Leguminous Plants from Some Areas of District Rajouri, J&K State, India. European J Med Plants 22(3):1–11

Bonkian LN, Yerbanga RS, Coulibaly MT, Lefèvre T, Sangaré I, Ouédraogo T et al (2017) Plants against malaria and mosquitoes in Sahel region of Burkina Faso: an ethno-botanical survey. Int J Herb Med 5

Borokini TI, Lawal IO (2014) Traditional medicine practices among the Yoruba people of Nigeria: A historical perspective. J Med Plants 2(6):20–33

Chang J-S, Chiang L-C, Chen C-C, Liu L-T, Wang K-C, Lin C-C (2001) Antileukemic activity of Bidens pilosa L. var. minor (Blume) Sherff and Houttuynia cordata Thunb. Am J Chin Med 29(02):303–312

Chang T-C, Wentzel EA, Kent OA, Ramachandran K, Mullendore M, Lee KH et al (2007) Transactivation of miR-34a by p53 broadly influences gene expression and promotes apoptosis. Mol Cell 26(5):745–752

Chekole G (2017) Ethnobotanical study of medicinal plants used against human ailments in Gubalafto District, Northern Ethiopia. J Ethnobiol Ethnomed 13:1–29

Chiang Y-M, Chuang D-Y, Wang S-Y, Kuo Y-H, Tsai P-W, Shyur L-F (2004) Metabolite profiling and chemopreventive bioactivity of plant extracts from Bidens pilosa. J Ethnopharmacol 95(2–3):409–419

Deba F, Xuan TD, Yasuda M, Tawata S (2008) Chemical composition and antioxidant, antibacterial and antifungal activities of the essential oils from Bidens pilosa Linn. var. Radiata. Food Control 19(4):346–352

El-Kamali HH (2009) Medicinal plants in east and central Africa: challenges and constraints. Ethnobotanical Leaflets 2009(2):12

Fokunang ET, Fonmboh DJ, Mballa RN, Nyuyki AB, Fokunang LB, Kaba N, et al. Pharmacovigilance of natural herbal medicines research for efficacy, safety and quality assurance of phytomedicine products. J Complement Altern Med Res 2020:21–37

Goudoum A, Ngamo Tinkeu LS, Madou C, Djakissam W, Mbofung CM (2016) Variation of some chemical and functional properties of Bambara groundnut (Voandzeia subterranean L. Thouars) during sort time storage. Food Sci Technol 36(2):290–295

Gurib-Fakim A (2006) Medicinal plants: traditions of yesterday and drugs of tomorrow. Mol Aspects Med 27(1):1–93

Horiuchi M, Seyama Y (2006) Antiinflammatory and antiallergic activity of Bidens pilosa L var radiata SCHERFF. J Health Sci 52(6):711–717

Kasali FM, Kadima JN, Peter EL, Mtewa AG, Ajayi CO, Tusiimire J et al (2021) Antidiabetic medicinal plants used in Democratic Republic of Congo: a critical review of ethnopharmacology and bioactivity data. Front pharmacol 12:757090

Kasika EL, Vasombolwa VK, Lejoly J (2014) Contribution to the knowledge of affinities of traditional medicine of Bantu of high and low lands in the territories of Beni and Lubero. J Med Plant Res 8(42):1245–1261

Kefalew A, Asfaw Z, Kelbessa E (2015) Ethnobotany of medicinal plants in Ada'a District, East Shewa Zone of Oromia regional state, Ethiopia. J Ethnobiol Ethnomed 11:1–28

Kimpouni V, Apani E, Motom M (2013) Analyse phytoécologique de la flore ligneuse de la Haute Sangha (République du Congo). Adansonia, sér 35(1):107–134

Kissanga R, Sales J, Moldão M, Alves V, Mendes H, Romeiras MM et al (2021) Nutritional and functional properties of wild leafy vegetables for improving food security in Southern Angola. Front sustain food syst 5:791705

Klotoé JR, Dougnon TV, Koudouvo K, Atègbo J, Loko F, Akoègninou A et al (2012) Ethnopharmacological survey on antihemorrhagic medicinal plants in South of Benin. European J Medl Plants 3(1):40–51
Kumari A, Kakkar P (2008) Screening of antioxidant potential of selected barks of Indian medicinal plants by multiple in vitro assays. Biomed Environ Sci 21(1):24–29
Kuo TF, Yang G, Chen TY, Wu YC, Tran Nguyen Minh H, Chen LS et al (2021) Bidens pilosa: nutritional value and benefits for metabolic syndrome. Food front 2(1):32–45
Lamorde M, Tabuti JR, Obua C, Kukunda-Byobona C, Lanyero H, Byakika-Kibwika P et al (2010) Medicinal plants used by traditional medicine practitioners for the treatment of HIV/AIDS and related conditions in Uganda. J Ethnopharmacol 130(1):43–53
Lima Silva F, Fischer DCH, Fechine Tavares J, Sobral Silva M, Filgueiras de Athayde-Filho P, Barbosa-Filho JM (2011) Compilation of secondary metabolites from Bidens pilosa L. Molecules 16(2):1070–1102
Lye KA, Bukenya-Ziraba R, Tabuti JR, Waako PJMHH (2008) Botanical-medicinal dictionary for east. Africa 2
Mahalakshmi S, Achala H, Ramyashree K, Prashith Kekuda T (2019) Rauvolfia tetraphylla L. (Apocynaceae)-a comprehensive review on its ethnobotanical uses, phytochemistry and pharmacological activities. Int J Pharm Bio Sci 9(2):664–682
Moshi MJ, Otieno DF, Mbabazi PK, Weisheit A (2010) Ethnomedicine of the Kagera region, north western Tanzania. Part 2: the medicinal plants used in Katoro Ward, Bukoba district. J Ethnobiol Ethnomed 6:1–5
Mtenga DV, Ripanda ASJH (2022) A review on the potential of underutilized blackjack (Biden Pilosa) naturally occurring in sub-Saharan. Africa 8(6):e09586
Ngezahayo J, Havyarimana F, Hari L, Stévigny C, Duez P (2015) Medicinal plants used by Burundian traditional healers for the treatment of microbial diseases. J Ethnopharmacol 173:338–351
Nguelefack T, Nana P, Atsamo A, Dimo T, Watcho P, Dongmo A et al (2006) Analgesic and anticonvulsant effects of extracts from the leaves of Kalanchoe crenata (Andrews). Haworth (Crassulaceae). J Ethnopharmacol 106(1):70–75
Nguta JM, Appiah-Opong R, Nyarko AK, Yeboah-Manu D, Addo PG (2015) Medicinal plants used to treat TB in Ghana. Int J Mycobacteriol 4(2):116–123
Njamen D, Mvondo MA, Djiogue S, Wanda GJMK, Nde CBM, Vollmer G (2013) Phytotherapy and women's reproductive health: the Cameroonian perspective. Planta Med 79(07):600–611
Nole T, Djeufack T, Lionel W (2016) Diabetes and arterial hypertension resorts of treatments and plants used for their treatments in three phytogeographic areas of Cameroon. Int J Tradit Complement Med 1(4):45–59
Novy JW (1997) Medicinal plants of the eastern region of Madagascar. J Ethnopharmacol 55(2):119–126
Nyamukuru A, Tabuti JR, Lamorde M, Kato B, Sekagya Y, Aduma PR (2017) Medicinal plants and traditional treatment practices used in the management of HIV/AIDS clients in Mpigi District, Uganda. J Herb Med 7:51–58
Orhan F (2016) Alleviation of salt stress by halotolerant and halophilic plant growth-promoting bacteria in wheat (Triticum aestivum). Braz J Microbiol 47(3):621–627
Orwa JA, Jondiko I, Minja RJ, Bekunda M (2008) The use of Toddalia asiatica (L) Lam.(Rutaceae) in traditional medicine practice in East Africa. J Ethnopharmacol 115(2):257–262
Otang-Mbeng W, Mashabela MN (2020) Industries R. a review of beneficial phytochemicals and postharvest studies on some indigenous leafy vegetables from the Mpumalanga province of South Africa. J Med Plant 12(4):533–544
Rahman A, Bajpai VK, Dung NT, Kang SC (2011) Antibacterial and antioxidant activities of the essential oil and methanol extracts of Bidens frondosa Linn. Int J Food Sci Technol 46(6):1238–1244
Rakotondrafara A, Rakotondrajaona R, Rakotoarisoa M, Ratsimbason M, Rasamison VE, Rakotonandrasana SR (2018) Ethnobotany of medicinal plants used by the Zafimaniry clan in Madagascar. J Phytopharmacol 7(6):483–484

Sharma A, Bargali K, Pande N (2009) The allelopathic potential of bryophyte extract on seed germination and seedling growth of Bidens biternata. Nat Sci Sleep 7:30–38

Shayanowako AIT, Morrissey O, Tanzi A, Muchuweti M, Mendiondo GM, Mayes S et al (2021) African leafy vegetables for improved human nutrition and food system resilience in Southern Africa: a scoping review. Sustainability 13(5):2896

Sundararajan P, Dey A, Smith A, Doss AG, Rajappan M, Natarajan S (2006) Studies of anticancer and antipyretic activity of Bidens pilosa whole plant. Afr Health Sci 6(1):27–30

Syamasamba MA, Kapiri MM, Muhesi KE, Mbayahi KE, Mavinga BM (2022) Ethnobotanical study of plants used by traditherapists for the treatment of malaria in the city of Butembo, North Kivu, East of the Democratic Republic of Congo. Indonesian J Innov Appl Sci (IJIAS) 2(3):219–233

Tabuti JR, Kukunda CB, Waako P (2010) Medicinal plants used by traditional medicine practitioners in the treatment of tuberculosis and related ailments in Uganda. J Ethnopharmacol 127(1):130–136

Tan PV, Dimo T, Dongo E (2000) Effects of methanol, cyclohexane and methylene chloride extracts of Bidens pilosa on various gastric ulcer models in rats. J Ethnopharmacol 73(3):415–421

Tsobou R, Mapongmetsem P-M, Van Damme P (2013) Medicinal plants used against typhoid fever in Bamboutos division, western Cameroon. Ethnobot Res Appl 11:163–174

Van Damme W (1996) Beauty in context: towards an anthropological approach to aesthetics: Brill

Woldeamanuel MM, Geda MK, Mohapatra S, Bastia TK, Rath P, Panda AK (2022) Ethnobotanical study of endemic and non-endemic medicinal plants used by indigenous people in environs of Gullele botanical garden Addis Ababa, central Ethiopia: a major focus on Asteraceae family. Front Pharmacol 13:1020097

Wu L, Nie L, Guo S, Wang Q, Wu Z, Lin Y et al (2022) Identification of medicinal Bidens plants for quality control based on organelle genomes. Front Pharmacol 13:842131

Wubetu M, Sintayehu M, Aeta MA (2018) Ethnobotany of medicinal plants used to treat various mental illnesses in Ethiopia: a systematic review. Asian J Plant Sci Res 8(1):9–33

Zahara K, Bibi Y, Tabassum S, Bashir T, Haider S, Araa A et al (2015) A review on pharmacological properties of Bidens biternata: a potential nutraceutical. Asian Pac J Trop Dis 5(8):595–599

Zahara K, Bibi Y, Masood S, Nisa S, Sher A, Ali N et al (2022a) Isolation and identification of bioactive compounds from bidens spp using HPLC-DAD and GC-MS analysis and their biological activity as anticancer molecules. Molecules 27(6):1927

Zahara K, Bibi Y, Arshad M, Kaukab G, Al Ayoubi S, Qayyum AJ (2022b) In-vitro examination and isolation of antidiarrheal compounds using five bacterial strains from invasive species Bidens bipinnata L. Saudi J Biol Sci 29(1):472–479

6 Genus *Bidens* in Indian Traditional Ayurvedic System of Medicine

Sehrish Imran, Muhammad Hasnain, Abdul Manan Yousaf, Yamin Bibi, Abdul Qayyum, and Farooq Ahmad

Abstract

Ayurveda, a traditional Indian system of medicine, is gaining increasing popularity, particularly for its effectiveness in treating chronic conditions. Typically, patients start conventional medications immediately after diagnosis, with Ayurvedic treatments often being used in conjunction with or following these conventional approaches. The genus *Bidens,* which includes a large number of medicinal plants, is well known for its role in ethnomedicine and is integral to the Indian system of medicine, Ayurveda. This study aims to document the *Bidens* species utilized in Ayurveda, along with their vernacular names and traditional uses. A total of five species within the *Bidens* genus, belonging to the Asteraceae family, have been documented. This work will serve as a valuable resource for students and researchers, providing easy access to information on Ayurvedic plant.

Keywords

Bidens · Ayurveda · Indian · Medicine · Conventional

S. Imran (✉) · M. Hasnain · A. M. Yousaf
Department of Botany, PMAS Arid Agriculture University, Rawalpindi, Pakistan

Y. Bibi
Department of Botany, Rawalpindi Women University, Rawalpindi, Punjab, Pakistan

A. Qayyum
Department of Agronomy, The University of Haripur, Haripur, Pakistan

F. Ahmad
Sustainable Development Study Centre, Government College University, Lahore, Pakistan

Y. Bibi et al. (eds.), *The Genus Bidens*,
https://doi.org/10.1007/978-981-96-4257-1_6

6.1 Introduction

Plants have played a significant role in human life, and their utilization for treating various diseases has a long history (Imran et al. 2023). The earliest documented records of using medicinal plants date back to 2600 BC, authored by the Sumerians and Akkadians. In ancient India, the knowledge of medicinal herbs and their application was described in the Rigveda and Atharvaveda (3500–1500 B.C.), paving the way for the alternative medical system of Ayurveda (Singh et al. 2020). In the present era, over 70% of people worldwide rely on plant-derived medicines to treat various illnesses and health conditions (Zahara et al. 2022).

Ayurvedic practice is around 3000 years old, with a long history of managing disease. The three basic principles, called doshas (vata, pitta, and kapha), are derived from five elements of Indian philosophy. Ayurveda's doshas can be identified as regulatory control factors for fundamental physiologic processes in living systems that maintain their identity throughout biologic history: vata and its subdoshas regulating input/output processes and motion; pitta and its subdoshas regulating throughput, turnover, and hence energy; and kapha and its subdoshas regulating storage, structure, and lubrication (Sukumaran and Parthiban 2014).

The genus *Bidens* belongs to the Asteraceae family, comprising about 230 species of flowering plants distributed across tropical and temperate regions worldwide. These plants are commonly known as beggarticks or burr marigolds. *Bidens* species are easily recognized by their distinctive seeds, which attach themselves to clothing and animal fur, aiding in seed dispersal (Norton 1984) (Xuan and Khanh 2016).

Despite the widespread traditional use of *Bidens* species, they remain relatively underexplored in mainstream Ayurvedic literature compared to other well-known medicinal plants. Given the global resurgence in interest in natural and complementary medicine, there is a significant opportunity to further investigate and document the Ayurvedic potential of *Bidens* species. By integrating traditional knowledge with modern scientific research, there is potential to develop new therapeutic applications and enhance the understanding of these plants within the Ayurvedic system. This chapter aims to provide a comprehensive overview of the genus *Bidens* within the context of Ayurvedic medicine (Table 6.1). It will cover the traditional uses, Ayurvedic properties, of key *Bidens* species. By the end of this chapter, readers will have a deeper understanding of the therapeutic potential of *Bidens* and its relevance in both traditional and modern medical practices.

6.2 Ayurvedic Application

6.2.1 Biden pilosa

Bidens pilosa L. is a particularly noteworthy species within this genus, distinguished by the wide range of naturally characterized products and the various biological activities associated with its extracts, fractions, and compounds (Arthur et al. 2012).

Table 6.1 Usage of genus *Bidens* in indigenous tribes of India

Tribe name	Location	*Bidens* species	Usage	References
Santhal	Jharkhand, West Bengal, and Odisha,	*Bidens pilosa*	Anti-inflammatory and antimicrobial properties	Soren and Jamir (2020)
Garo	Northeastern state of Meghalaya	*Biden* spp	Treat wounds and infections	Nakhro (2022)
Bhils	Rajasthan and Madhya Pradesh	*Bidens tripartita*	Diuretic and detoxifying effects, particularly in treating urinary tract infections	Sharma (2023)
Mishing	Assam	*Bidens pilosa*	Treating digestive issues and infections	Young (2022)
Khasis	Meghalaya	*Biden plants*	Skin conditions and respiratory issues	Sruthi and Mukherjee (2020)
Toda	Nilgiri Hills of Tamil Nadu	*Bidens plants*	For their anti-inflammatory and analgesic properties, often in the form of poultices for pain relief	Nair and Sunny (2022)
Konyak	Nagaland	*Bidens* spp	Treating fevers and digestive issues	Konyak et al. (2021)
Irula	Tamil Nadu and Kerala	*Bidens* spp	To treat skin conditions and infections	Stanley and Muthukumar (2021)
Baiga	Madhya Pradesh and Chhattisgarh	*Bidens* spp	Anti-inflammatory and analgesic properties often in remedies for pain relief and inflammation	Soni and Soni (2022)
Lushai	Mizoram	*Bidens*pilosa	called vawkpuithal and treating infections, rheumatism, diarrhoea, ear, eyes tooth ache , skin ailments	Sharma et al. (2001)
Lepcha	Sikkim and West Bengal	*Bidens* spp	Treat respiratory problems and infections	Bentley (2021)
Korku	Madhya Pradesh and Maharashtra	*Bidens* spp	Anti-inflammatory and analgesic properties Often in remedies for pain and inflammation	Choudhary and Abdul Azeez (2020)
Mikir	Ri-Bhoi	*Bidens* pilosa	Treating digestive issues and skin conditions	Bhardwaj et al. (2024)
Gond	Central India	*Bidens* spp	Anti-inflammatory and analgesic properties Often in remedies for pain relief and wound healing	Soni and Soni (2022)
Siddi	Karnataka, Gujarat, and Maharashtra	*Bidens* spp	To treat infections and inflammation	Singh and Dalal (2020)

Species of the *Bidens* genus and their varieties are often given vernacular names that reflect their distinctive characteristics. For example, due to their sticky achenes, *Bidens* species are commonly known by names such as Spanish needles, beggar's ticks, devil's needles, cobbler's pegs, broomstick, pitchforks, and farmer's friends in English and other languages. In Chinese, they are sometimes referred to as xian feng cao, meaning "all bountiful grass," in recognition of their vigorous growth (Bartolome et al. 2013). The Ayurvedic properties of *Bidens pilosa*, referred to as "Rasa Panchaka," include its taste (Rasa), potency (Virya), post-digestive effect (Vipaka), and inherent qualities (Gunas). These properties interact with the body's doshas (Dosas) in a distinct way, influencing their balance and function. All parts of the *Bidens pilosa* plant—including the whole plant, aerial parts (such as leaves, flowers, seeds, and stems), and roots, whether fresh or dried—are commonly utilized in traditional folk medicine. These components are often prepared in various forms, such as dry powder, decoctions, macerations, or tinctures (Redl and Erisman 1994). *B. pilosa* is used as a herb and as an ingredient in teas and herbal remedies. Its branches and leaves, whether dried or fresh, are used in sauces and teas (Rybalchenko et al. 2010). In the 1970s, the United Nations Food and Agriculture Organization (FAO) advocated the growth of *B. pilosa* in Africa because it is easy to cultivate, edible, tasty, and safe (Zhu et al. 2021). Generally, this plant is applied as a dry powder or tincture when used topically and as a powder, maceration, or decoction when used internally. *Bidens pilosa*, whether used as a whole plant or in parts, has been reported to be effective in treating over 40 different conditions, including inflammation, immune system disorders, digestive issues, infections, cancers, metabolic syndrome, wounds, and more (see table 6.1) (Bartolome et al. 2013). *Bidens pilosa* is typically ingested, but it can also be applied externally. For example, fresh *B. pilosa* is used to treat snake bites and wounds (Owoyemi and Oladunmoye 2017), while in Trinidad and Tobago, an aqueous solution made from its leaves is used for bathing babies and children (Hall and Nguyen 2021). *Bidens pilosa* is sometimes used on its own but is also frequently included in medicinal mixtures with other plants, such as *Aloe vera*, *Plectranthus mollis*, *Valeriana officinalis*, and *Cissus sicyoides*, among others (Ayyanar and Ignacimuthu 2005). It is not yet clear whether the commonly used mixtures of *Bidens pilosa* or the combinations of its compounds produce synergistic effects, and this requires further investigation. However, one study has indicated that different varieties of *B. pilosa* share similar phytochemical compositions and may be interchangeable in use (Chien et al. 2009).

6.2.2 Biden biternata

Bidens biternata is known by various vernacular names across different regions. In India, it is referred to as Phutjom, Kerrari, and Kuro. In Thailand, it is called Chonchum, while in Indonesia, it is known as Ketul and Agedi. In China, the plant is referred to as Jinzhan or Yinpin. In Pakistan, it goes by the names Dipmal and Phutium (Zahara et al. 2015). This plant is commonly found, especially in the Western Ghats regions of Kerala, India. The Paniya and Kattunaayika tribes of

Wayanad District in Kerala use it both as a leafy vegetable and as a remedy for ailments such as hepatitis, colds, coughs, dysentery, and other conditions (Sukumaran and Parthiban 2014). *Bidens biternata* is a highly medicinal plant, traditionally used to treat nearly 26 different ailments, including serious skin and lung infections (Ayyanar and Ignacimuthu 2005; Zahara et al. 2019).

6.2.3 Bidens tripartita L.

Bidens tripartita L. is widely distributed across the European part of the Transcaucasia, Siberia, Central Asia (excluding Turkmenistan), and the southern regions of the Far East. Its range also extends to North Africa and North America. In Kazakhstan, this species is commonly found throughout various regions (Tomczykowa et al. 2011). In traditional medicine, *Bidens tripartita* has long been used in the form of water infusions and decoctions, often combined with baths, to address various ailments such as scrofula, rickets, exudative diathesis, and skin conditions like acne and boils. These preparations are also applied in the treatment of gout, arthritis, and joint rheumatism. Additionally, they are recommended for improving appetite and digestion and for treating liver and spleen disorders, colds, bronchitis, and diabetes mellitus (Orhan et al. 2016; Oproshanska et al. 2021). The extract of *Bidens tripartita* is widely used to treat various skin conditions, including psoriasis, seborrhea, urticaria, diathesis, acne, pimples, wounds, ulcers, and minor skin cracks. Its positive effects on the skin are largely attributed to the presence of tannins, which also help reduce excessive sweating in the armpits and feet. As a result, *B. tripartita* is often used in baths, lotions, and wipes to treat microbial eczema of the feet and epidermophytosis (Karazhan 2013). Additionally, masks made from this plant have been shown to reduce oily skin, tone the skin, and provide a rejuvenating effect. Wiping the face with a decoction of *B. tripartita* has been found to help reduce acne (Kumari and Kakkar 2008). For diathesis, an infusion made from 10 to 30 g of *B. tripartita* herbs is commonly added to bathwater (Kaskoniene et al. 2020). The infusion of *Bidens tripartita* herb exhibits diuretic, diaphoretic, choleretic, and bactericidal properties while also aiding digestion and normalizing metabolic imbalances. Additionally, the oil extract from this plant is known for its anti-inflammatory and wound-healing effects (Kuzmishyna 2024).

6.2.4 Bidens bipinnata

Bidens bipinnata is well known for anticancerous activity. Whole plant of *B. bipinnata* was traditionally used to cure a variety of ailments, including laryngitis, acute nephritis, dysentery, snake bite, chronic diarrhea, and malaria (Yang et al. 2012). Dried aerial parts of *Bidens bipinnata*. Are added in a cup of hot water for 10–15 minutes. Strained and drink. This infusion is often used for its anti-inflammatory and diuretic properties (Zahara et al. 2015). Fresh or dried *Bidens bipinnata* leaves are boiled, strained, and consume. This decoction is traditionally used to treat respiratory issues and digestive problems. Dried *Bidens bipinnata* leaves are ground into a fine powder. It can be mixed with honey or warm water and taken orally to boost immunity and manage diabetes.

6.2.5 Bidens gardneri

Bidens gardneri Bak., a member of the Asteraceae family, is a plant that grows between 30 and 100 cm tall and produces yellow flowers. In Brazil, it is commonly referred to as **"picão-preto."** This plant thrives in tropical and subtropical regions, from sea level up to altitudes of 3000 m. It is widely used in traditional medicine, particularly for treating liver disorders such as hepatitis and jaundice. The infusion of *B. gardneri* is commonly used for its anti-inflammatory, diuretic, antirheumatic, and antidiabetic properties (Felix et al. 2020). However, the full biological effects of its indiscriminate use as an herbal medicine remain not fully understood.

6.3 Conclusion

The genus *Bidens* holds significant Ayurvedic potential due to its wide range of medicinal properties. In India, this plant is used as a wild vegetable with high nutritional value. Its use aligns with the holistic principles of Ayurveda, aiming to promote overall health and balance within the body.

References

Arthur G, Naidoo K, Coopoosamy RJ (2012) Bidens pilosa L.: Agricultural and pharmaceutical importance. J Med Plants Res 6(17):3282–3287

Ayyanar M, Ignacimuthu S (2005) Traditional knowledge of Kani tribals in Kouthalai of Tirunelveli hills, Tamil Nadu, India. J Ethnopharmacol 102(2):246–255

Bartolome AP, Villaseñor IM, Yang W-C (2013) Bidens pilosa L. (Asteraceae): botanical properties, traditional uses, phytochemistry, and pharmacology. Evid Based Complement Alternat Med 2013(1):340215

Bentley J (2021) Protectors of the land and water: citizenship, territory and vulnerability among the Lepcha in Sikkim and West Bengal. Asian Ethnicity 22(2):330–352

Chien S-C, Young PH, Hsu Y-J, Chen C-H, Tien Y-J, Shiu S-Y et al (2009) Anti-diabetic properties of three common Bidens pilosa variants in Taiwan. Phytochemistry 70(10):1246–1254

Choudhary HR, Abdul Azeez EP (2020) The Korku tribe of Melghat region in India and their current state: an ethnographic study. The Oriental Anthropologist 20(1):97–112

Owoyemi OO, Oladunmoye MK (2017) Phytochemical screening and antibacterial activities of Bidens pilosa L. and Tridax procumbens L. on skin pathogens. Int J Mod Biol Med 8(1): 24-46.

Felix JM, Bertoncello LA, do Amaral Alves MC, de Miranda Pedroso TF, Dos Santos RC, Formagio ASN et al (2020) Effects of Bidens gardneri baker leaves aqueous extract in glucose homeostasis of rats. Biomed Pharmacother 127:110185

Hall D, Nguyen TA (2021) Radical narratives in the American jobs plan (AJP) and the global potential of 'Biden plans'

Imran S, Bibi Y, Munawar T, Yousaf AM, Hasnain M (2023) A panoramic review on ethnomedicinal, therapeutic, phytochemical, and advance attributes, of the genus Ziziphus Mill., native to Pakistan. Ethnobot Res Appl. 25: 1-32.

Karazhan N (2013) Sravnitel'nyj analiz komponentnogo sostava spirtovyh izvlechenij iz travy vidov cheredy metodom zhidkostnoj hromatografii [the comparative analysis of components composition of the spirit extracts from Bidens species herb by liquid chromatography method]. 4:49–56

Kaskoniene V, Bimbiraite-Surviliene K, Kaskonas P, Tiso N, Cesoniene L, Daubaras R et al (2020) Changes in the biochemical compounds of Vaccinium myrtillus, Vaccinium vitis-idaea, and forest litter collected from various forest types. Turk J Agric For 44(6):557–566

Konyak PA, Semy K, Puro N (2021) Neon-timber forest products as a means of livelihood in Mon district, Nagaland, India. Curr Sci 121(6):837–840

Kumari A, Kakkar P (2008) Screening of antioxidant potential of selected barks of Indian medicinal plants by multiple in vitro assays. Biomed Environ Sci 21(1):24–29

Kuzmishyna II (2024) Medicinal plants and medicinal raw materials: theory and practice

Nair KL, Sunny G (2022) Cultural documentation and collection development: Toda tribes of Nilgiris. Chitrolekha J Art Design 6(2)

Nakhro K (2022) A study on nutritional status among the adult Garos' of west Garo hills, north-East India (Meghalaya). Glasnik Antropoloskog drustva Srbije 55:19–25

Norton WT (1984) Isolation and characterization of myelin. Springer, Myelin, pp 161–199

Oproshanska T, Mishchenko O, Shapoval O, Khvorost O, Koiro O, Bondariev Y (2021) Study of membrane-stabilizing and anti-inflammatory activity of tincture from herba Bidens Tripartita L PharmacologyOnline. 3:1863-1869

Orhan N, İçöz ÜG, Altun L, Aslan M (2016) Anti-hyperglycaemic and antioxidant effects of Bidens tripartita and quantitative analysis on its active principles. Iran J Basic Med Sci 19(10):1114

Redl R, Erisman BP (ed) (1994) Reducing distortion in peak-current-controlled boost power-factor correctors. In: Proceedings of 1994 IEEE applied power electronics conference and exposition-ASPEC'94. IEEE

Rybalchenko N, Prykhodko V, Nagorna S, Volynets N, Ostapchuk A, Klochko V et al (2010) In vitro antifungal activity of phenylheptatriyne from Bidens cernua L. against yeasts. Fitoterapia 81(5):336–338

Sharma HK, Chhangte L, Dolui AK (2001) Traditional medicinal plants in Mizoram, India. Fitoterapia 72(2): 146-161

Sharma R (2023) An ecological study of the Bhil tribe in Rajasthan. Innov Res Thoughts 9(4):158–164

Singh K, Dalal S (2020) Socio-economic dynamics and livelihood patterns in the siddi community. Annals NAGI 40(1):151-162

Singh YD, Jena B, Ningthoujam R, Panda S, Priyadarsini P, Pattanayak S, et al. Potential bioactive molecules from natural products to combat against coronavirus. Adv Tradit Med 2020:1–12

Soni LN, Soni A (2022) Gonds in a Chhattisgarh village: economic pattern and type of subsistence. Tribal People of Central India: Problems and Prospects,274

Bhardwaj Y, Pulicherla Y, Nagayya S, Cheemanapalli S, Meher SK, Taboh M, Bhuyan B (2024) Exploration of medicinal plants diversity of Ri-Bhoi district, Meghalaya. Journal of Drug Research in Ayurvedic Sciences 9(4): 222-246.

Soren P, Jamir WJS (2020) Chapter-5 the Santhals: their culture and traditions. In: Trends in sociology, psychology and anthropology, p 79

Sruthi P, Mukherjee S (2020) Women, nature and culture: an ecofeminist reading of the matrilineal culture of the Khasis, Jaintia and Garo Tribes of Meghalaya. Cultura de los Cuidados 24:165–177

Stanley FS, Muthukumar P (2021) Health status and living conditions of Irulas in Coimbatore district of Tamil Nadu, India. Elementary Educ Online 20(4)

Sukumaran S, Parthiban BJBD (2014) Vascular plant diversity of Udayagiri fort, Kanyakumari district, Tamilnadu, India. Bioscience Discovery 5(2):204–217

Tomczykowa M, Leszczyńska K, Tomczyk M, Tryniszewska E, Kalemba D (2011) Composition of the essential oil of Bidens tripartita L. roots and its antibacterial and antifungal activities. J Med Food 14(4):428–433

Xuan TD, Khanh TD (2016) Chemistry and pharmacology of Bidens pilosa: an overview. J Pharm Investig 46:91–132

Yang X-W, Huang M-Z, Jin Y-S, Sun L-N, Song Y, Chen H-S (2012) Phenolics from Bidens bipinnata and their amylase inhibitory properties. Fitoterapia 83(7):1169–1175

Young KN, 'The wait for the flood is also the wait for fish!': Water commons among the Mishing on Majuli Island (Assam), India. Millenn J Hum Soc Sci. 3(4): 2022

Zahara K, Bibi Y, Tabassum S, Bashir T, Haider S, Araa A et al (2015) A review on pharmacological properties of Bidens biternata: a potential nutraceutical. 5(8):595–599

Zahara K, Bibi Y, Qayyum A, Nisa S (2019) Investigation of antimicrobial and antioxidant properties of Bidens Biternata. J SCI TECHNOL TRANS A SCI. 43:725–734

Zahara K, Bibi Y, Masood S, Nisa S, Sher A, Ali N et al (2022) Isolation and identification of bioactive compounds from bidens spp. using HPLC-DAD and GC-MS analysis and their biological activity as anticancer molecules. Molecules 27(6):1927

Zhu YJ, Li CH, Yu SH, Jiang HQ, Tian ZH, Wang WQ et al (2021) Two new polyacetylene glucosides and a new caffeoyl derivative with angiogenic activity from Bidens parviflora Willd. Phytochem Lett 42:82–86

Nutraceutical Potential of *Bidens* Species 7

Zakir Ullah, Banzeer Ahsan Abbasi, Javed Iqbal, Shumaila Ijaz, Mamoona Munir, Tabassum Yaseen, Sobia Kanwal, Tariq Mahmood, Sayed Afzal Shah, Imran Khan, and Shobana Sampath

Abstract

The *Bidens* species, sometimes called bur marigolds or beggarticks, gained significant interest owing to their diverse pharmacological and biomedical applications in traditional medicine and nutraceuticals around the globe. The scientific research on the phytochemistry and therapeutic applications of *Bidens* species is systematically examined in this chapter, accenting the species' geographic range and therapeutic potentials. As determined by phytochemical investigation, numerous *Bidens* plants contain bioactive substances, including phenolic acids, terpenoids, polyacetylenes, and flavonoids. The pharmacological activity of these phytoconstituents is diverse and includes hepatoprotective, anticancer, anti-inflammatory, antibacterial, and antioxidant properties. For several generations, people from many cultures have used the species of *Bidens* in traditional

Zakir Ullah and Banzeer Ahsan Abbasi authors have equally contributed to this chapter.

Z. Ullah
Department of Plant Sciences, Faculty of Biological Sciences, Quaid-i-Azam University Islamabad, Islamabad, Pakistan

B. A. Abbasi (✉) · M. Munir
Department of Botany, Rawalpindi Women University, Rawalpindi, Pakistan
e-mail: banzeer.abbasi@f.rwu.edu.pk; mamoona.munir@f.rwu.edu.pk

J. Iqbal (✉) · T. Yaseen
Department of Botany, Bacha Khan University, Charsadda, Khyber Pakhtunkhwa, Pakistan
e-mail: tabassumyaseen@bkuc.edu.pk

S. Ijaz · T. Mahmood
School of Biomedical Engineering, Shenzhen University Medical School, Shenzhen University, Shenzhen, China
e-mail: tmahmood@qau.edu.pk

Y. Bibi et al. (eds.), *The Genus Bidens*,
https://doi.org/10.1007/978-981-96-4257-1_7

medicine. Infections, gastrointestinal issues, and inflammatory problems are just a few of the disorders for which *Bidens pilosa* has been used throughout Asia. Skin diseases, fever, and malaria have all been treated using *Bidens* species in African traditional medicine. *Bidens* plants have also been used for their analgesic, wound-healing, and anti-inflammatory qualities. Furthermore, a growing body of research validates the potential of *Bidens* species in functional foods and nutraceutical formulations. In preclinical research, extracts and isolated chemicals from *Bidens* plants have shown promise in treating chronic illnesses like diabetes, cancer, and cardiovascular conditions. Even yet, more investigation is necessary to clarify the molecular mechanisms of action, improve extraction methods, standardize formulations, and assess the safety profiles of products based on *Bidens*. To sum up, *Bidens* species are an important source of bioactive substances with a variety of pharmacological characteristics and medical uses. The production of nutraceuticals and pharmaceutical products derived from *Bidens* can be enhanced through the integration of traditional knowledge with modern scientific approaches for the betterment of world health and well-being.

Keywords

Phytochemicals · Anti-inflammatory · Antioxidant · Traditional medicine · Bioactive compounds · Pharmacological properties

7.1 Introduction

Bidens is a genus of the Asteraceae family and has a global distribution with over 230 species. The taxonomy and systematics of the genus are still complicated and disjointed despite their widespread distribution (Fig. 7.1) (Antoniak et al. 2023). Beggarticks, blackjack, burr marigolds, cobbler's pegs, Spanish needles, stickseeds, tickseeds, and tickseed sunflowers are some of the common names for *Bidens* species. These names are derived from the fruits' characteristic bristles and barbed skin

S. Kanwal
Department of Biology and Environmental Sciences, Allama Iqbal Open University, Islamabad, Pakistan
e-mail: sobia.kanwal@aiou.edu.pk

S. A. Shah
Department of Biological Sciences (PWD Campus), National University of Medical Sciences (NUMS), Rawalpindi, Pakistan
e-mail: afzal.shah@numspak.edu.pk

I. Khan
Department of Food Science and Technology, The University of Haripur, Haripur, Pakistan
e-mail: imrankhan572@uoh.edu.pk

S. Sampath
Department of Biotechnology, Vel Tech Rangarajan Dr. Sagunthala R&D Institute of Science and Technology, Chennai, Tamil Nadu, India

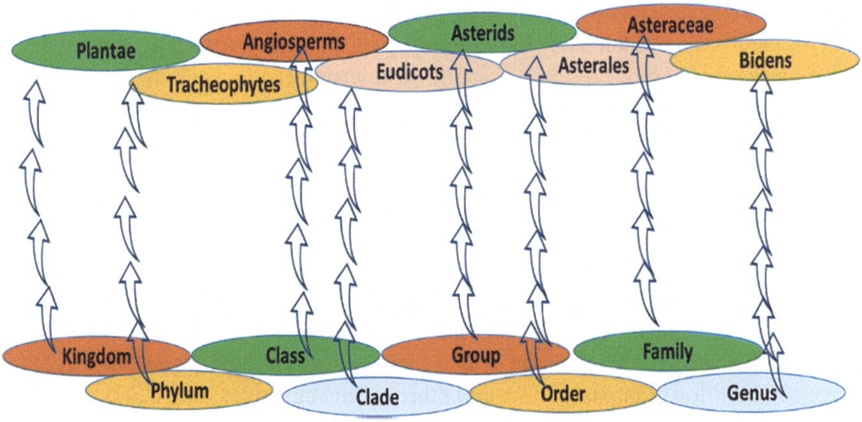

Fig. 7.1 Description of medicinally important *Bidens* genus with many biologically assorted species

(Shandukani et al. 2018). The Latin words bis (meaning "two") and dens (meaning "tooth") are the source of the genus name *Bidens,* which refers to the distinctive fruits with two teeth (Pozharitskaya et al. 2010). The dominant species including the *Bidens* genus include *Bidens cernua* (Nodding beggarticks), *Bidens alba* (L.) DC. (Romerillo), etc. (Horiuchi and Seyama 2006). The Americas, Africa, Polynesia, Europe, and Asia are home to these species (Yang et al. 2015). The ability of *Bidens* species to disperse by zoochory, in which seeds stick to clothes, fur, or feathers to make their way to other environments, is one of their most important traits (Mendel et al. 2020). *Bidens* species have established themselves in various habitats, including several marine islands, because of their successful dispersion mechanism (Korkmaz et al. 2011). However, some *Bidens* species are restricted to minimal geographic ranges, leading to several species being threatened with extinction. This is especially notable in the Hawaiian Islands, where the unique biogeography and the absence of native mammals have influenced the evolution of certain *Bidens* species (Pradeesh 2018). On these islands, some taxa have developed reduced burrs, which appear to be an adaptation for wind dispersal rather than attachment to animals. This evolutionary shift highlights the genus's adaptability to different dispersal vectors in response to local ecological conditions (Fig. 7.1) (Redl et al. 1994).

7.2 Nutritional Values of *Bidens* Species

As mentioned earlier, *Bidens* species, often known as beggarticks or bur marigolds, are plants with various applications in traditional medicine and as ornamental plants (Chiang et al. 2004). The nutritional values of *Bidens* species can vary depending on the specific species and part of the plant being analyzed (Chauhan et al. 2019).

Though there are some general aspects often reported, the young leaf of *B. biternata* has high quantities of crude fiber (0.55%), moisture (0.03%), sugar (0.08 mg/g), total carbohydrate (16.06 mg/g), total protein (22.1 mg/g), and vitamins A and E. The stem has a notable concentration of lipids and free fatty acids (Zahara et al. 2019). Because *B. biternata* has a wealth of micronutrients, such as iron, magnesium, manganese, aluminum, zinc, and copper, it is frequently served as an appetizer in many different cultures. Compared to other plant components including the mature leaf, immature leaf, and stem, the root has the largest concentrations of manganese, iron, copper, and zinc (Zahara et al. 2022).

Human nutrition cannot exist without plant-based proteins, and *B. biternata* is well known for having a high total protein content, especially in its leaves (22.1 mg/g) (47). Compared to wheat (22–35 mg/g), rice (30–36 mg/g), and maize (28–42 mg/g), this protein content is similar (48). Amino acid content is high in *B. biternata's* young leaf and stem. Methionine is one of them that is more prevalent in young leaves. In addition, compared to several other green vegetables, like *Cassia tora* and *Cissus quadrangularis*, *B. biternata* has higher quantities of proteins, total reducing sugar, carbs, amino acids, and crude fiber (Zahara et al. 2019).

Bidens pilosa has been acknowledged for its potential in traditional Chinese medicine and as a functional meal for the treatment of hypertensive disorders. Using a bioactivity-guided fractionation approach, bioactive compounds with renin- and angiotensin I–converting enzyme (ACE)-inhibitory effects were discovered from the ethanol extract of *B. pilosa* in this work (Dzoyem et al. 2013). Compound identification was achieved using chemometrics analysis based on liquid chromatography-mass spectrometry (LC–MS). The fraction that included polyphenols showed the strongest dual inhibitions against ACE and renin activity. By using LC–MS/MS, a total of eighteen marker metabolites were positively or provisionally identified. Of them, five compounds demonstrated strong dual inhibitory effects against ACE and renin in vitro (Mokganya and Tshisikhawe 2019). This work is the first to show that *B. pilosa's* active ingredients can control the renin-angiotensin system (RAS) in vitro. The described procedure offers a quick and efficient way to distinguish bioactive ingredients from nutraceuticals and functional meals.

B. pilosa is an edible plant known for its high nutritional value. It contains significant proteins, carbohydrates, fibers, vitamins, and minerals. The plant has been traditionally used to manage metabolic syndrome due to its rich phytochemical profile, which includes flavonoids, polyacetylenes, and phenolic compounds (Horiuchi and Seyama 2006). Similarly, *Bidens bipinnata* is a nutrient-rich wild plant that contains minerals, fiber, proteins, carbohydrates, vitamins, including β-carotene, and at least 17 types of amino acids (Kumar et al. 2024). This species provides essential nutrients and has potential health benefits due to its rich nutritional profile.

Moreover, *Bidens frondosa* has shown significant nutritional value with a composition that includes minerals, vitamins, fiber, and essential fatty acids. Its leaves contain appreciable percentages of moisture, ash, crude protein, crude lipid, crude fiber, and carbohydrates (Kumar et al. 2024). The plant also provides essential minerals such as sodium, potassium, calcium, magnesium, iron, zinc, phosphorus, copper, manganese, and nitrogen (Bhagawan and Kusumawati 2020). Furthermore, *Bidens alba* is recognized for its nutritional content, including proteins, carbohydrates, fibers, vitamins, and minerals (Torres-Fajardo et al. 2020). This species has been traditionally used in various regions as a food and medicinal plant, providing essential nutrients and health benefits (Zahara et al. 2015).

Likewise, *Bidens hawaiensis*, a species endemic to the Hawaiian Islands, has been studied for its genomic properties, which contribute to its nutritional and medicinal value (Pradeesh n.d.). This species is part of an adaptive radiation event, showcasing significant genetic diversity and potential health benefits (Pradeesh n.d.). These *Bidens* species exhibit a wide range of nutritional components, making them valuable for both traditional and modern dietary uses (Kamaraj et al. 2022). Their rich profiles of proteins, carbohydrates, vitamins, minerals, and other bioactive compounds highlight their potential as nutraceuticals, contributing to overall health and well-being (Singh et al. 2023).

7.3 Phytochemical Constituents and Pharmacological Activity

Various *Bidens* species, including *Bidens pilosa*, *Bidens hawaiensis*, *Bidens alba*, *Bidens frondosa*, and *Bidens bipinnata*, have garnered interest due to their traditional medicinal uses and recognized bioactive ingredients (Table 7.1) (Atawodi et al. 2014). *B. pilosa* contains 301 phytochemicals categorized into phytosterols,

Table 7.1 Nutraceutical constituents in *Bidens* species

Nutrient	*Bidens* species	Constituents	References
Proteins	*Bidens pilosa*, *Bidens bipinnata*	Crude protein, total proteins	Malik et al. (2018)
Carbohydrates	*Bidens bipinnata*	Total carbohydrates, total soluble sugars, starch	Malik et al. (2018)
Vitamins	*Bidens pilosa*, *Bidens bipinnata*	β-carotene, various vitamins	Sanoussi et al. (2015)
Minerals	*Bidens pilosa*, *Bidens bipinnata*	Calcium, iron, magnesium, potassium, zinc, phosphorus, copper, manganese	Omara et al. (2020)
Fats	*Bidens pilosa*, *Bidens bipinnata*	Crude fat, free fatty acids	Asif (2015)
Fibers	*Bidens bipinnata*	Crude fiber	Kumari et al. (2022)
Flavonoids	*Bidens bipinnata*, *Bidens pilosa*	Quercetin, kaempferol, luteolin, apigenin, various glucosides	Joshi et al. (2018)
Phenolic compounds	*Bidens pilosa*, *Bidens bipinnata*	Caffeic acid, gallic acid	Kumari et al. (2022)
Saponins	*Bidens bipinnata*	Various saponins	Asif (2015)
Terpenoids	*Bidens bipinnata*, *Bidens pilosa*	Various terpenoids	Omara et al. (2020)
Steroids and sterols	*Bidens bipinnata*, *Bidens pilosa*	β-sitosterol, stigmasterol, daucosterol	Joshi et al. (2018)
Anthraquinones	*Bidens bipinnata*	Various anthraquinones	Atawodi et al. (2014)

phenolic acids, polyacetylenes, fatty acids, flavonoids, terpenes, and pheophytins (Omara et al. 2020). Key active compounds include polyacetylenes, flavonoids, triterpenes, and essential oils, which contribute to the plant's diverse pharmacological effects (Table 7.2).

Table 7.2 Different species of *Bidens* genus and their biological activities

Biological activity	*Bidens* species	Description	References
Antimalarial	*Bidens frondosa* *Bidens pilosa,*	Ethanol extracts showed up to 90% inhibition of *Plasmodium falciparum* growth in vitro; acetylene and flavonoid compounds were identified as active substances	Kumari et al. (2022)
Antioxidant	*Bidens pilosa, B. subalternans*	Aqueous infusion showed protective effects on hemolysis induced by radicals and increased cytokine production, indicating antioxidant and immunomodulatory activity	Joshi et al. (2018)
Anti-inflammatory	*Bidens pilosa*	Extracts showed inhibition of inflammatory mediators and enzymes	Alamgir (2017)
Antimicrobial	*Bidens pilosa*	Ethanol extracts and fractions showed broad-spectrum antibacterial activity, with ethyl acetate and petrol fractions being the most effective	Sharma et al. (2009)
Antitumor	*Bidens pilosa*	Chloroform fraction and hydroalcoholic extract significantly reduced tumor volume and increased lifespan in the Ehrlich ascites carcinoma mouse model	de Almeida et al. (2019)
Herbicide Resistance	*Bidens* spp.	High genetic variability within populations and resistance to ALS-inhibiting herbicides were observed	Kumar et al. (2024)
Cadmium Accumulation	*Bidens pilosa*	Different ecotypes of *B. pilosa* showed varied capabilities for cadmium accumulation and tolerance, with higher accumulation linked to higher antioxidant enzyme activities	Sharma et al. (2023)
Reproductive Characteristics	*Bidens frondosa*	High seed production, versatile mating system (self- and cross-pollination), and high germination rate contribute to invasiveness	Sharma et al. (2009)
Symbiotic Relationships	*Bidens pilosa*	Symbiotic relationship with AM fungus *Glomus mosseae* enhances drought resistance by adjusting the antioxidant system and reducing cell membrane damage	Sharma et al. (2023)
Predation and Biological Control	*B. subalternans,*	Molecular ecology techniques identified key predators feeding on *C. bidens*, including *Chrysoperla externa* and *Harmonia axyridis*, which can aid in biological control strategies	Al-Snafi (2016)

Phytochemical constituents significantly influence the pharmacological activity of *Bidens* species. The total phenolic content, an indicator of antioxidant properties, is highest in the stems of *B. biternata* (0.16 mg/g), followed by the root (0.15 mg/g), mature leaf (0.042 mg/g), and young leaf (0.04 mg/g) (Omara et al. 2020). *B. biternata* has low phytic acid content (0.35), reducing the risk of nutrient absorption inhibition (Omara et al. 2020). Tannic acid, found in low concentrations in mature and young leaves, negatively affects growth rate, feed intake, feed efficiency, and protein digestibility (Joshi et al. 2018).

Bidens species are rich in flavonoids, which contribute to their medicinal properties. Some of the flavonoids identified in different *Bidens* species include luteolin is known for its anti-inflammatory, antioxidant, and anticancer properties (Atawodi et al. 2014). Another common flavonoid in *Bidens* species, quercetin, has strong antioxidant properties and has been studied for its potential benefits in reducing inflammation, allergies, and blood pressure (Malik et al. 2018). Moreover, apigenin is known for its anti-inflammatory, antioxidant, and anticancer activities; apigenin is another flavonoid present in some *Bidens* species.

Thirty-three phenolic compounds have been identified in various parts of its plants, including common acids-like compounds found in leaves, stems, and roots. Twelve derivatives of p-coumaric and caffeoylquinic acids are present throughout the plant. *Bidens* species exhibit significant terpene diversity, with 99 distinct molecules, including triterpenes, sesquiterpenes, diterpenes, and monoterpenes.

Twenty flavonoid glycosides have been identified, with compounds in the leaves and other compounds throughout the plant. Sixteen okanin chalcone glycosides are found in the flowers, leaves, and aerial parts, with compounds in the leaves (70). Ten flavonoids, including quercetin structures, while other compounds, found in leaves and aerial parts, have a unique structure. Compounds like okanin structures (Joshi et al. 2018). These phytochemicals highlight the potential of *Bidens* species for various therapeutic applications due to their antioxidant, anti-inflammatory, antimicrobial, and anticancer properties (Fig. 7.2).

Bidens species exhibit a broad spectrum of biological and pharmacological activities, including anti-inflammatory, antimicrobial, anti-allergy, antihypertensive, anticarcinogenic, and smooth muscle relaxant effects (Asif 2015).

Bidens bipinnata shows significant antiplasmodial activity, with its flavonoid and phenolic-rich extracts reducing parasitemia and alleviating malaria symptoms (Hu et al. 2018). *Bidens pilosa*, *Bidens bipinnata*, and *Bidens alba* have antimicrobial properties, whereas *Bidens bipinnata* and *Bidens cernua* possess cytotoxic effects and may influence tumor growth through apoptosis and cell cycle regulation (Fig. 7.2) (Kviecinski et al. 2011).

In vitro studies confirm that *Bidens* extracts have a vasorelaxant effect on isolated rat aortas, likely due to their ability to block Ca^{+2} channels. Further studies reveal that *Bidens* extracts can restore normal blood pressure and hypertriglyceridemia in rats fed a high-fructose diet, primarily through vascular actions rather than changes in glucose or insulin levels (Nole et al. 2016).

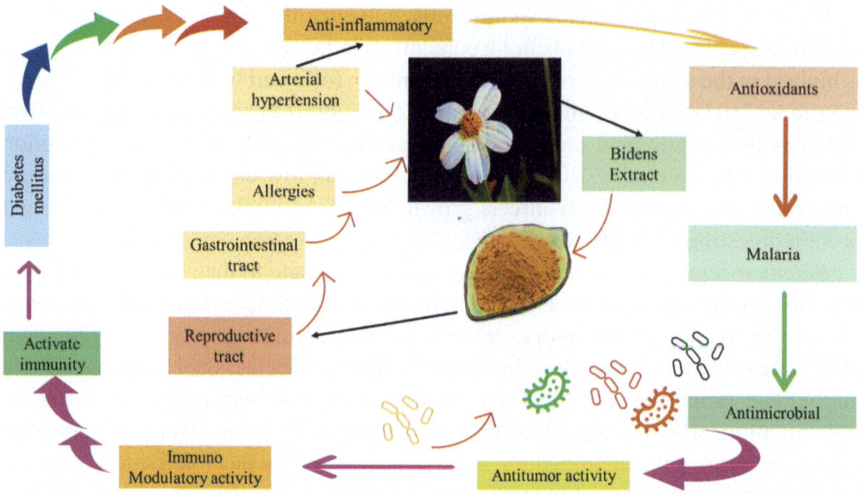

Fig. 7.2 Different biological and biomedical potentials of *Bidens* species for the treatment of different diseases

7.4 Experimental and Human Toxicology

Toxicology evaluations usually entail testing on both humans and experimental animals; however, there are currently insufficient studies with large enough sample sizes to conclusively categorize *Bidens* as safe and nontoxic (Ong et al. 2008). However, when mice were injected with a dosage of 1 g per kg of body weight, several early experimental tests did not reveal any signs of toxicity (Kviecinski et al. 2011). Additionally, oral infusions of crushed powder from *Bidens* aerial parts at a concentration of 100 mg/mL did not exhibit any toxicity when given to rats for 28 days at a dose limit of 2000 mg/kg (134). Rats exposed to ethanol or water-based extracts of *Bidens* leaves have demonstrated negligible harmful effects in vivo (135.136). Moreover, neither erythema nor skin edema were brought on by repeated dosages in subsequent studies (Ezuruike and Prieto 2014).

However, as tea prepared from *Bidens* aerial parts by distillation and decoction has shown genotoxic impacts in vitro, care should be taken when using these infusions. *Bidens* infusions at a dosage of 40 μl/ml culture media and 2 mg/ml extract are not advised to be used (Uysal et al. 2018). Furthermore, it is doubtful where some studies' sample collections came from because growing medicinal plants in contaminated places may be dangerous. According to recent studies, *Bidens* not only accumulate cadmium (Cd) and other metals but also eliminate arsenic (As), which makes it a potentially hazardous food for humans but an efficient environmental bioremediation of As and Cd (Ong et al. 2008). As and Cd co-contamination causes the concentrations of these elements to build up in *Bidens* tissues to rise in proportion to increased soil levels. There are no reports on this plant's ability to chelate human tissue currently. Because of this, caution should be used while collecting and harvesting *Bidens* for medical uses. Additionally, the plant material should be analyzed to see whether its origins raise any safety issues. Furthermore, the origin of herbal material should always be considered because environmental factors can have a considerable impact on the pharmacological activities of plants, including weather, soil type, and harvest time. The commercial anti-allergic product ClearGuard™, which contains *Bidens*, is safe for use in humans and is like prescription medications like loratadin (Fokunang et al. 2020). It is safe to consume 2 g of dried *Bidens* species twice a day (de Almeida et al. 2019). More human study is required even though *B. pilosa* has a long history of traditional usage, and no significant negative effects have been reported, suggesting that it is probably safe.

7.5 Conclusion

The genus *Bidens* encompasses a diverse array of species that have been extensively utilized in traditional and modern medicine across the globe. Rich in phytochemicals, these plants exhibit a wide range of bioactive compounds that confer significant health benefits. Their medicinal properties include anti-inflammatory, antioxidant, antimicrobial, and immunomodulatory effects, making them valuable in the treatment and prevention of various ailments. Ethnobotanical studies highlight the importance of *Bidens* species in traditional medicine, where they are used to treat conditions such as infections, inflammation, and digestive disorders. Modern scientific research has corroborated many of these traditional uses, demonstrating the potential of *Bidens* extracts and compounds in developing new nutraceuticals and pharmaceuticals. The global distribution of *Bidens* species, coupled with their adaptability to different environments, underscores their significance not only in natural ecosystems but also in medicinal and nutritional applications. As research continues to uncover the pharmacological potential of these plants, *Bidens* species are poised to play a crucial role in the advancement of natural health products and therapies. In conclusion, the nutraceutical and medicinal applications of *Bidens* species represent a promising frontier in both traditional and modern medicine. Their rich phytochemical profile and demonstrated therapeutic effects offer valuable opportunities for health promotion and disease management, warranting further exploration and utilization in the fields of healthcare and wellness.

References

Alamgir A (2017) Therapeutic use of medicinal plants and their extracts: volume 1. Springer

Al-Snafi AE (2016) Medicinal plants with antimicrobial activities (part 2): Plant based review. Sch Acad J Pharm 5(6):208–239

Antoniak K, Studzińska-Sroka E, Szymański M, Dudek-Makuch M, Cielecka-Piontek J, Korybalska K (2023) Antiangiogenic, Anti-Inflammatory and Antioxidant Properties of Bidens tripartite Herb, *Galium verum* Herb, and *Rumex hydrolapathum* Root. Molecules 28(13):4966

Asif M (2015) Chemistry and antioxidant activity of plants containing some phenolic compounds. Chem Int 1(1):35–52

Atawodi SE, Olowoniyi OD, Adejo GO, Liman ML, Dubey N (2014) Review of the antioxidant potential of African medicinal and food plants. Plants as a source of natural antioxidants 34

Bhagawan WS, Kusumawati D (ed) (2020) Ethnobotanical medicinal plant study of Tengger tribe in Ranu Pani Village, Indonesia. Proceedings of the 3rd International Conference on Education & Social Science Research (ICESRE)

Chauhan BS, Ali HH, Florentine S (2019) Seed germination ecology of *Bidens pilosa* and its implications for weed management. Sci Reports 9(1):16004

Chiang Y-M, Chuang D-Y, Wang S-Y, Kuo Y-H, Tsai P-W, Shyur L-F (2004) Metabolite profiling and chemopreventive bioactivity of plant extracts from *Bidens pilosa*. J Ethnopharmacol 95(2–3):409–419

de Almeida RF, Trevisan MTS, Thomaziello RA, Breuer A, Klika KD, Ulrich CM et al (2019) Nutraceutical compounds: Echinoids, flavonoids, xanthones and caffeine identified and quantitated in the leaves of *Coffea arabica* trees from three regions of Brazil. Food Res Int 15:493–503

Dzoyem JP, Tshikalange E, Kuete V (2013) Medicinal plants market and industry in Africa. In: Medicinal plant research in Africa. Elsevier, pp 859–890

Ezuruike UF, Prieto J (2014) The use of plants in the traditional management of diabetes in Nigeria: Pharmacological and toxicological considerations. J Ethnopharmacol 155(2):857–924

Fokunang ET, Fonmboh DJ, Mballa RN, Nyuyki AB, Fokunang LB, Kaba N, et al. Pharmacovigilance of natural herbal medicines research for efficacy, safety and quality assurance of phytomedicine products J. Complement. Altern. Med. Res 2020:21–37

Horiuchi M, Seyama Y (2006) Antiinflammatory and antiallergic activity of *Bidens pilosa* L var radiata SCHERFF. J Health Sci 52(6):711–717

Hu H-M, Bai S-M, Chen L-J, Hu W-Y, Chen G (2018) Chemical constituents from *Bidens bipinnata* Linn. Biochem Syst Ecol 79:44–49

Joshi S, Ballabh B, Negi P, Dwivedi S (2018) Diversity, distribution, use pattern and evaluation of wild edible plants of Uttarakhand, India. Defence Life Sci J 3(2):126–135

Kamaraj C, Ragavendran C, Kumar RCS, Ali A, Khan SU, Luna-Arias JP et al (2022) The antiparasitic potential of Asteraceae plants: A comprehensive review on therapeutic and mechanistic aspects for biocompatible drug discovery. Phytomed plus 2(4):100377

Korkmaz H, Mumcu Ü, Alkan S (2011) A new alien species record for the flora of Turkey: *Bidens bipinnata* (Asteraceae). Biological Diversity Conservation 4(2):49–54

Kumar M, Saini RV, Gupta M, Singh RJBC (2024) Green synthesis of silver nanoparticles (Cha-AgNPs) using Chenopodium album extract and evaluation of their antifungal potential against pathogenic fungi. Bomass Conversion Biorefinery:1–12

Kumari N, Kumar M, Mekhemar M, Lorenzo JM, Pundir A, Devi KB et al (2022) Therapeutic uses of wild plant species used by rural inhabitants of Kangra in the western Himalayan region. South African J Bot 148:415–436

Kviecinski M, Benelli P, Felipe K, Correia J, Pich C, Ferreira S et al (2011) SFE from *Bidens pilosa* Linné to obtain extracts rich in cytotoxic polyacetylenes with antitumor activity. The J Supercritical Fluids 56(3):243–248

Malik K, Ahmad M, Bussmann RW, Tariq A, Ullah R, Alqahtani AS et al (2018) Ethnobotany of anti-hypertensive plants used in northern Pakistan. Frontiers in pharmacol 9:789

Mendel M, Chłopecka M, Latek U, Karlik W, Tomczykowa M, Strawa J et al (2020) Evaluation of the effects of *Bidens tripartita* extracts and their main constituents on intestinal motility–An ex vivo study. J ethnopharmacol 259:112982

Mokganya M, Tshisikhawe M (2019) Medicinal uses of selected wild edible vegetables consumed by Vhavenda of the Vhembe District Municipality, South Africa. S Afr J Bot 122:184–188

Nole T, Djeufack T, Lionel W (2016) Diabetes and arterial hypertension resorts of treatments and plants used for their treatments in three phytogeographic areas of Cameroon. Int J Trad Complement Med 1(4):45–59

Omara T, Kagoya S, Openy A, Omute T, Ssebulime S, Kiplagat KM et al (2020) Antivenin plants used for treatment of snakebites in Uganda: ethnobotanical reports and pharmacological evidences. Tropical Med Health 48:1–16

Ong P-L, Weng B-C, Lu F-J, Lin M-L, Chang T-T, Hung R-P et al (2008) The anticancer effect of protein-extract from *Bidens alba* in human colorectal carcinoma SW480 cells via the reactive oxidative species-and glutathione depletion-dependent apoptosis. Food chemical toxicol 46(5):1535–1547

Pozharitskaya O, Shikov A, Makarova M, Kosman V, Faustova N, Tesakova S et al (2010) Anti-inflammatory activity of an HPLC-fingerprinted aqueous infusion of aerial part of *Bidens tripartita* Phytomedicine 17(6):463–46

Pradeesh S (2018) In vitro anticancer activity of bidens biternata (LOUR.) MERR. & SHERIFF–An Ethno Medicinal Plant Of Kerala. J Medicinal Aromatic Plant Sci 3(1):15–20

Pradeesh S (n.d.) Preliminary phytochemical screening and antimicrobial activity of bidens biternata (LOUR.) MERR. & SHERIFF-OF SOUTHERN WESTERN GHATS

Redl K, Breu W, Davis B, Bauer R (1994) Anti-inflammatory active polyacetylenes from *Bidens campylotheca*. Planta Med 60(01):58–62

Sanoussi F, Ahissou H, Dansi M, Hounkonnou B, Agre P, Dansi A et al (2015) Ethnobotanical investigation of three traditional leafy vegetables [*Alternanthera sessilis* (L.) DC. *Bidens pilosa* L. *Launaea taraxacifolia* Willd.] widely consumed in southern and central Benin. J Biodiv Env Sci 6(2):187–198

Shandukani PD, Tshidino SC, Masoko P, Moganedi KM (2018) Antibacterial activity and in situ efficacy of *Bidens pilosa* Linn and Dichrostachys cinerea Wight et Arn extracts against common diarrhoea-causing waterborne bacteria. BMC Complement Altern Med 18:1–10

Sharma A, Bargali K, Pande N (2009) The allelopathic potential of bryophyte extract on seed germination and seedling growth of *Bidens biternata*. Nat Sci Sleep 7:30–38

Sharma A, Kaur A, Kohli RK, Singh HP, Batish D (2023) *Bidens pilosa* (Asteraceae) invasion reshapes the pattern of plant communities and edaphic properties across the north-western Himalayan landscape. Eco Inform 77:102281

Singh A, Kumar R, Sharma S (2023) Natural products and hypertension: scope and role in antihypertensive therapy. Asian J Nursing Educ Res 13(2):162–166

Torres-Fajardo RA, González-Pech PG, Sandoval-Castro CA, JFdJ T-A (2020) Small ruminant production based on rangelands to optimize animal nutrition and health: Building an interdisciplinary approach to evaluate nutraceutical plants. Animals (Basel) 10(10):1799

Uysal S, Ugurlu A, Zengin G, Baloglu MC, Altunoglu YC, Mollica A et al (2018) Novel in vitro and in silico insights of the multi-biological activities and chemical composition of *Bidens tripartita* L. Food Chemical Toxicol 111:525–536

Yang W, Tien Y, Chung C, Chen Y, Chiou W, Hsu S et al (2015) Effect of Bidens pilosa on infection and drug resistance of Eimeria in chickens. Res veterinary sci 98:74–81

Zahara K, Bibi Y, Tabassum S, Bashir T, Haider S, Araa A et al (2015) A review on pharmacological properties of *Bidens biternata*: A potential nutraceutical. 5(8):595–599

Zahara K, Bibi Y, Qayyum A, Nisa S (2019) Investigation of antimicrobial and antioxidant properties of *Bidens Biternata*. Iran J Sci Technol Trans A: Sci 43:725–734

Zahara K, Bibi Y, Arshad M, Kaukab G, Al Ayoubi S, Qayyum A (2022) In-vitro examination and isolation of antidiarrheal compounds using five bacterial strains from invasive species *Bidens bipinnata* L. Saudi J Biol Sci 29(1):472–479

Part III

Chemistry

Phenolic Compounds Diversity of Genus *Bidens*

8

Bisma Javed, Mamoona Munir, Sadia Abdullah,
Mushtaq Ahmad, Banzeer Ahsan Abbasi,
Sumreen Dawood, and Lixin Zhang

Abstract

The genus *Bidens* of Asteraceae family comprises around 230 species. Beggarticks, Spanish needles, blackjack, burr marigolds, stickseeds, tickseeds, and tickseed sunflowers are some of names of the plants. The genus has a wide distribution, but the systematics and taxonomy have been described as complicated and disjointed. Plants are annual or perennial shrubs or herbs, known for their diverse phenolic compounds that significantly influence their pharmacological properties. These compounds include flavonoids, phenolic acids, and tannins, which have strong antibacterial, antioxidant, and anti-inflammatory properties. *Bidens* species contain flavonoids with potent anti-inflammatory and anticancer effects, such as luteolin and glycosides. Phenolic acids, such as caffeic acid and its esters, help protect plants from stress-induced diseases. *Bidens* has little research on its tannins, which have astringent and antibacterial properties and have medicinal potential. The bioactive components of *Bidens* species have the potential to be therapeutically useful, and this chemical heterogeneity supports this possibility (diverse genus of weedy herbs). Popular names used for the fruiting bodies of these plants include beggarticks, blackjack, burr marigolds,

B. Javed · M. Munir (✉) · S. Abdullah · B. A. Abbasi
Department of Botany, Rawalpindi Women University, Rawalpindi, Pakistan
e-mail: mamoona.munir@f.rwu.edu.pk

M. Ahmad
Department of Plant Sciences, Quaid-e-Azam University, Islamabad, Rawalpindi, Pakistan

S. Dawood
Department of Botany, University of Mianwali, Mianwali, Pakistan

L. Zhang
College of Life Sciences, Northwest A&F University, Yangling, Shaanxi, People's Republic of China

Y. Bibi et al. (eds.), *The Genus Bidens*,
https://doi.org/10.1007/978-981-96-4257-1_8

cobbler's pegs, Spanish needles, stickseeds, tickseeds, and tickseed sunflowers. The majority of the fruits have bristles and barbs on them. The derived name *Bidens*, which refers to the relative distinguishing traits, comes from the Latin concepts bis ("two") and dens ("tooth").

Keywords

Bidens · Bio-activities · Traditional uses · Phenolic · Tannins · Flavonoids

8.1 Introduction

Asteraceae (daisy) is a family of flowering plants that includes the genus *Bidens* (Gaspar et al. 2002). (Diverse genus of weedy herbs). Popular names used for the fruiting bodies of these plants include beggarticks, blackjack, burr marigolds, cobbler's pegs, Spanish needles, stickseeds, tickseeds, and tickseed sunflowers. The majority of the fruits have bristles and barbs on them. The derived name *Bidens*, which refers to the relative distinguishing traits, comes from the Latin concepts bis ("two") and dens ("tooth") (Hall 1967). Herbaceous plants that range in height from 10 to 150 cm are found in the vast genus *Bidens*. These plants can be annual or perennial. *Bidens* blooms are star-shaped, with several tiny tubular flowers arranged in a center. It comprises a ring of five to seven broad ray florets that are colored yellow, white, orange, red, or pink. Corolla is absent in several species. The so-called achene fruits are lengthy, dark-shelled seeds that develop after pollination. Many of the common names for *Bidens* originate from their pappus, an appendage utilized for propagation (Tsuda et al. 1994).

The genus comprises approximately 230 species that are found all over the world in various warm-temperate and tropical regions; Asia, Europe, Polynesia, Africa, and the Americas are home to various species (Gaspar et al. 2002; Sun and Ganders 1988). Blackjack can be found growing wild from sea level to 2400 m above sea level, or it can be found growing as a weed in crops like cotton, pulses, and cereals, in sunny or slightly shaded areas. For germination, temperature of 25 to 30 °C is ideal. Lower than 15 °C and higher than 45 °C are not ideal temperatures. It can withstand very high salinity levels of up to 100 mM NaCl and grows best in soils with a pH range of 4–9. Flooding inhibits the sprouting of seeds (Ganders et al. 2000). The seeds of *Bidens* are zoochorous, and they spread to other locations by attaching themselves to clothing, fur, or feathers. Because of this, they have been able to settle in a large area, including numerous marine islands. Many of these species, particularly those found in the Hawaiian Islands, are currently in danger of going extinct, and some are found only in extremely small geographic areas (Arthur et al. 2012).

8.2 Phenolic Compounds

Plants produce phenolic compounds as secondary metabolites when they break down phenylpropanoid in pentose phosphate and shikimic acid (Randhir et al. 2004). They vary in complexities from basic phenolic molecules to intricate polymerizing compounds, which are made up of benzene rings containing one or more hydroxyl substituents (Velderrain-Rodríguez et al. 2014). Phenols are among the most prominent secondary metabolites in plants, and they occur throughout the whole biochemical cycle. These phenolic compounds, which are often referred to as polyphenols, consist of a variety of substances, such as complex flavonoids, phenolic acids, simple flavonoids, and colored anthocyanins (Babbar et al. 2014). Usually, these phenolic compounds are related to the defense systems of the plant. On the other hand, phenolic metabolites are essential for other functions, such as adding pigments for defense against herbivores, accelerating pollination, and having antimicrobial and antifungal qualities (Alasalvar et al. 2001; Acamovic and Brooker 2005; Edreva et al. 2008).

In plants, these phenolics are broken down by the rhizosphere's plant-growth-promoting microorganisms, which also improve the fertility of the soil. Additionally, they increase the porosity of the soil, which helps plants absorb more nutrients, and they help the minerals and other components of the soil to chelate (Bais et al. 2006). The sub-epidermal tissues of plants frequently pick up these phenolic chemicals during times of stress and pathogen invasion. Plant physiology, age, stage of growth, environment, and kind of pathogen attack are only a few examples of the numerous internal and external variables that affect the production process and level of the phenolic compounds that have been collected (Ozyigit et al. 2007).

Studies indicate that fruits, vegetables, and beverages are the main dietary sources of phenolic compounds. As dietary antioxidants, plant polyphenols may guard against oxidative damage in human health and illness (Balasundram et al. 2006). Plant-based foods and beverages that are vital to health and nutrition contain phenolic compounds, which are naturally occurring antioxidants. Several studies have shown that phenolic compounds are the most prevalent dietary antioxidants in the typical human diet among all other antioxidants (Table 8.1). Phenolic compounds have garnered significant attention in recent times as the reports show that they may be able to prevent a number of human ailments (Fiorentino et al. 2008; Hooper and Cassidy 2006; Pu et al. 2013).

8.3 Species-Wise Distribution of Phenolic Compounds in Genus *Bidens*

Bidens is well known for its ethno-medicinal uses. Studies have indicated that a significant number of species of *Bidens* contain flavonoids, acetylenes, lactones, and sesquiterpenes. However, the majority of prevalent ones include C17-, C14-, and C13-polyacetylenes, thiophenes, carotene, vitamin C, and coumarins (umbelliferon, scopoletin, and aesculetin) (Matos et al. 2015) (Table 8.2).

Table 8.1 Classification of phenolic compounds

Class	Subclass	Examples					
Simple phenols	Hydroxybenzoic acid	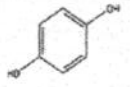 (Kevan and Baker 1983)		Hydroquinone $C_6H_4(OH)_2$			
Phenolic acids	Hydrobenzoic acid	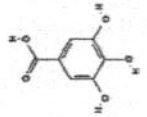 (Spjut 1994)	Gallic acid $C_7H_6O_5$	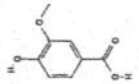 (Sánchez et al. 2023)	Vanillic acid $C_8H_8O_4$	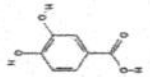 (Wagatsuma et al. 2021)	Protocatechuic acid
	Hydrocinnamic acid	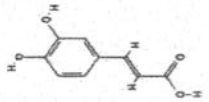 (Kompała-Bąba et al. 2023)	Caffeic acid $C_9H_8O_4$	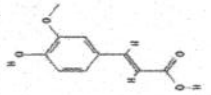 (Kompała-Bąba et al. 2023)	Ferulic acid $C_{10}H_{10}O_4$	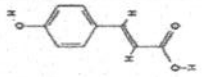 (Kuo et al. 2021)	p-coumaric acid $C_9H_8O_3$
Flavonoids	Flavonols	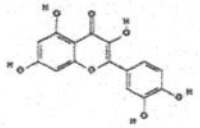 (Kompała-Bąba et al. 2023)	Quercetin $C_{15}H_{10}O_7$	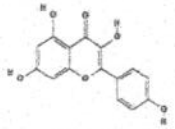 (Hall and Nguyen 2021)	Kampherol $C_{15}H_{10}O_6$	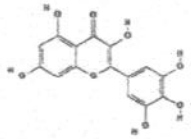 (Zahara et al. 2015)	Myricetin $C_{15}H_{10}O_8$
	Flavones	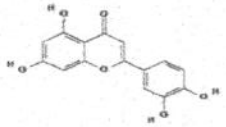			Luteolin $C_{15}H_{10}O_6$ (Bargielski 2020)		
	Isoflavones	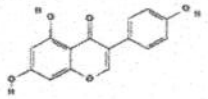 (Douglas et al. 2021)	Genistein $C_{15}H_{10}O_5$		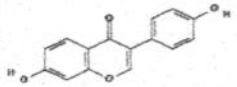 (Fokunang et al. 2020)	Daidzein $C_{15}H_{10}O_4$	

	Flavanone	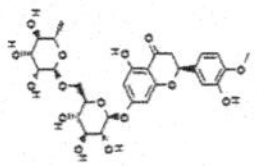 (Kissanga et al. 2021)	Hesperidin $C_{28}H_{34}O_{15}$	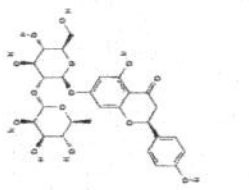 (Gurib-Fakim 2006)	Naringin $C_{27}H_{32}O_{14}$
	Flavanols	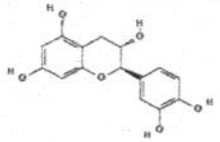 (Zahara et al. 2022)	Catechin $C_{15}H_{14}O_6$	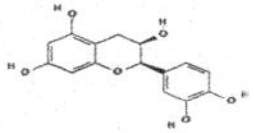 (Lima Silva et al. 2011)	Epicatechin $C_{15}H_{14}O_6$
	Anthocyanins	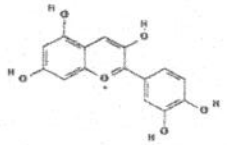 (Mtenga and Ripanda 2022)	Cyanidin $C_{15}H_{11}O_6^+$	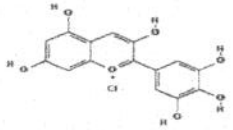 (Bartolome et al. 2013a)	Delphinidin $C_{15}H_{11}ClO_7$
Stilbenes		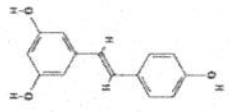 (Arthur et al. 2012)	Resveratrol $C_{14}H_{12}O_3$		
Chalcones		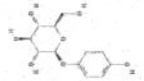 (Kissanga et al. 2021)	Arbutin $C_{12}H_{16}O_7$		
Lignans		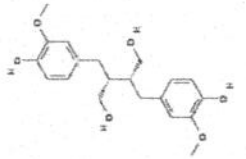 (Lye et al. 2008)	Secoisolar $C_{20}H_{26}O_6$		

(continued)

Table 8.1 (continued)

Class	Subclass	Examples			
Tannins	Hydrolyzable tannins	 (Shayanowako et al. 2021)	Tannic acid $C_{76}H_{52}O_{46}$	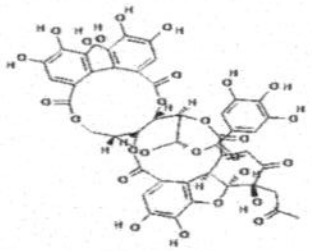 (Otang-Mbeng et al. 2020)	Ellagitannins $C_{44}H_{32}O_{27}$
	Condensed tannins	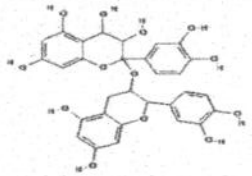 (Shayanowako et al. 2021)	Procyanidin $C_{30}H_{26}O_{13}$		
Coumarins		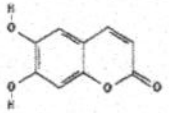 (Moshi et al. 2010)	Esculetin $C_9H_6O_4$		
Xanthones		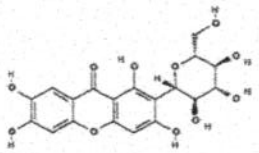 (Sharma et al. 2009)	Mangiferin $C_{19}H_{18}O_{11}$		
Aurones		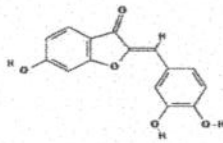 (Orwa et al. 2008)	Sulfuretin (2Z)-2-[(3,4-dihydroxyphenyl)methylidene]-6-hydroxy-1-benzofuran-3-one $C_{15}H_{10}O_5$		

Table 8.2 Diversity of phenolic compounds in genus *Bidens*

Phenolic compound	Type	Bioactivities	References
Phenolic compounds in *Bidens pilosa*			
Flavonoids	Luteolin	Antiallergic, antioxidant, antiproliferative, immunosuppressive, anti-inflammatory	Bartolome et al. (2013a), Lima Silva et al. (2011), Xuan and Khanh (2016), Chiang et al. (2004), Tewtrakul et al. (2003), and Wu et al. (1997)
	Cytopiloin		
	Centaureidin,		
	Centaurein		
	Quercetin	Antioxidant, hypertension, vasodilation, obesity, hypercholesterolemia, and atherosclerosis.	
Phenolics	Rutin	Antioxidant activity	
	Chlorogenic acid	Neuroprotective, DNA-protective, and antidiabetic.	
	Gallic acid	Anti-angiogenic, anti-inflammatory, anticancer, and antioxidant.	
	Caffeic acid	Anti-inflammatory and anticancer	
Tannins	Complex phenolics	Antimicrobial, antibacterial, antiviral, cardioprotective, antidiabetic and antiobesity	
Phenolic compounds in *Bidens tripartite*			
Flavonoids	Aurones	Antitumor, antimicrobial, antiviral, antioxidant, anti-Alzheimer, antitumor, antidiabetic, anti-Parkinson's, anti-inflammatory, and anti-nociceptive	Constantinescu and Lungu (2021), Mustafa and Mostafa (2020), Fu et al. (2020), Osipova et al. (2020), Cao et al. (2020), Zhou et al. (2020), Rammohan et al. (2020), Maliyakkal et al. (2022), Hsieh et al. (2000), and de Campos-Buzzi et al. (2006)
	Chalcones		
	Flavones		
Tannins	Present	Antimicrobial, antibacterial, antiviral, cardioprotective, antidiabetic, and antiobesity	
Coumarins	Likely present	Anti-fungal, antibacterial, anti-inflammatory, antioxidant, anti-cholinesterase, antidiabetic, antineoplastic, anti-HIV, anticoagulant, and anticancer properties	Jung et al. (2001)
Phenolics	Caffeic acid	Anti-inflammatory and anticancer	Bartolome et al. (2013a), Lima Silva et al. (2011), Xuan and Khanh (2016), Chiang et al. (2004), Tewtrakul et al. (2003), and Wu et al. (1997)
	Luteolin	Antiallergic, antioxidant, antiproliferative, immunosuppressive, anti-inflammatory	

(continued)

Table 8.2 (continued)

Phenolic compound	Type	Bioactivities	References
Phenolic compounds in *Bidens cernua*			
Flavonoids	Luteolin	Antiallergic, antioxidant, antiproliferative, immunosuppressive, anti-inflammatory	Bartolome et al. (2013a), Lima Silva et al. (2011), Xuan and Khanh (2016), Chiang et al. (2004), Tewtrakul et al. (2003), and Wu et al. (1997)
	Flavanomarein	Antioxidant	
	Tilianin	Anti-inflammatory	Gao et al. (2019)
	Maritimetin	Radical scavenger and plant metabolite	
	Cynaroside	Cardio protective, inhibitor for H_2O_2-induced apoptosis, antioxidant, anti-inflammatory, antibacterial, antifungal, and anticancer.	
	Cosmosiin	Immunomodulatory and antioxidant effects	
Phenolic compounds in *Bidens bipinnata*			
Flavonoids	Hyperin	Antioxidant, analgesic, antiallergic, anti-inflammatory properties, inhibit enzymes	Jung et al. (2001)
Flavonoids coumarins	Catechin	Antiviral, anti-inflammatory, antiallergenic, antimicrobial, and anticancer	
	Isoquercitrin	Antioxidant, chemoprotective effects against allergies, diabetes, cancer, oxidative stress	
	Astragalin	Anti-inflammatory and anti-atopic dermatitis	
	Quercetin	Hypertension, vasodilation, obesity, hypercholesterolemia, and atherosclerosis	Shen et al. (2015)
	Biscoumaric acid	Anti-tuberculosis, anti-HIV, anti-inflammatory, antibacterial, anticancer, and antioxidant	
Phenolic acids	Present but not studied in detail	Anti-inflammatory and anticancer actions, neuroprotective, DNA-protective, and antidiabetic, antimicrobial activities, antioxidant and anti-inflammatory	

(continued)

Table 8.2 (continued)

Phenolic compound	Type	Bioactivities	References
Ceramides	Present- 9 in number	Function as auxiliary messengers in the processes of cell signaling, perform variety of biological roles, structural components of cellular membranes, function is in skin's structure	Figueroa Valverde et al. (2008)
Phenolic compounds in *Bidens frondosa*			
Flavonoids	Flavones	Antitumor, antimicrobial, antiviral, antioxidant, anti-Alzheimer, antitumor, antidiabetic, anti-Parkinson's, anti-inflammatory, and anti-nociceptive	Constantinescu and Lungu (2021), Mustafa and Mostafa (2020), Fu et al. (2020), Osipova et al. (2020), Cao et al. (2020), Zhou et al. (2020), Rammohan et al. (2020), Maliyakkal et al. (2022), Hsieh et al. (2000), and de Campos-Buzzi et al. (2006)
Flavonoid	Chalcones		
	Aurones		
Phenolic compounds in *Bidens ferulifolia*			
Flavonoids	Flavones	Antitumor, antimicrobial, antiviral, antioxidant, anti-Alzheimer, antitumor, antidiabetic, anti-Parkinson's, anti-inflammatory, and anti-nociceptive	Constantinescu and Lungu (2021), Mustafa and Mostafa (2020), Fu et al. (2020), Osipova et al. (2020), Cao et al. (2020), Zhou et al. (2020), Rammohan et al. (2020), Maliyakkal et al. (2022), Hsieh et al. (2000), and de Campos-Buzzi et al. (2006)
Flavonoids	Anthocyanidins	Reduction of cardiovascular disease incidence, control of obesity, the mitigation of diabetes, and anti-inflammatory and anti-carcinogenic	
	Anthochlors		Varro (1999), Hu et al. (2017), Mustafa et al. (2020), and Sun et al. (2024)
	Dihydroflavonols	Anti-inflammatory activity	
	Flavonols	Cardiovascular disease, antioxidant	
Phenolic compounds in *Bidens aurea*			
Flavonoids	Aurones	Antitumor, antimicrobial, antiviral, antioxidant, anti-Alzheimer, antitumor, antidiabetic, anti-Parkinson's, anti-inflammatory, and anti-nociceptive	Constantinescu and Lungu (2021), Mustafa and Mostafa (2020), Fu et al. (2020), Osipova et al. (2020), Cao et al. (2020), Zhou et al. (2020), Rammohan et al. (2020), Maliyakkal et al. (2022), Hsieh et al. (2000), and de Campos-Buzzi et al. (2006)
	Chalcones		

(continued)

Table 8.2 (continued)

Phenolic compound	Type	Bioactivities	References
Phenolic compounds in *Bidens parviflora wild*			
Flavonoids	Flavones	Antitumor, antimicrobial, antiviral, antioxidant, anti-Alzheimer, antitumor, antidiabetic, anti-Parkinson's, anti-inflammatory, and anti-nociceptive	Constantinescu and Lungu (2021), Mustafa and Mostafa (2020), Fu et al. (2020), Osipova et al. (2020), Cao et al. (2020), Zhou et al. (2020), Rammohan et al. (2020), Maliyakkal et al. (2022), Hsieh et al. (2000), and de Campos-Buzzi et al. (2006)
Phenolic acids	Flavonones	And antifungal	Mustafa and Mostafa (2020)
	Glycosides	Antimicrobial, antioxidant, immunomodulatory, antimicrobial activity, and estrogenic activity	
	Chalcones	Antitumor, antimicrobial, antiviral, antioxidant, anti-Alzheimer, antitumor, antidiabetic, anti-Parkinson's, anti-inflammatory, and anti-nociceptive	Constantinescu and Lungu (2021), Mustafa and Mostafa (2020), Fu et al. (2020), Osipova et al. (2020), Cao et al. (2020), Zhou et al. (2020), Rammohan et al. (2020), Maliyakkal et al. (2022), Hsieh et al. (2000), and de Campos-Buzzi et al. (2006)
	Aurones		
	Neolignan glucosides		
	Caffeoylquinic acid derivatives	Anti-inflammatory and antioxidant activities	Figueroa Valverde et al. (2008)
Phenolic compounds in *Bidens alba*			
Flavonoid		Antioxidant and anti-inflammatory activity	Figueroa Valverde et al. (2008), Wang et al. (2006a), Yang et al. (2009), and Ramabulana et al. (2020)
Phenolics	Caffeic acid	Antioxidant and anti-inflammatory activity	

8.3.1 *Bidens pilosa*

The bioactivities of *B. pilosa* are linked to a variety of phytochemicals; approximately 201 metabolites have been defined, some of which have been identified as aromatics, flavonoids, phenylpropanoids, aliphatics, terpenoids, alkaloids, porphyrins (Bartolome et al. 2013a, b; Borges et al. 2013; MU/Chipaca Muysc Cubun—Lengua Muisca n.d.; Lima Silva et al. 2011), tannins, and cardiac glycosides. It has been demonstrated that the flavonoids and polyenes, including luteolin, cytopiloin

(Cp), centaureidin, and centaurein, have antiallergic, antioxidant, antiproliferative, immunosuppressive, and anti-inflammatory qualities (Bartolome et al. 2013a; Lima Silva et al. 2011; Xuan and Khanh 2016; Chiang et al. 2004; Tewtrakul et al. 2003; Wu et al. 1997). Phenolic compounds were observed in the plant samples, such as gallic acid (36.37–152.75 mg/100 g), chlorogenic acid (45.15–503.21 mg/100 g), and rutin (24.23–323.44 mg/100 g), and caffeic acid was recognized first (Yan et al. 2020). Almost 20 different plants contain quercetin, a significant bioflavonoid. Its widely recognized benefits include those against inflammation, oxidative stress, hypertension, vasodilation, obesity, hypercholesterolemia, and atherosclerosis. The discovery of this flavonoid in *B. pilosa* supports the plant's purported anti-inflammatory qualities and raises the possibility of novel pharmaceutical applications involving quercetin (David et al. 2016). Fatty acids are abundant in *B. pilosa* and have been linked to anti-inflammatory benefits by regulating oxidative stress and the expression of genes that produce inflammatory mediators (Quaglio et al. 2020). The anticancer effect of *B. pilosa* has been shown in a few preclinical investigations. It has been found that *B. pilosa* contains extra flavonoids and polyphenols that may have antioxidant and preventive properties against damages caused by free radicals (Rasool Hassan 2012). According to studies, *B. pilosa* extracts or components have long been utilized as phytomedicines exhibiting antidiabetic properties. Worldwide, *B. pilosa* is utilized for a wide range of applications, such as phytoremediation of heavy metal-polluted soils (Sun et al. 2009; Yan et al. 2020), weed control (Xuan et al. 2005), food and traditional medicine, and plant biopesticide (Li et al. 2021).

8.3.2 *Bidens tripartita*

Among *B. tripartite*, the biologically active substances (BAS) are polysaccharides, phenolcarboxylic acids, flavonoids, tannins, coumarins, aromatic hydroxy aldehydes, and poly-acetylenes (Rasool Hassan 2012). Studies on the phytochemical composition of bur-marigold plant revealed the presence of coumarins, carotenoids, minor levels of vitamin C, aurones, chalcones, and flavones. The principal constituents of *B. tripartita* were determined to be luteolin-7-glucoside, luteolin, derivatives of caffeic acid, and an unidentified luteolin glycoside (Pagani et al. 1972). While amounts of cosmene and eugenol were found in the flower heads of *B. tripartita*, the green sections of the plant include polyacetylenic substances, linoleic acid, and ocimene (Varro 1999). *Bidens tripartita* contains caryophyllene, which inhibits inflammatory cytokines, chemokines, and eicosanoids while also altering several signaling cascades to protect tissues. Thus, caryophyllene by molecular docking studies an active substance that can reduce inflammation, alter immunity, and fight against coronavirus infection (disease in 2019).

In *Bidens tripartita*, the methyl ester of 9,12-octadecadienoic acid is an important fatty acid that has antibacterial properties. Plants contain a linear diterpene called phytol have both antibacterial and anti-inflammatory properties. Certain dietary supplements, such as gamma and alpha-tocopherol (vitamin E), have antiradical qualities. *Bidens tripartita* contains phytosterols, such as stigmasterol and campesterol, which have characteristics similar to those of animal cholesterol. These phytosterols can lower heart disease risk and have the ability to lower cholesterol levels (la Casa García 1995).

8.3.3 *Bidens cernua*

Previous findings on *B. cernua* phytochemistry have shown the existence of secondary plant metabolites, including flavonoids, polyacetylenes, and sesquiterpenes. Six flavonoids have been identified through phytochemical analysis of the aerial portions of *Bidens cernua* L. luteolin, flavanomarein, tilianin, maritimetin, cynaroside, and cosmosiin. The quantitative study of flavonoids revealed that the contents of flavonoids in flowers were 0.81%, while the contents in the herb were 1.62%, comparable to hyperoside. Tannin content overall was found to be 7.87% in flowers and 5.83% in herbs. The essential oil extracted from *B. cernua* herb was found to have a chemical makeup of 92.2%, or 99 components. The essential oil's antimicrobial activity was assessed on fungi, bacteria, and Gram-positive and Gram-negative organisms. Strong antibacterial and antifungal properties were shown by the oil, particularly against strains of Gram-positive bacteria (Li et al. 2008).

8.3.4 *Bidens bipinnata*

Not much research has been done on the chemistry of *Bidens bipinnata*. According to Li et al., acetylenic glucosides are present (Li et al. 2004). *Bidens bipinnata* is known to contain the primary flavonoids hyperin, catechin, isoquercitrin, astragalin, and quercetin (Shen et al. 2015). A phytochemical investigation was then conducted by Hu et al., who discovered the presence of nine ceramides, 13 flavonoids, 1 triterpenoid, and 1 polyacetylene. The bioactivities of these isolated compounds have not, however, been studied. 16-pregnenolone and 9-octadecenoic acid (Z)-methyl ester were identified as the two antibacterial compounds (Hu et al. 2018). With an oxo group at position 20 and a beta-hydroxy group at position 3, the steroid 16-pregnenolone is a 20-oxo derivative of pregnene. Numerous additional steroids have already been found in plants, such as *Oryza sativa* L. (Macías et al. 2006). Nevertheless, no reports of this chemical belonging to the genus *Bidens* exist. According to a study by Valverde et al., pregnenolone seems to be a potent antibacterial agent (Figueroa Valverde et al. 2008).

8.3.5 *Bidens frondosa*

According to earlier research *B. frondosa* was found to include flavones, terpenes, chalcones, aurones, polyenic D-glucosides (Brandão et al. 1997; Pagani et al. 1972), and acetylene (Brandão et al. 1997).

8.3.6 *Bidens ferulifolia*

The primary flavonoid pigments that were found were flavones, anthocyanidins, and anthochlors. Low levels of dihydroflavonols and flavonols were also seen in a few genotypes. Often, the pigment concentrations in the basal sections were higher than those in the outer part. The color of *B. ferulifolia* flowers is determined by four principal pigment classes: carotenoids, flavones, anthocyanins, and anthochlors. The first three have a shared metabolic pathway, which makes the production of anthocyanins, anthochlors, and flavones competitive (Varro 1999).

8.3.7 *Bidens aurea*

An abundance of flavonoid chemicals, primarily aurones and chalcones, is present in the species *Bidens aurea* (Aiton) Sherff (la Casa García 1995).

8.3.8 *Bidens parviflora* Wild

This species has previously been found to include flavones (Li et al. 2008), flavonones (Hoffmann and Hölzl 1988a), flavonoid glycosides (Li et al. 2008; Hoffmann and Hölzl 1988b), flavonol glycosides (De Tommasi et al. 1998; Sashida et al. 1991), chalcones (Redl et al. 1992; Bauer et al. 1992), aurones (Sashida et al. 1991; Redl et al. 1992), sterols, polyacetylene glucosides (Hoffmann and Hölzl 1988a; Bauer et al. 1992), and monoterpenes (Hoffmann and Hölzl 1988a; Liu et al. 2020). Previous research on this plant produced the following compounds: sucrose esters (Wang et al. 2003), phenolic acids (Wang et al. 2006a), polyacetylene glucosides (Wang et al. 2001), monoterpene glycosides (Wang et al. 2007), neolignan glucosides (Wang et al. 2006b), phenolic glucosides (Wang et al. 1994), and derivatives of caffeoylquinic acid (Wang et al. 2006c).

8.3.9 *Bidens alba*

The nutrient-dense wild plant is comparable to that of kale, or leaf cabbage, which is cultivated for its edible leaves and is rich in fiber, protein, carotenes, folate, and magnesium. Indians, Zulus, and South Africans boil the fresh or dried leaves for consumption (Zeevaart and Creelman 1988).

8.4 Future Recommendations

1. Use advanced techniques such as GC–MS, NMR spectroscopy, and LC–MS/MS to identify and quantify phenolic compounds.

2. To comprehend the variations in phenolic profiles within the genus, compare species and populations.
3. Use transcriptomics and genomic methods to look at the manufacture of phenolic chemicals in the genus *Bidens*.
4. Verify through in vivo and clinical trials the therapeutic benefits of phenolic compounds produced from the *Bidens* species.
5. Use ecologically friendly farming practices to boost the amount and level of phenolic compounds in *Bidens*.
6. Examine the effects of different agronomic methods on the phenolic profile of different *Bidens* species.
7. Use biotechnological methods to enhance the synthesis of specific phenolic compounds.
8. Look into the potential for using plant cell cultures or microbial fermentation to produce phenolic chemicals from *Bidens* species.
9. In order to classify and identify species of *Bidens*, use phenolic profiles as chemotaxonomic indicators.
10. Analyze the connection between speciation and ecological adaptation. Also consider the significance of phenolic compound diversity in terms of evolution within the species.
11. Consider the potential applications of phenolic compounds from the *Bidens* species in the production of nutraceuticals, cosmetics, and functional foods.
12. Investigate whether phenolic compounds can be widely extracted and purified for use in the commercial sector.

8.5 Conclusion

The main groups of phenolic compounds found in the genus *Bidens* are tannins, coumarins, phenolic acids, and flavonoids. The unique chemical structures and distinctive features of each of these groups contribute to the overall biological activity of the plant. These compounds have potential health benefits and medicinal uses. Flavonoids, which are widely recognized for their anti-inflammatory, antioxidant, and anticancer properties, are abundant in *Bidens*, i.e., quercetin, luteolin, and glycosides. Gallic acid, chlorogenic acid, and caffeic acid are phenolic acids that are also widely present and enhance the antioxidant activity of the plant. *Bidens* contains phenolic chemicals that reduce oxidative stress and scavenge free radicals, which are linked to serious diseases like cancer and cardiovascular disorders. Phenolic compounds like quercetin and luteolin inhibit the inflammatory response. Phenolic chemicals in *Bidens* have a wide range of biological actions, which makes them an attractive option for therapeutic development. Natural phenolic compounds have less toxicity and fewer adverse effects than synthetic medicines, which makes them a safer option. Finding additional phenolic compounds in *Bidens* species may help develop new medicinal products to treat cancer, infections, and inflammatory diseases.

References

Acamovic T, Brooker J (2005) Biochemistry of plant secondary metabolites and their effects in animals. Proc Nutr Soc 64(3):403–412

Alasalvar C et al (2001) Comparison of volatiles, phenolics, sugars, antioxidant vitamins, and sensory quality of different colored carrot varieties. J Agric Food Chem 49(3):1410–1416

Arthur G, Naidoo K, Coopoosamy R (2012) Bidens pilosa L: agricultural and pharmaceutical importance. J Med Plants Res 6(17):3282–3287

Babbar N et al (2014) Influence of different solvents in extraction of phenolic compounds from vegetable residues and their evaluation as natural sources of antioxidants. J Food Sci Technol 51:2568–2575

Bais HP et al (2006) The role of root exudates in rhizosphere interactions with plants and other organisms. Annu Rev Plant Biol 57:233–266

Balasundram N, Sundram K, Samman S (2006) Phenolic compounds in plants and Agri-industrial by-products: antioxidant activity, occurrence, and potential uses. Food Chem 99(1):191–203

Bargielski RC (2020) Fields brook superfund site: race, class, and environmental justice in a blasted landscape. University of South Florida, Tampa

Bartolome AP et al (2013a) Bidens pilosa L. (Asteraceae): botanical properties, traditional uses, phytochemistry, and pharmacology. Evid Based Complement Alternat Med 2013(1):340215

Bartolome AP, Villaseñor IM, Yang WC (2013b) Bidens pilosa L. (Asteraceae): botanical properties, traditional uses, phytochemistry, and pharmacology. Evid Based Complement Alternat Med 2013:1–51. https://doi.org/10.1155/2013/340215

Bauer R, Redl K, Davis B (1992) Four polyacetylene glucosides from Bidens campylotheca. Phytochemistry 31(6):2035–2037

Borges C et al (2013) Bidens pilosa L. (Asteraceae): traditional use in a community of southern Brazil. Revi Bras Plant Med 15:34–40

Brandão M et al (1997) Antimalarial activity of extracts and fractions from Bidens pilosa and other Bidens species (Asteraceae) correlated with the presence of acetylene and flavonoid compounds. J Ethnopharmacol 57(2):131–138

Cao Y et al (2020) Licochalcone B, a chalcone derivative from Glycyrrhiza inflata, as a multifunctional agent for the treatment of Alzheimer's disease. Nat Prod Res 34(5):736–739

Chiang Y-M et al (2004) Metabolite profiling and chemopreventive bioactivity of plant extracts from Bidens pilosa. J Ethnopharmacol 95(2–3):409–419

Constantinescu T, Lungu CN (2021) Anticancer activity of natural and synthetic chalcones. Int J Mol Sci 22(21):11306

David AVA, Arulmoli R, Parasuraman S (2016) Overviews of biological importance of quercetin: a bioactive flavonoid. Pharmacogn Rev 10(20):84

de Campos-Buzzi F et al (2006) Antinociceptive effects of synthetic chalcones obtained from xanthoxyline. Arch Pharm 339(7):361–365

De Tommasi N, Piacente S, Pizza C (1998) Flavonol and chalcone ester glycosides from Bidens andicola. J Nat Prod 61(8):973–977

Douglas MR et al (2021) Are populations of economically important bonefish and queen conch'open'or'closed'in the northern caribbean basin? Mar Ecol 42(2):e12639

Edreva A et al (2008) Stress-protective role of secondary metabolites: diversity of functions and mechanisms. Gen Appl Plant Physiol 34(1–2):67–78

Figueroa Valverde L et al (2008) Synthesis and antibacterial activity of pregnenolone-vitamin B1 conjugate. J Mex Chem Soc 52(2):130–135

Fiorentino A et al (2008) Isolation and structure elucidation of antioxidant polyphenols from quince (Cydonia vulgaris) peels. J Agric Food Chem 56(8):2660–2667

Fokunang ET et al (2020) Pharmacovigilance of natural herbal medicines research for efficacy, safety and quality assurance of phytomedicine products. J Complement Altern Med Res 12:21–37

Fu Y et al (2020) New chalcone derivatives: synthesis, antiviral activity and mechanism of action. RSC Adv 10(41):24483–24490
Ganders FR, Berbee M, Pirseyedi M (2000) ITS base sequence phylogeny in Bidens (Asteraceae): evidence for the continental relatives of Hawaiian and Marquesan Bidens. Syst Bot 25(1):122–133
Gao F, Xiao J, Huang G (2019) Current scenario of tetrazole hybrids for antibacterial activity. Eur J Med Chem 184:111744
Gaspar T et al (2002) Auxins in the biology of roots. In: Waisel Y, Eshel A, Kafkafi U (eds) Plant roots: the hidden half, 3rd edn. CRC Press, Boca Raton, pp 587–616
Gurib-Fakim A (2006) Medicinal plants: traditions of yesterday and drugs of tomorrow. Mol Aspects Med 27(1):1–93
Hall GW (1967) A biosystematic study of the North American complex of the genus Bidens (Compositae). Indiana University, Bloomington
Hall D, Nguyen TA (2021) Radical narratives in the American jobs plan (AJP) and the global potential of 'Biden plans'
Hoffmann B, Hölzl J (1988a) Weitere acylierte chalkone aus Bidens pilosa. Planta Med 54(05):450–451
Hoffmann B, Hölzl J (1988b) A methylated chalcone glucoside from Bidens pilosa. Phytochemistry 27(11):3700–3701
Hooper L, Cassidy A (2006) A review of the health care potential of bioactive compounds. J Sci Food Agric 86(12):1805–1813
Hsieh HK et al (2000) Synthesis and anti-inflammatory effect of chalcones. J Pharm Pharmacol 52(2):163–171
Hu X-L et al (2017) Recent developments of coumarin hybrids as anti-fungal agents. Curr Top Med Chem 17(29):3219–3231
Hu H-M et al (2018) Chemical constituents from Bidens bipinnata Linn. Biochem Syst Ecol 79:44–49
Jung J-C, Jung Y-J, Park O-S (2001) A convenient one-pot synthesis of 4-hydroxycoumarin, 4-hydroxythiocoumarin, and 4-hydroxyquinolin-2 (1 H)-one. Synth Commun 31(8):1195–1200
Kevan P, Baker HG (1983) Insects as flower visitors and pollinators. Annu Rev Entomol 28(1):407–453
Kissanga R et al (2021) Nutritional and functional properties of wild leafy vegetables for improving food security in Southern Angola. Front Sustain Food Syst 5:791705
Kompała-Bąba A et al (2023) Taxonomic diversity and selection of functional traits in novel ecosystems developing on coal-mine sedimentation pools. Sustainability 15(3):2094
Kuo TF et al (2021) Bidens pilosa: Nutritional value and benefits for metabolic syndrome. Food Front 2(1):32–45
Li S et al (2004) New acetylenic glucosides from Bidens bipinnata L INNE. Chem Pharm Bull 52(4):439–440
Li Y-L et al (2008) Flavonoids and a new polyacetylene from Bidens parviflora Willd. Molecules 13(8):1931–1941
Li B-L et al (2021) Polyacetylenes from Bidens pilosa Linn., promising insecticides with anti-inflammatory effects on HaCaT cells. Ind Crop Prod 171:113929
Lima Silva F et al (2011) Compilation of secondary metabolites from Bidens pilosa L. Molecules 16(2):1070–1102
Liu Y et al (2020) Plant diversity in herbal tea and its traditional knowledge in Qingtian County, Zhejiang Province, China. Plant Divers 42(6):464–472
Lye KA et al (2008) Botanical-medicinal dictionary for East Africa, Makerere herbarium handbook 2. Department of Botany, Makerere University, Kampala
Macías FA et al (2006) Bioactive steroids from Oryza sativa L. Steroids 71(7):603–608
Maliyakkal N et al (2022) Ameliorative effect of ethoxylated chalcone-based MAO-B inhibitor on behavioural predictors of haloperidol-induced parkinsonism in mice: evidence of its antioxidative role against Parkinson's diseases. Environ Sci Pollut Res 29(5):7271–7282

Matos M et al (2015) Coumarins – an important class of phytochemicals chapter. In: Phytochemicals-Isolalation, characterisation and role in human health, agricultural and biological sciences. IntechOpen, London

Moshi MJ et al (2010) Ethnomedicine of the Kagera region, north western Tanzania. Part 2: the medicinal plants used in Katoro Ward, Bukoba district. J Ethnobiol Ethnomed 6:1–5

Mtenga DV, Ripanda AS (2022) A review on the potential of underutilized blackjack (Biden Pilosa) naturally occurring in sub-Saharan Africa. Heliyon 8(6):e09586

MU/Chipaca Muysc Cubun—Lengua Muisca (n.d.) Available online: http://muysca.cubun.org/MU/chipaca. Accessed on 27 Apr 2021

Mustafa M, Mostafa YA (2020) A facile synthesis, drug-likeness, and in silico molecular docking of certain new azidosulfonamide–chalcones and their in vitro antimicrobial activity. Monatshefte für Chemie-Chemical Monthly 151:417–427

Mustafa YF, Mohammed ET, Khalil RR (2020) Antioxidant and antitumor activities of methanolic extracts obtained from red delicious and granny smith apples' seeds. Syst Rev Pharm 11(4):570–576

Orwa JA et al (2008) The use of Toddalia asiatica (L) Lam. (Rutaceae) in traditional medicine practice in East Africa. J Ethnopharmacol 115(2):257–262

Osipova V et al (2020) Synthesis and antioxidant activity of new hydroxy derivatives of chalcones. Russ Chem Bull 69(3):504–509

Otang-Mbeng W, Mashabela MN, Industries R (2020) A review of beneficial phytochemicals and postharvest studies on some indigenous leafy vegetables from the Mpumalanga Province of south Africa. J Med Plants 12(4):533–544

Ozyigit II, Kahraman MV, Ercan O (2007) Relation between explant age, total phenols and regeneration response in tissue cultured cotton (Gossypium hirsutum L.). Afr J Biotechnol 6(1):3

Pagani F, Romussi G, Bohlmann F (1972) Notiz uber die struktur des polyin-glucosids aus Bidens frondosa L. Chem Ber 105:3126

Pu F, Ren X-L, Zhang X-P (2013) Phenolic compounds and antioxidant activity in fruits of six Diospyros kaki genotypes. Eur Food Res Technol 237:923–932

Quaglio AE et al (2020) Bidens pilosa (Black Jack) standardized extract ameliorates acute TNBS-induced intestinal inflammation in rats. Planta Med 86(05):319–330

Ramabulana AT, Steenkamp P, Madala N, Dubery IA (2020) Profiling of chlorogenic acids from Bidens Pilosa and differentiation of closely related positional isomers with the aid of UHPLC-QTOF-MS/MS-based in-source collision-induced dissociation. Metabolites 2020, 10. https://doi.org/10.3390/METABO10050178

Rammohan A et al (2020) Design, synthesis, docking and biological evaluation of chalcones as promising antidiabetic agents. Bioorg Chem 95:103527

Randhir R, Lin Y-T, Shetty K (2004) Stimulation of phenolics, antioxidant and antimicrobial activities in dark germinated mung bean sprouts in response to peptide and phytochemical elicitors. Process Biochem 39(5):637–646

Rasool Hassan B (2012) Medicinal plants (importance and uses). Pharm Anal Acta 3(10):2153–2435

Redl K, Davis B, Bauer R (1992) Chalcone glycosides from Bidens campylotheca. Phytochemistry 32(1):218–220

Sánchez J et al (2023) Arid and semi-arid environments: their relationship with the dispersion and germination of species. Revista mexicana de ciencias forestales 14(75):35–67

Sashida Y et al (1991) New aurone glucosides and new phenylpropanoid glucosides from Bidens pilosa. Chem Pharm Bull 39(3):709–711

Sharma A, Bargali K, Pande N (2009) The allelopathic potential of bryophyte extract on seed germination and seedling growth of Bidens biternata. Nat Sci Sleep 7:30–38

Shayanowako AIT et al (2021) African leafy vegetables for improved human nutrition and food system resilience in Southern Africa: a scoping review. Sustainability 13(5):2896

Shen A-Z et al (2015) Total flavonoids of Bidens bipinnata L. ameliorate experimental adjuvant-induced arthritis through induction of synovial apoptosis. BMC Complement Altern Med 15:1–9

Spjut RW (1994) A systematic treatment of fruit types. Garden, Bronx

Sun M, Ganders FR (1988) Mixed mating systems in Hawaiian Bidens (Asteraceae). Evolution 42(3):516–527

Sun YB, Zhou Q, Liu W, An J, Xu ZQ, Wang L (2009) Joint effects of arsenic and cadmium on plant growth and metal bioaccumulation: a potential Cd-hyperaccumulator and As-excluder Bidens Pilosa L. J Hazard Mater 165:1023–1028. https://doi.org/10.1016/j.jhazmat.2008.10.097

Sun M et al (2024) Enhanced anti-inflammatory activity of Tilianin based on the novel amorphous nanocrystals. Pharmaceuticals 17(5):654

Tewtrakul S et al (2003) HIV-1 integrase inhibitory substances from Coleus parvifolius. Phytother Res 17(3):232–239

Tsuda T et al (1994) Antioxidative pigments isolated from the seeds of Phaseolus vulgaris L. J Agric Food Chem 42(2):248–251

Varro ET (1999) Phytomedicines: back to the future. J Nat Prod 62(11):1589–1592

Velderrain-Rodríguez G et al (2014) Phenolic compounds: their journey after intake. Food Funct 5(2):189–197

Wagatsuma T et al (2021) Burr marigold (Bidens tripartita L) roots directly and immediately scavenge rhizosphere methane with highly exuded hydrogen peroxide via a rhizosphere Fenton reaction. Plant Soil 459:289–313

Wang J et al (1994) Phenolic glucosides from Bidens parviflora and their anti-histamine activities. Chin Tradit Herb Drug 38:647–649

Wang N et al (2001) Antiallergic agents from natural sources. 3. Structures and inhibitory effects on nitric oxide production and histamine release of five novel polyacetylene glucosides from Bidens parviflora WILLD. Chem Pharm Bull 49(8):938–942

Wang N et al (2003) Bioactive sucrose esters from Bidens parviflora. Phytochemistry 62(5):741–746

Wang J, Wang N, Yao X (2006a) Structures and anti-histamine activities of phenolic acid derivatives from Bidens parviflora Willd. Chin J Med Chem 16(3):168

Wang NL et al (2006b) Two neolignan glucosides and antihistamine release activities from Bidens parviflora W ILLD. Chem Pharm Bull 54(8):1190–1192

Wang J et al (2006c) Caffeoylquinic acid derivatives from Bidens parviflora and their antihistamine release activates. Chin Tradit Herb Drug 24:966–970

Wang N-L et al (2007) Two new monoterpene glycosides and a new (+)-jasmololone glucoside from Bidens parviflora Willd. J Asian Nat Prod Res 9(5):449–455

Wu H et al (1997) Clinical therapeutic effect of drug-separated moxibustion on chronic diarrhea and its immunologic mechanisms. J Traditi Chin Med Chung i tsa Chih Ying wen pan 17(4):253–258

Xuan TD, Khanh TD (2016) Chemistry and pharmacology of Bidens pilosa: an overview. J Pharm Investig 46:91–132

Xuan TD et al (2005) Biological control of weeds and plant pathogens in paddy rice by exploiting plant allelopathy: an overview. Crop Prot 24(3):197–206

Yan A et al (2020) Phytoremediation: a promising approach for revegetation of heavy metal-polluted land. Front Plant Sci 11:359

Yang X et al (2009) Simultaneous determination of myricitrin, hyperin, quercitroside, and quercetin in folium Rhododendri micranthi by RP-HPLC. J Chromatogr Sci 47(8):714–717

Zahara K et al (2015) A review on pharmacological properties of Bidens biternata: A potential nutraceutical. Asian Pac J Trop Dis 5(8):595–599

Zahara K et al (2022) Isolation and identification of bioactive compounds from Bidens spp. using HPLC-DAD and GC-MS analysis and their biological activity as anticancer molecules. Molecules 27(6):1927

Zeevaart JA, Creelman RA (1988) Metabolism and physiology of abscisic acid. Ann Rev Plant Physiol Plant Mol Biol 39(1):439–473

Zhou W et al (2020) Design, synthesis and anti-tumor activity of novel benzimidazole-chalcone hybrids as non-intercalative topoisomerase II catalytic inhibitors. Molecules 25(14):3180

Essential Oil Diversity of *Bidens* Species 9

Aleeza Nawab

Abstract

Bidens belongs to the family Asteraceae and is known for its essential oil (EO). They are called by multiple vernacular names, e.g., daisies, blackjacks, and beggarticks depending upon shape and color of the flower. The EO is present in varying amounts in different parts of the plant. Variety of extraction methods have been employed that affect the quality and composition of essential oil. The genus *Bidens* has been studied for their essential oil around the globe, e.g., North America, China, Hawaii, Kenya, Colombia, Australia, Poland, and Brazil. Essential oils of *Bidens* are loaded with polyacetylenes, phenolic acids, terpenoids, glycosides, saponins, polysaccharides, and alkaloids. The chemical constituents present in essential oil of *Bidens* species have unique applications in food, cosmetics and pharmaceutical industry. The chapter explores the essential oil constituents of different *Bidens* species, their phytochemistry, and potential uses. These chemical constituents may be extracted and used to treat cancer, asthma, anxiety, and other infectious diseases. But due to limited research on its anticancerous potential. There is a need to fully explore the potential of the essential oil in treating cancer and asthma. The details of the essential oil constituents are added separately for each species.

Keywords

Essential oil · *Bidens* · Asteraceae · Phytochemistry

A. Nawab (✉)
Department of Botany, Pir Mehr Ali Shah Arid Agriculture University, Rawalpindi, Pakistan

Y. Bibi et al. (eds.), *The Genus Bidens*,
https://doi.org/10.1007/978-981-96-4257-1_9

9.1 Introduction

Essential oils (EO) are concentrated extracts of plant that contain the aromatic compounds from numerous plant parts, i.e., leaves, flowers, bark, or roots. These oils are utilized for their medicinal properties, fragrance, and other benefits. They play an essential role in several aspects of well-being. These concentrated plant extracts propose benefits, i.e., aromatherapy, natural medicine, and even as natural cleaning products.

Genus *Bidens* is a diverse genus of Asteraceae family, including around 280 species. The Essential Oils contents from dried leaves, stem, and flowers of *Bidens* species have been studied in Iran, Nigeria, Cameroon, Japan, USA, and China for better preservation of food qualities and development of medicinal products. Various techniques were used and shown a variety of compounds from different parts of *Bidens* One prominent species is *Bidens pilosa*, normally known as "hairy beggar-tick." Hydro distillation of dried leaves of *Biden pilosa* was carried out resulting in 27 identified compounds, which accounted 97.57% of total oil content (Goudoum et al. 2016; Sakuda 1988; Deba et al. 2008; Riahi et al. 2008; Ogunbinu et al. 2009).

9.2 Essential Oil Diversity in Key Species of Genus *Bidens*

Essential Oils of genus *Bidens* have been traditionally used in Africa and South America since ancient times. It is reported to have various pharmacological and biological activities, e.g., antimalarial, antioxidant, anti-inflammatory, antidiabetic, antihyperglycemic, antileukemic, antimicrobial, antitumor, and antibacterial.

9.2.1 *Bidens pilosa*

Essential oils from different species of *Bidens* have been studied in recent years due to immense diversity of chemical compounds using different extraction techniques. *Biden pilosa* contains about 300 chemical constituents from different parts that exhibit antimicrobial, anti-inflammatory, anticandidal, antitumor, antioxidant, antibacterial, and antiangiogenic properties. These chemical compounds are polyacetylenes, polyacetylene glycosides, fatty acids, terpenes, okanin glycosides, phytosterols, phenolic acids, aurones, flavonoids, and flavonoids glycosides (Deba et al. 2008; Kuo et al. 2021; Xuan and Khanh 2016).

Due to its wide medicinal properties, the plant extracts of blackjacks have been proven to treat different diseases. Due to the increasing use of synthetic drugs, antibiotic resistance has developed inside human body due to its great tolerance to the variety of sentimental conditions. In this case, the herbal extracts have been proved effective as an alternative treatment due to their strong antibacterial properties (Ashafa and Afolayan 2009). Blackjacks release allelochemicals that contain insecticidal, allelopathic, fungicidal, and nematocidal activities, respectively (Balah et al. 2024) (Table 9.1).

Table 9.1 Chemical constituents of essential oil from *Bidens pilosa*

	Compound name	Chemical formula	Plant part
1	Palmitic acid	$C_{16}H_{32}O_2$	Seeds
2	p-Hydroxybenzoic acid	$C_7H_6O_3$	Roots and leaves
3	Limonene	$C_{10}H_{16}$	Leaves and flowers
4	Carvone	$C_{10}H_{14}O$	Leaves
5	Caffeic acid	(HO)2C6H3CH=CHCO2H	Leaves
6	α-Pinene	$CH_3(C{=}CH_2)\ C(CH_3){=}CH_2$	Leaves
7	α-Pinene oxide	$CH_3(C{=}CH_2)\ C(CH_3){=}CH_2O$	Leaves
8	β-Pinene	$CH_3(CH_2)\ C(CH_3){=}CH_2$	Leaves
9	Ferulic acid	$C_{10}H_{10}O_4$	Stem and leaves
10	Vanillic acid	$C_8H_8O_4$	Stem and leaves
11	Salicylic acid	$C_7H_6O_3$	Stem and leaves
12	p-Coumaric acid	$HOC_6H_4CH{=}CHCO_2H$	Leaves
13	Quercetin	$C_{15}H_{10}O_7$	Leaves and flowers
14	Epoxy-myrcene	$C_{10}H_{16}O$	Flowers
15	Limonene dioxide	$C_{10}H_{16}O_2$	Flowers
15	Myrcene	$C_{10}H_{16}$	Flowers
16	Octadecanoic acid	$C_{18}H_{36}O_2$	Flowers
17	9-Octadecenoic acid (Z)-, oxiranylmethyl ester	$C_{19}H_{34}O_3$	Flowers
18	Octadecenoic acid (E)-	$C_{18}H_{32}O_2$	Flowers
19	Glycidyl palmitate	$C_{19}H_{36}O_4$	Flowers
20	n-Hexadecenoic acid	$C_{16}H_{32}O_2$	Flowers
21	(Z)-15-Octadecenoic acid methyl ester	$C_{19}H_{36}O_2$	Flowers
22	Hexadecenoic acid, methyl ester	$C_{17}H_{34}O_2$	Flowers
23	Isoquinaldonitrile,3-methyl-, 2-oxide	$C_{11}H_{10}N_2O$	Flowers
24	1,1,1,3,5,5,5-Heptamethyltrisiloxane	$C_9H_{24}OSi$	Flowers

9.2.2 *Bidens tripartita*

Bidens tripartita is commonly called bur-marigold and is traditionally used by local communities for a variety of diseases. It is highly known for treating problems of kidney and gallbladder stones, bleeding, and ruptured blood vessels. Medicinal properties are due to the presence of polyacetylenes, polysaccharides, and polyphenolic compounds. *Bidens tripartita* possess anti-inflammatory, antibacterial, immunostimulant, antifungal, hepatoprotective, and hyposensitive activities (Kotov et al. 2020). In addition, methylene chloride extract of *Bidens tripartita* can inhibit cancer cells reported by one study. Studies on oil extraction from different parts of *Bidens tripartita* have been studied but rarely reported studies on roots.

The essential oils are highly volatile and contain antioxidants, flavonoids, lactones, carotenoids, mineral elements, coumarins, polysaccharides, and amines. Essential oil from roots, leaves, and aerial parts have been extracted via cold extraction, hydro distillation, steam distillation, and superficial CO_2 extraction have been carried out and gas chromatography and mass spectrophotometry was used for detection of chemical compounds. The presence of chemical compounds in specific plant part varies. The chemical compounds detected by GC–MS analysis of EO of *Bidens tripartita* are α-fellandren, p-cymene, β-linalool, β-ocimene, (Z)-cymen, phellandrene, and cubebene only present in aerial parts, while limonene, vitamin E, methyl palmitoleate, phytol, spathulenol, 9,12-octadecadienoic acid methyl ester, N-butylbenzene sulfonamide, Gamma stigmasterol, tocopherol, and campesterol in leaves or caryophyllene, α-pinene, β-elemene p-cymene are present in both leaves as well as aerial parts of plants. Whereas β-ocimene, α-pinene, β-cymene, and β-elemene have been reported in roots (Tomczykowa et al. 2011). Other chemical compounds are also present, but their percentage composition is not known. The name of compounds, % composition, and chemical formulas have been listed in Table 9.2.

9.2.3 *Bidens frondosa*

Bidens frondosa contains commonly called "devil beggarticks." It is highly known for its essential oil. The EO of *Bidens frondosa* contain diverse chemical compounds showing anticancerous, antidiabetic, anti-inflammatory, and antioxidant properties. The composition of essential oil depends upon edaphic and climatic factors. The leaf is rich in alpha and beta pinene, whereas flower is rich in germacrene D and caryophyllene oxide. The root contains high amounts of eugenol and methyl eugenol. These chemical compounds have the potential to treat asthma and cancer. The chemical constituents are present in varying amounts in various parts of plants, e.g., borneol (17.66%), β-cadre, and 4-terpineol (17.26%). Other compounds present in *Bidens frondosa* are 8,3′,4′-trihydroxyflavone-7-O-(6″-O-p-coumaroyl)-β-d-glucopyranoside, (3E,5E,11E)-tridecatriene-7,9-diyne-1,2,13-triol-2-O-β-d-glucopyranoside, caryophyllene oxide, 2′-butoxyethylconiferin, spathulenol, chalcones, or α-terpineol, respectively (Li et al. 2017) (Table 9.3). In addition, more research is needed to fully explore its treatment for cancer.

Table 9.2 Chemical compounds present in EO of *Bidens tripartita* with their molecular formula

1	Gamma tocopherol	$C_{20}H_{40}O_2$
2	Spathulenol	$C_{15}H_{24}$
3	Methyl palmitoleate	$C_{19}H_{36}O_2$
4	9,12-Octadecadienoic acid methyl ester	$C_{20}H_{36}O_2$

Table 9.3 Chemical compounds present in essential oil of *Bidens frondosa* with their molecular formula

Sl. no.	Name of compound	Percentage	Molecular formula
1	(3E,5E,11E)-tridecatriene-7,9-diyne-1,2,13-triol-2-O-β-d-glucopyranoside	1.23–1.83%	$C_{19}H_{24}O_8$
2	2′-Butoxyethylconiferin	0.1–0.5	$C_{22}H_{34}O_9$
3	8,3′,4′-Trihydroxyflavone-7-O-(6″-O-p-coumaroyl)-β-d-glucopyranoside	0.85–1.45%	$C_{30}H_{26}O_{13}$
4	Caryophyllene oxide	20.50%	$C_{15}H_{24}O$
5	Borneol	17.66%	$C_{10}H_{18}O$
6	Spathulenol	1.5–4.5%	$C_{15}H_{24}O$
7	α-Terpineol	0.2–1.5%	$C_{10}H_{18}O$
8	β-Cedrene	6.94%	$C_{15}H_{24}$
9	4-Terpineol	17.26%	$C_{10}H_{18}O$
10	Chalcones	0.5–1%	$C_{15}H_{12}O$
11	Aurones	1–2%	$C_{15}H_{10}O_5$
12	Polysaccharides	10–20%	$(C_6H_{10}O_5)n$
13	Flavonoids	5–10%	$C_{15}H_{10}O_6$-$C_{16}H_{12}O_8$
14	Luteolin-7-glycoside	2–5%	$C_{21}H_{20}O_{10}$
15	Luteolin-7-O-glycoside	1–3%	$C_{21}H_{20}O_{10}$

9.2.4 *Bidens bipinnata*

Bidens bipinnata is extensively studied for its antimicrobial properties in the recent years. It possesses unique medicinal properties and is highly known for treating dysentery, acute nephritis, and malaria (Kulsoom Zahara et al. 2022). Its essential oil consists of ceramides, flavonoids, steroids, phenylpropanoids, and pyrimidines (Hu et al. 2018). It is widely utilized among local communities for treating sore throats and malaria. Pharmacological investigations showed its antiliver fibrosis and antihypertensive properties. The study on *Bidens bipinnata* showed inhibition against alpha amylase activity that showed its antidiabetic potential (Yang et al. 2012) (Table 9.4).

9.2.5 *Bidens segetum*

Bidens segetum is commonly called corn marigold due to its flower appearance. It is native to Brazil, and its decoction is traditionally used among local communities as hepatoprotective, anti-inflammatory, and blood pressure–lowering properties. The EO of *Bidens* segetum contain volatile compounds, e.g., myrcene, *sabinene, β-phellandrene, germacrene D, E-β-ocimene, bicyclogermacrene, and α-pinene* (Silveira et al. 2009). It possess high antioxidant properties due to the presence of high amount of coumarins (Nascimento et al. 2015). Due to these volatile

Table 9.4 Chemical constituents of *Bidens bipinnata* with their molecular formula

	Name of compound	Molecular formula
1	Isokanin	$C_{16}H_{10}O_6$
2	(2S, 3S, 4R, 8E)-2-(2′(R), 3′(R)-dihydroxytriocosanoylamino]-8-octadecene-1, 3, 4-triol	$C_{51}H_{100}N_2O_{11}$
3	(2S, 3S, 4R, 8E)-2-[(2′R)-2′-hydroxytriocosanoylamino]-10-octadecene-1, 3, 4-triol	$C_{58}H_{124}N_2O_{11}$
4	(2S, 3S, 4R, 8E)-2-[(2′R)-2′-hydroxytetracosanoylamino]-10-octadecene-1, 3, 4-triol	$C_{64}H_{126}N_2O_{11}$
5	(2S) isookanin-4′-methoxy-7-O-β-d-glucopyranoside	$C_{21}H_{24}O_{11}$
6	3,6,8-Trichloro-5, 7, 3′, 4′-tetrahydroxyflavone	$C_{15}H_7Cl_3O_6$
7	8,3′,4′-Trihydroxyflavone-7-O-β-d-glucopyranoside	$C_{21}H_{20}O_{11}$
8	(2R) isookanin-4′-methoxy-7-O-β-d-glucopyranoside	$C_{21}H_{24}O_{11}$
9	(3E,11E)-Tridecadiene-6,8,10-atriyne-1,13-diol-2-O-β-D-glucopyranoside	$C_{21}H_{28}O_9$
10	4-O-(2″, 3″-O-diacetyl-6″-O-p-coumaroyl-β-d-glucopyranosyl)-p-coumaric acid	$C_{28}H_{28}O_{12}$.
11	(6E,12E)-Tetradecadiene-8,10-diyne-1,14-diol-3-O-β-D-glucopyranoside	$C_{19}H_{24}O_8$
12	(6E,12E)-3-Oxo-tetradecadiene-8,10-diyne-14-hydroxyl-1-O-β-D-glucopyranoside	$C_{19}H_{26}O_7$
13	(6E,12E)-Tetradecadiene-8,10-diyne-1,3,14-triol	$C_{14}H_{18}O_3$
14	(2S) (5E,11E)-Tridecadiene-7,9-diyne-1,2,13-triol	$C_{13}H_{18}O_3$
14	(3E,11E)-Tridecadiene-6,8,10-atriyne-1,2,13-triol	$C_{13}H_{14}O_3$
15	9-Octadecenoic acid (Z)-, methyl ester	$C1_9H_{36}O_2$
16	16-Pregnenolone	$C_{21}H_{30}O_3$

Table 9.5 Essential oil composition of *Bidens segetum*

Sl. no.	Compounds	Percentage	Molecular formula
1	β-Phellandrene	18–45%	$C_{10}H_{14}$
2	Myrcene	11.2%	$C_{10}H_{16}$
3	Germacrene D	6–32%	$C_{15}H_{24}$
4	δ-Cadinene	3.2%	$C_{15}H_{24}$
5	E-β-Ocimene	11.6%	$C_{10}H_{16}O$
6	α-Pinene	7.3%	$C_{10}H_{16}$
7	Bicyclogermacrene	5.7%	$C_{15}H_{24}$
8	Sabinene	6.7%	$C_{10}H_{16}$

constituents, it showed high activity against HSV1 and ACVr uncovering its medicinal potential (Table 9.5).

9.2.6 *Bidens gardenii*

It is a traditional medicinal plant widely distributed in region Pantanal in Brazil. Few studies have been reported on EO composition of *gardenii*. The study reported the presence of some sesquiterpenes, e.g., α-guaiene, α-himachalane, alloaromadendrene, methyl linoleate, β-caryophyllene, germacrene D, bicyclogermacrene, α-amorphene, and β-chamigrene The analysis of volatile constituents is done by SPME that enhance the accuracy and is completely solvent free. *Bidens gardenii* decoction, tea, tincture, and poultice are traditionally used as Indian and Chinese medicine to treat eczema, pain relief, and to treat respiratory problems (Table 9.6).

9.2.7 *Bidens sulphurea*

It is commonly called sulfur cosmos due to resemblance of its flower with sulfur. It is native to Southern America, Central America, and Mexico. Its phytoconstituents possess anticancerous, antibacterial, and antipyretic properties. Due to the presence of highly volatile phytoconstituents, this specie is reported as weed control (Da Silva et al. 2017). Despite its role as weed control, due to a variety of chemical constituents, e.g., bicyclogermacrene, bicyclolemene, germacrene D, β-caryophyllene, and α-copaene, *Bidens sulphurea* can treat inflammatory bowel disease (IBD), gastric disorders, and control fever (Table 9.7).

Table 9.6 Essential oil constituents of *Bidens gardenii*

	Compounds	Percentage	Molecular formula
1	α-Himachalane	2.8%	$C_{15}H_{24}$
2	α-Guaiene	0.4%	$C_{15}H_{24}$
3	Methyl linoleate	0.9%	C19H34O2
4	β-Caryophyllene	88.6% in aerial parts	$C_{15}H_{24}$
		88.2% in flowers	
		95.9% in fruit	
5	Alloaromadendrene	0.2%	$C_{15}H_{24}$
6	Germacrene D	24.3% in aerial parts	
		43.1% in fruits	
		33.2% in flowers	
7	Bicyclogermacrene	5.7% in aerial parts	$C_{15}H_{24}$
		2.7% in fruits	
		1.6% in flowers	
8	β-Chamigrene	2.7%	$C_{15}H_{24}$
9	α-Amorphene	0.2%	$C_{15}H_{24}$

Table 9.7 Essential oil constituents of *Bidens sulphurea*

Sesquiterpenes ($C_{15}H_{24}$)				
Sl. no.	Compounds	Aerial parts	Flowers	Fruits
1	Bicyclogermacrene	5.7%	1.6%	2.7%
2	β-Caryophyllene	77.4	90.5%	77.4%
3	Germacrene D	61.7%	44.3%	17.8%
4	α-Copaene	Not specified		
5	Bicyclolemene	Not specified		

Table 9.8 Essential oil composition of *Bidens sulphurea*

	Sesquiterpenes ($C_{15}H_{24}$)	
Sl. no.	Compounds	Percentage
1	Bicyclogermacrene	44.3%
2	β-Caryophyllene	77.4%
4	β-Chamigrene	2.7%
5	Alloaromadendrene	0.2%
6	α-Amorphene	0.2%
7	Germacrene D	61.7%
8	α-Guaiene	0.4%
9	α-Himachalane	2.8%
10	Methyl linoleate	0.9%

9.2.8 *Bidens graveolens*

Bidens graveolens is commonly called "Stinking Begger ticks." The essential oil of *B. graveolens* have been extracted via hydro distillation identified by GC/MS analysis that led to identification of some volatile constituents, e.g., bicyclogermacrene (44.3%), β-caryophyllene (77.4%), β-chamigrene (2.7%), alloaromadendrene (0.2%), α-amorphene (0.2%), germacrene D(61.7%), α-guaiene (0.4%), α-himachalane (2.8%), and methyl linoleate (0.9%). The highest percentage of germacrene 61.7% constitute EO of *Bidens graveolens* (Silva et al. 2020) (Table 9.8). Due to the presence of these chemical constituents, they can be used to treat diabetes, respiratory problems, gastric ulcer, diabetes, inflammation, and skin infections (Delgado-Paredes et al. 2021).

9.2.9 *Bidens parviflora*

Bidens parviflora is commonly called small, flowered beggartick due to distinctive small flower as compared to other species of *Bidens*. The EO has been extracted from *B. parviflora* that resulted in the identification of 16 chemical constituents. The chemical constituents extracted from the whole parts of the *B. parviflora* are apigenin, luteolin, quercetin, rhamnetin, rhamnazin, tricin, 3′-O-methylorobol,

Table 9.9 Chemical constituents of essential oils of *Bidens parviflora*

	Chemical compounds	Percentage	Molecular formula
1	Apigenin	$C_{15}H_{10}O_5$	1.09%
2	Quercetin	$C_{15}H_{10}0_7$	1.03%
3	Luteolin	$C_{15}H1_0O_6$	0.96%
4	Isorhamnetin	$C_{16}H_{12}O_7$	0.82%
5	Rhamnetin	$C_{16}H_{12}O_7$	0.52%
6	Rhamnazin	$C_{16}H_{12}O_7$	0.68%
7	Tricin	$C_{17}H_{14}O_7$	0.64%
8	3′-O-Methylorobol	$C_{16}H_{12}O_6$	0.71%
9	5,7,4′-Trihydroxy-6,8,3′-trimethoxyflavone	$C_{18}H_{16}O_8$	0.35%
10	7,3′-Dimethylquercetin	$C_{17}H_{12}O_7$	0.45%
11	8-(β-D-glucopyranosyl)-3-hydroxy-1-methoxy-1,6,9-nonatriyne	$C_{16}H_{22}O_7$	
12	3′-O-Methylquercetin	$C_{17}H_{12}O_7$	0.59%
13	5,7,3′,4′-tetrahydroxy-6,8-dimethoxyflavone-7-O-β-D-glucopyranoside	$C_{23}H_{24}O_{14}$	0.31%
14	3,5,3′,4′-Tetrahydroxy-6,7,8-trimethoxyflavone	$C_{18}H_{16}O_9$	0.42%
15	5,7,3′,4′-Tetrahydroxy-6,8-dimethoxyflavone	$C_{17}H_{14}O_8$	0.48%

5,7,4′-trihydroxy-6,8,3′-trimethoxyflavone, 7,3′-dimethylquercetin, 5,7,3′,4′-tetrahydroxy-6,8-dimethoxyflavone-7-O-β-D-glucopyranoside, 3,5,3′,4′-tetrahydroxy-6,7,8-trimethoxyflavone, and 5,7,3′,4′-tetrahydroxy-6,8-dimethoxyflavone. The highest percentage of the compound is apigenin that is about 1.09% reported in EO of *B. parviflora* (Table 9.9) (Li et al. 2008).

9.2.10 *Bidens procera*

Bidens procera is commonly called tall beggarticks widely distributed in Madagascar, tropical Africa, and Asia. The essential oil is rich in polyacetylenes, flavonoids, and phenolic acids (Table 9.10). The essential oil is medicinally important that possess anti-inflammatory, wound healing, antimicrobial, and antioxidant properties due to the presence of various terpenes, chlorogenic acid, quercetin, apigenin, and luteolin (Cai et al. 2022). It is also used to treat respiratory tract infections. Essential oil from *Bidens procera* has been extracted in recent years. The chemical compounds present in EO of *B. procera* are listed in table along with their molecular formulas.

Table 9.10 Essential oil constituents of *Bidens procera*

	Chemical compounds	Molecular formula
1	Farnisin	$C_{16}H_{18}O_5$
2	Luteolin	$C_{15}H_{10}O_6$
3	Esculetin	$C_9H_6O_4$
4	Methyl hematinate	$C_{16}H_{16}O_8$
5	Maritimein	$C_{15}H_{14}O_5$
6	(7′R,8′R)-Sapiumin B	$C_{16}H_{14}O_7$
7	Okanin 4′-O-β-D-glucoside	$C_{20}H_{20}O_9$
8	6″-Acetylmaritimein	$C_{16}H_{16}O_6$
9	2,3-Diacetyl butanetetrol	$C_{10}H_{14}O_5$
10	(Z)-6,7,3′,4′-Tetrahydroxyaurone	$C_{15}H_{10}0_6$
11	Okanin 4′-O-(6″-E-p-coumaroyl)-β-D-glucopyranoside	$C_{22}H_{22}O_{11}$
12	3-Methylquercetin	$C_{16}H_{10}O_7$
13	Methyl 4,5-di-O-caffeoyl quinate	$C_{19}H_{20}O_{10}$
14	Methyl 4-O-caffeoyl-2-C-methyl-D-erythronate	$C_{16}H_{16}O_8$
15	3-O-Caffeoyl-2-C-methyl-D-erythrono-1,4-lactone	$C_{17}H_{18}O_7$
16	Methyl 2-O-caffeoyl-2-C-methyl-D-erythronate	$C_{16}H_{16}O_8$

Table 9.11 Essential oil constituents of *Bidens cernua*

	Chemical compound	Molecular formula
1	1-Phenylhepta1,3,5-triyne	57.1%
2	2-epoxide	3.1%
3	α-selinene	1.1%
4	(E, E)-1,3,11-Tridecatriene5,7,9-triyne	1.8%
5	Humulene-1	3.0%

9.2.11 *Bidens cernua*

Bidens cernua is commonly called nodding beggartick, which is native of North America. It is traditionally used in treating various ailments including respiratory tract infections, digestive issues, skin conditions, and anti-inflammatory issues. The *Bidens cernua* contains pharmacologically important chemical constituents, holds ethnopharmacological significance, and can be used in the preparations. The liquid oil obtained through hydrodistillation of blooming aerial parts of *Bidens cernua* is lighter than water with strong pungent smell due to the presence of thiophene derivatives. The oil was rich in polyacetylenes and deficient in oxidized compounds (Table 9.11). The phenylheptatriynes is a chemotaxonomic marker and is nonvariable.

9.3 Conclusion

The essential oils obtained from *Bidens* species have been reported to contain a large number of bioactive compounds, i.e., alkaloids, phenols, terpenes, etc. The therapeutic potential of their essential oils is contributed by these compounds. The essential oils derived from this genus are widely utilized for management of skin-related problems by local communities, all over the world. They are also reported to have anti-inflammatory properties. But there is a need to fully explore their molecular mechanism for revealing its potential applications and developing therapeutic drugs for human well-being.

References

Ashafa A, Afolayan AJ (2009) Screening the root extracts from Biden pilosa L. var. radiata (Asteraceae) for antimicrobial potentials. J Med Plants Res 3(8):568–572

Balah MA, Al-Andal A, Radwan AM, Donia AEM (2024) Unveiling allelopathic dynamics and impacts of invasive Erigeron bonariensis and Bidens pilosa on plant communities and soil parameters. Sci Rep 14(1):10159

Cai F-J, Wang L, Zhao W, Tian J-L, Kong D-G, Liu Q et al (2022) Phytochemical and chemotaxonomic investigations on the whole herbs of Bidens procera LC Xu ex XW Zheng. Biochem Syst Ecol 101:104395

Da Silva BP, Nepomuceno MP, Varela RM, Torres A, Molinillo JM, Alves PL et al (2017) Phytotoxicity study on *Bidens sulphurea* Sch. Bip. as a preliminary approach for weed control. J Agric Food Chem 65(25):5161–5172

Deba F, Xuan TD, Yasuda M, Tawata S (2008) Chemical composition and antioxidant, antibacterial and antifungal activities of the essential oils from Bidens pilosa Linn. var. Radiata. Food Control 19(4):346–352

Delgado-Paredes G, Delgado-Rojas P, Rojas-Idrogo C (2021) Peruvian medicinal plants and cosmopolitan plants with potential use in the treatment of respiratory diseases and COVID-19. Int J Plant Anim Environ Sci 11:295–321

Goudoum A, Abdou AB, Ngamo LST, Ngassoum MB, Mbofung CMF (2016) Antioxidant activities of essential oil of Bidens pilosa (Linn. Var. Radita) used for the preservation of food qualities in North Cameroon. Food Sci Nutr 4(5):671–678

Hu H-M, Bai S-M, Chen L-J, Hu W-Y, Chen GJ (2018) Chemical constituents from Bidens bipinnata Linn. Biochem Syst Ecol 79:44–49

Kotov S, Kotov E, Bezruk I, Gontova T, Kotov A (2020) The study of the extraction dynamics of biologically active substances from the Bidens tripartita L. herb and antioxidant activity of the obtained extracts. EUREKA, Health Sci 6:95–101

Kulsoom Zahara YB, Arshad M, Kaukab G, Al Ayoubi S, Qayyum A (2022) In-vitro examination and isolation of antidiarrheal compounds using five bacterial strains from invasive species Bidens bipinnata L. Saudi J Biol Sci 29(1):472–479

Kuo TF, Yang G, Chen TY, Wu YC, Tran Nguyen Minh H, Chen LS et al (2021) Bidens pilosa: nutritional value and benefits for metabolic syndrome. Food Front 2(1):32–45

Li Y-L, Li J, Wang N-L, Yao X-S (2008) Flavonoids and a new polyacetylene from Bidens parviflora Willd. Molecules 13(8):1931–1941

Li HY, Chen XB, Liu QZ, Liu ZL (2017) Chemical composition and insecticidal properties of the essential oil of Bidens frondosa L (Asteraceae) against booklice (Liposcelis bostrychophila). Trop J Pharm Res 16(1):171

Nascimento AL, Raggi L, Young MCM, Moreno PR (2015) Chemical characterization of the volatile compounds of the flowers of Bidens segetum Martius ex Colla (Asteraceae). J Essent Oil Res 27(1):70–75

Ogunbinu AO, Flamini G, Cioni PL, Adebayo MA, Ogunwande IA (2009) Constituents of Cajanus cajan (L.) Millsp., Moringa oleifera Lam., Heliotropium indicum L. and Bidens pilosa L. from Nigeria. Nat Prod Commun 4(4):1934578X0900400427

Riahi S, Ganjali MR, Pourbasheer E, Norouzi P (2008) QSRR study of GC retention indices of essential-oil compounds by multiple linear regression with a genetic algorithm. Chromatographia 67:917–922

Sakuda Y (1988) Constituents of essential oils from Bidens pilosa L. and Ambrosia trifida L. Kochi Joshi Daigaku Kiyo, Shizen Kagakuhen 36:1–5

Silva AC, Bizzo HR, Vieira RF, Bringel JB Jr, Azevedo DA, Uekane TM, Rezende CM (2020) Characterization of volatile and odor-active compounds of the essential oil from Bidens graveolens Mart. (Asteraceae). Flavour Fragr J 35(1):79–87

Silveira CS, Martins FO, Costa CS, Romanos MTV, Kaplan MAC, Boylan F (2009) In vitro cytotoxic, antioxidant and antiviral effects of Pterocaulon alopecuroides and Bidens segetum extracts. Rev Bras Farmacogn 19:343–348

Tomczykowa M, Leszczyńska K, Tomczyk M, Tryniszewska E, Kalemba D (2011) Composition of the essential oil of Bidens tripartita L. roots and its antibacterial and antifungal activities. J Med Food 14(4):428–433

Xuan TD, Khanh TD (2016) Chemistry and pharmacology of Bidens pilosa: an overview. J Pharm Investig 46:91–132

Yang X-W, Huang M-Z, Jin Y-S, Sun L-N, Song Y, Chen H-S (2012) Phenolics from Bidens bipinnata and their amylase inhibitory properties. Fitoterapia 83(7):1169–1175

Aliphatic Natural Products Isolated from *Bidens* Species

10

Syeda Sobia Gilani

Abstract

According to the United Nations World Health Organization (UNWHO), 80% of the world's population obtains their primary medical treatment from medicinal plants. From ancient medicinal practices to modern scientific investigations, *Bidens* species have captivated the attention of scientific researchers. *Bidens*, commonly known as beggarticks or bur-marigolds, is a genus of traditional medicinal plants enriched in phytochemical composition. The advancement in phytochemical analysis and bioassay techniques has made it easier to identify and characterize the variety of secondary metabolites found in *Bidens* species, such as terpenoids, poly-acetylenes and flavonoids, and aliphatic chemicals. This chapter is a guide on aliphatic natural products, their extraction, structural elucidation, and possible pharmaceutical applications. Aliphatic compounds found in *Bidens* include fatty acids (i.e., saturated, unsaturated fatty acids, i.e., palmitic acid, stearic acid, oleic acid, and linoleic acid). These compounds have diverse pharmacological activities, including anti-inflammatory, antimicrobial, and antioxidant properties. In addition to fatty acids, *Bidens* species contain a remarkable array of esters, which are produced when fatty acids condense with alcohols. These esters not only contribute to the distinct aroma and flavor of flowers and fruits but also have potential applications in the perfume and flavoring industries. In addition, the polyketides isolated from *Bidens* include chalcones, flavonoids, and anthocyanins, which exhibit antioxidant, anti-inflammatory, and anticancer activities. Poly-acetylenes from *Bidens* plants possess diverse biological activities, including antimicrobial, antifungal, and cytotoxic properties, indicating their potential as pharmacological and agricultural candidates. Despite their ubiquity, *Bidens* plants remained relatively understudied until recent years, when

S. S. Gilani (✉)
Department of Botany, Arid Agriculture University, Rawalpindi, Pakistan

Y. Bibi et al. (eds.), *The Genus Bidens*,
https://doi.org/10.1007/978-981-96-4257-1_10

their pharmacological, agricultural, and ecological significance began to emerge. By unraveling the biosynthetic pathways and biological activities of aliphatic compounds from *Bidens* species, researchers can unlock new opportunities for drug discovery, crop improvement, and environmental sustainability. This will result in an improved understanding of the medicinal plants and their wonders.

Keywords

Bidens · Aliphatic compounds · Pharmaceutical applications · Biological activities

10.1 Introduction

Hydrocarbons are molecules made entirely of carbon and hydrogen. In the field of organic chemistry, they are divided into two main categories: aliphatic compounds and aromatic compounds. Aliphatic compounds can exist in either unsaturated forms, such as hexene and hexyne, or saturated forms, like hexane, where all the carbon–carbon bonds are single and the structure is "saturated" with hydrogen atoms. Open-chain compounds, which lack any kind of ring, always fall into the aliphatic category and can exhibit straight or branched structures. Additionally, cyclic molecules that are not aromatic can also be classified as aliphatic (Kumada et al. 1968). As was previously established, there are several structural variations among aliphatic molecules. They may be hexane is a linear molecule with a structure like a snake (hexane), the branched molecule (3-methyl hexane), and cyclic molecule (cyclohexane) (Albertsson and Varma 2002).

Aliphatic compounds are a class of organic compounds that can exhibit unsaturation, characterized by double bonds (alkenes) or triple bonds (alkynes), or saturation, with single bonds (alkanes) (Bartolome et al. 2013). However, if other elements (heteroatoms) are bonded to the carbon chain, the compound is no longer considered a hydrocarbon and is therefore not classified as an aliphatic compound. Common heteroatoms that can be bonded to the carbon chain include oxygen, nitrogen, sulfur, and chlorine. It's important to note that compounds with heteroatoms bonded to the carbon chain may still be referred to as aliphatic, if the hydrocarbon component of the molecule is aliphatic, such as in the case of aliphatic amines. This distinction is made to differentiate them from aromatic amines. Methane (CH4), for instance, is the most basic and basic aliphatic molecule; it is made up of one carbon atom bound to four hydrogen atoms.

The *Bidens* genus in the Asteraceae family are known for their rich chemical diversity, particularly in aliphatic natural products. The genus *Bidens* encompasses over 200 species, distributed worldwide, especially in tropical and sub-tropical regions (Bo et al. 2012). One notable species, *Bidens pilosa*, is extensively studied for its pharmacologically active constituents (Bartolome et al. 2013). The aliphatic compounds in *Bidens* include fatty acids, alcohols, aldehydes, and ketones, which contribute to the plant's bioactivity. These compounds are primarily synthesized through the acetate-malonate pathway, a common biosynthetic route in plants (Alamgir 2017). Fatty acids, such as linoleic and oleic acid, are predominant in *Bidens* species and possess anti-inflammatory properties (Wahjudi et al. 2024). In addition to fatty acids, *Bidens* species produce significant quantities of aliphatic alcohols like hexadecanol and octadecanol (Xuan and Khanh 2016). These alcohols are known to act as precursors for the biosynthesis of waxes, which are essential for protecting the plant's surface (Chen et al. 2016). Furthermore, aliphatic aldehydes and ketones, such as nonanal and hexanone, are involved in plant defense mechanisms against herbivores and pathogens (Mutuma et al. 2022).

The isolation and characterization of these aliphatic compounds from *Bidens* involve various analytical techniques, including gas chromatography–mass spectrometry (GC–MS) and nuclear magnetic resonance (NMR) spectroscopy. These methods allow for precise identification and quantification of the compounds, providing insights into their functional roles. The studies have employed high-performance liquid chromatography (HPLC) to separate and purify individual aliphatic constituents from *Bidens* extracts (Singh et al. 2020). The biological activities of *Bidens* aliphatic compounds have garnered attention for their potential therapeutic application. For instance, linoleic acid isolated from *Bidens pilosa* has demonstrated significant antimicrobial activity against various bacterial strains (Deba et al. 2008). Moreover, hexadecanol and octadecanol exhibit antioxidant properties, which could be beneficial in preventing oxidative stress-related diseases (Ravichandran et al. 2018). Despite the promising bioactivities, the mechanisms underlying the actions of *Bidens* aliphatic compounds remain largely unexplored. Future research should focus on elucidating these mechanisms through molecular and cellular studies. Additionally, the ecological roles of these compounds in plant–plant and plant–insect interactions warrant further investigation (Rosenthal and Berenbaum 2012). Understanding the chemistry of *Bidens* and their aliphatic natural products not only contributes to the field of natural product chemistry but also offers potential avenues for the development of new pharmacological agents. Integrating chemical analysis with biological studies can pave the way for discovering novel compounds with unique bioactivities. As research advances, the comprehensive exploration of *Bidens* aliphatic compounds could lead to significant breakthroughs in medicine and agriculture (Gagare et al. 2024).

10.2 Structural Diversity of Aliphatic Natural Products in *Bidens*

The genus *Bidens* comprises over 230 species found across tropical and sub-tropical regions. These species are well known for their rich phytochemical profiles, particularly their aliphatic compounds. Aliphatic compounds, characterized by their linear or branched hydrocarbon chains, include a variety of bioactive molecules, such as fatty acids, terpenoids, and polyacetylenes. These compounds have been found to possess significant pharmacological properties, making *Bidens* species valuable in traditional and modern medicine. This chapter explores the diversity of aliphatic compounds in the *Bidens* genus and their therapeutic applications, providing a comprehensive overview of Biden species and their bioactive compounds (Zhang et al. 2020). Since the genus *Bidens* was first identified in 1753, there has been a rise in interest in both basic research and practical applications. This is mostly because of how widely it is used in foods, beverages, and medications. A total of 116 publications have detailed Biden's exploitation and therapeutic application. As of now, 201 chemicals have been discovered from this plant, including 70 aliphatics, 60 flavonoids, 8 porphyrins, 25 terpenoids, 19 phenyl-propanoids, 13 aromatics, and 6 miscellaneous compounds. Future studies should concentrate on the relationship between phytochemicals and their bioactivities, as not yet entirely established. In this chapter, we discuss the significance of the identified compounds in light of the intricacy of their biological activities (Bartolome et al. 2013).

Aliphatic natural products encompass a wide range of chemical structures, spanning from simple saturated hydrocarbons to complex polyketides and polyacetylenes. Within *Bidens* species, these compounds manifest in various forms, each with its unique chemical signature and biological significance (Gagare et al. 2024). One prominent group of aliphatic compounds found in *Bidens* is fatty acids, which serve as essential building blocks for lipid biosynthesis and membrane structure. These consist of unsaturated and saturated fatty acids like oleic and linoleic acid, and palmitic and stearic acid, respectively. Fatty acids from *Bidens* plants have a variety of pharmacological actions, in addition to their roles as metabolic intermediates, such as anti-inflammatory, antioxidant, and antibacterial qualities (Idris et al. 2023). In addition to fatty acids, *Bidens* species produce a remarkable array of esters, which result from the condensation of fatty acids with alcohols. These esters contribute to the characteristic aroma and flavor profiles of *Bidens* flowers and fruits while also possessing potential applications in perfumery and flavoring industries (Lima Silva et al. 2011). Polyketides represent another class of aliphatic natural products abundant in *Bidens* plants. These compounds are synthesized through the iterative condensation of acetyl-CoA units by polyketide synthase enzymes, leading to the formation of structurally diverse molecules with pharmacological relevance. Examples of polyketides isolated from *Bidens* include chalcones, flavonoids, and anthocyanins, which exhibit antioxidant, anti-inflammatory, and anticancer activities (Dembitsky et al. 2018). Polyacetylenes, characterized by carbon–carbon triple bonds in their backbone, represent a unique class of aliphatic compounds found in certain *Bidens* species. These compounds are biosynthesized via the polyketide pathway, followed by desaturation and cyclization reactions mediated by specific enzymes. Polyacetylenes from *Bidens* plants possess diverse biological activities, including antimicrobial, antifungal, and cytotoxic properties, making them promising candidates for pharmaceutical and agricultural applications (Chiang et al. 2004).

10.3 Aliphatic Natural Compounds from Key Species of Genus *Bidens*

10.3.1 *Bidens pilosa*

Of all the species in the genus, *Bidens pilosa* has been investigated the most. From the various parts of this plant, major chemical constituents (including 301 compounds) belonging to the following groups have been identified or isolated: phenolic acids, terpenes, pheophytins, fatty acids, aurones, chalcones, okanin glycosides, flavonoids, flavone glycosides, and phytosterols. Numerous of them are thought to be bioactive substances that may be responsible for some of the pharmacological effects. This plant is known to possess a variety of biological and pharmacological properties, including antimalarial, antiallergy, antihypertensive, smooth muscle relaxant, anticancerogenic, antidiabetic, anti-inflammatory, antimicrobial, and antioxidant properties (Xuan and Khanh 2016). In TCM, polyacetylene substances are typically regarded as an effective building block. Numerous investigations have demonstrated the various pharmacological benefits of naturally occurring polyacetylenes, including their antibacterial, antifungal, antiproliferative, and anti-inflammatory properties (Deba et al. 2008). Important inflammatory enzymes involved in the inflammatory response, such as cyclooxygenase-2 (COX-2) and inducible nitric oxide synthase (iNOS), are inhibited by polyacetylenes. Furthermore, these substances have demonstrated efficacy against many bacterial and fungal diseases, endorsing their application in the management of infections. In addition, *B. pilosa* has aliphatic acids, which support its antioxidant qualities by scavenging free radicals and lowering oxidative stress. These diverse medicinal effects highlight *B. pilosa*'s potential as a source of bioactive aliphatic chemicals for use in medicine (Xuan and Khanh 2016).

The *B. pilosa* has 24 different terpene compounds, including monoterpenes, sesquiterpenes, diterpenes, triterpenes, and tetraterpenes. Their chemical structures are represented by compounds 1–24. 28 monoterpenes (C10), 58 sesquiterpenes (C15), 6 diterpenes (C20), and 6 triterpenes (C25) are found among them; the remaining constituents are various forms of terpenoid derivatives. Both hydrocarbons and oxygenated molecules are present in them. There are eight different types of monoterpenes: acyclic monoterpenes, which include compounds 1–4, and monocyclic monoterpenes, which include compounds 5, 6, and 7. In addition, the bicyclic compounds 8, 9, 10, and 11 have been found. Except for two molecules, 12 and 13, which are linear sesquiterpenes, the majority of sesquiterpenes are monocyclic, bicyclic, or tricyclic. Sesquiterpene's unique structures differ significantly from one another. The majority of chemicals, known as monoterpenes and sesquiterpenes, are present in essential oils that are extracted from different sections of *B. pilosa*. Compounds 14–17 acyclic diterpenes were extracted from the entire *B. pilosa* plant. The chemicals 18, which are tricyclic diterpenes, are components of the essential oils that are extracted from the shoots. Whole plants have also been shown to contain the single acyclic polyterpene compound 19 as well as the pentacyclic polyterpenes (triterpenes) components 20–23. The tetraterpene, component 24, was the sole compound discovered in the leaf (Xuan and Khanh 2016).

10.3.2 *Bidens tripartita*

Bidens tripartita, commonly known as three-lobe beggarticks, is another species with significant therapeutic potential. The plant *B. tripartita* is rich in many biologically active substances, such as minerals like essential oil, chlorophylls, flavonoids, cinnamic acid, tannins, polyphenol content, polysaccharides, carotenoids, and ascorbic acid. There are triterpenes, unsaturated aliphatic hydrocarbons, fatty acid, and sterol ethers, mostly consisting of stigmasterol in the *B. tripartita* petroleum (Rodin et al. 2021). The aliphatic compounds in *B. tripartita,* such as essential fatty acids and terpenoids, have been shown to possess hypoglycemic properties, making them beneficial in managing diabetes (Berganayeva et al. 2023). The aliphatic compounds identified from this plant are p-cymene (6.0%), hexanal (5.7%), p-cymene-9-ol (3.4%), and β-element (2.6%) (Tomczykowa et al. 2011).

10.3.3 *Bidens frondosa*

Bidens frondosa, known as devil's beggarticks, contains aliphatic compounds with potent antimalarial properties. These compounds inhibit the growth of *Plasmodium falciparum,* the parasite responsible for malaria, by disrupting its life cycle. In addition to their antimalarial activity, the aliphatic compounds in *B. frondosa* also exhibit significant anti-inflammatory and analgesic properties, which are beneficial in managing pain and inflammation. These substances show antioxidant action, lower the risk of chronic illnesses, and shield cells from oxidative stress, which enhances their therapeutic potential.

The aliphatic compounds isolated from this plant include a new polyacetylene glucoside called (3 E,5 E,11 E)-tridecatriene-7,9-diyne-1,2,13-triol-2-O-β-d-glucopyranoside. This compound exhibits significant anti-inflammatory activity. Another compound identified is phenylpropanoid glucoside and 2′-butoxyethylconiferin. It belongs to the phenylpropanoid class and contributes to the plant's bioactivity (Ahn and Kim 2016).

10.3.4 *Bidens bipinnata*

Bidens bipinnata, or Spanish needles, is notable for its anticancer properties. The aliphatic compounds identified from this plant include E-2-hexenyl O-β-D-glucopyranoside, n-hexyl O-β-D-glucopyranoside, isopentyl O-β-D-glucopyranoside, n-butyl O-α-D-fructofuranoside, and n-butyl-O-β-D-fructopyranoside. The aliphatic compounds isolated from *B. bipinnata* have been shown to inhibit the growth of cancer cells and induce apoptosis, making them promising candidates for cancer treatment. These compounds modulate cell signaling pathways involved in cell proliferation and survival, thereby exerting their anticancer effects. The anti-inflammatory properties of its aliphatic compounds help in reducing inflammation and pain, which are beneficial in managing inflammatory conditions. The diverse pharmacological activities of *B. bipinnata* highlight its potential as a valuable medicinal plant for developing new treatments for cancer, infections, inflammation, and oxidative stress-related diseases. The four aliphatics were among the 38 chemicals that were isolated and identified as a result of the phytochemical study conducted on *Bidens bipinnata Linn.* (Asteraceae). Based on information from NMR and MS spectroscopy, all chemical structures were determined (Hu et al. 2018).

10.3.5 *Bidens parviflora*

Bidens parviflora, commonly known as small flower beggarticks, has been traditionally used for its antifungal properties. The whole herb of *Bidens parviflora Willd.* was extracted in ethanol to yield two new polyacetylene glucosides: (10S)-tridecane-2E,12-diene-4,6,8-triyne-1-ol-10-O-β-D-glucopyranoside and (12R)-tridecane-2E,8E-diene-4,6-diyne-1,14-diol-12-O-β-D-glucopyranoside. The aliphatic compounds isolated from *B. parviflora* have demonstrated potent antifungal activity by disrupting the cell membrane integrity of fungi, leading to their death. In addition to its antifungal effects, *B. parviflora* exhibits anti-inflammatory properties, which are beneficial in reducing inflammation and pain. The antioxidant activity of its aliphatic compounds helps in protecting cells from oxidative damage, contributing to their overall health benefits. Furthermore, *B. parviflora* is a possible source of novel antimicrobial agents for the treatment of bacterial and fungal illnesses due to the antibacterial qualities of these compounds. The therapeutic applications of *B. parviflora* highlight its potential as a valuable medicinal plant with diverse health benefits (Zhu et al. 2021).

10.3.6 *Bidens alba*

Bidens alba, also known as shepherd's needle, is another species with significant medicinal potential. The aliphatic compounds, i.e., the polyacetylene phenylhepta-1,3,5-triyne identified (Grombone-Guaratini et al. 2004) and quantitated by high-pressure liquid chromatography in *B. alba* have shown promising anti-inflammatory and antimicrobial activities, making them useful in treating a variety of ailments. These compounds inhibit key inflammatory pathways, thereby reducing inflammation and pain. The antimicrobial activity of its aliphatic compounds supports their use in treating infections caused by bacteria and fungi. The diverse pharmacological activities of *B. alba* underscore its potential as a novel therapeutic agent for treating inflammation, infections, and oxidative stress-related conditions (Norton 1984).

10.3.7 *Bidens odorata*

Bidens odorata, commonly known as the Spanish needle, contains aliphatic compounds with notable pharmacological activities. The composition of *Bidens odorata* Cav. was determined by GC–MS, and following 1–16 aliphatic compounds were found (Scogin and Zakar 1976). *B. odorata*'s antibacterial activity makes it a viable treatment option for a range of bacterial and fungal illnesses. Additionally, the antioxidant properties of its aliphatic compounds help in mitigating oxidative stress-related damage, which is implicated in various chronic diseases. The hepatoprotective activity of these compounds further contributes to liver health by protecting against toxins and oxidative damage. The multifaceted therapeutic applications of *B. odorata* highlight its potential as a valuable medicinal plant with diverse health benefits (Zavala et al. 2017).

10.3.8 *Bidens biternata*

Bidens biternata, known as the twin fruit beggarticks, is another species with significant medicinal potential. The alkaloids, tannins, terpenoids, reducing sugar, glycosides, flavonoids, anthraquinones, phlorotannins, iridoids, and steroids were among the phytochemicals that were derived. Estimates were also made for several nutritional parameters, including total proteins, total carbs, total reducing sugar, various amino acids, free fatty acids, crude fiber, lipids, total moisture content, vitamins, etc. (Zahara et al. 2015). The aliphatic compounds, i.e., tris-(2,4-di-tert-butylphenyl); 4-hydroxy-2,4′-dimethoxychalcone; and 2,4-di-tert-butylphenol are known for their biological activities. These compounds inhibit key inflammatory pathways, thereby reducing inflammation and pain. Additionally, *B. biternata* exhibits antimicrobial properties that help in treating infections caused by bacteria and fungi. The antioxidant activity of aliphatic components enables their application in defending cells against oxidative stress and lowering the danger of chronic illnesses.

10.3.9 *Bidens connata*

Bidens connata, commonly known as the connate beggarticks, contains aliphatic compounds with notable pharmacological activities. By preventing the generation of proinflammatory cytokines, these substances have shown important anti-inflammatory qualities. The chemical composition of the essential oil isolated from the aerial blooming parts of *Bidens cernua* was analyzed. The principal compound was identified as 1-phenylhepta-1,3,5-triyne, followed by sesquiterpenes caryophyllene oxide and humulene-1,2-epoxide (Chalchat et al. 2009). Additionally, the antioxidant properties of its aliphatic compounds help in mitigating oxidative stress-related damage, which is implicated in various chronic diseases. The hepatoprotective activity of these compounds further contributes to liver health by protecting against toxins and oxidative damage. The multifaceted therapeutic applications of *B. connata* highlight its potential as a valuable medicinal plant with diverse health benefits (Crowe and Parker 1981).

10.3.10 *Bidens procera*

Bidens procera, known as the nodding beggarticks, is another species with significant medicinal potential. The aliphatic compounds in *B. cerenua* may be helpful in treating a range of illnesses because of their encouraging antioxidant and anti-inflammatory properties. These substances lend support to their use in lowering the risk of chronic illnesses and shielding cells from oxidative stress. The promise of *B. cerenua* as a source of novel therapeutic compounds for the treatment of infections, oxidative stress-related diseases, and inflammation is highlighted by its various pharmacological activity (Tomczykowa et al. 2017). Two new polyacetylenes, tridecane-2E-monoene-4,6,8-triyntylen-1,13-diol-12-O-β-glucoside (1) and tetradecane-2E,8E-diene-4,6-diyne-1,14-diol-13-O-β-glucoside (2), were discovered through a phytochemical analysis of *Bidens procera.* By comparing the published data with NMR and MS spectra, their chemical structures were clarified (Cai et al. 2022).

The genus *Biden* is a rich source of aliphatic compounds with diverse pharmacological activities. The therapeutic potential of these compounds, demonstrated in various species, underscores their significance in traditional and modern medicine. The anti-inflammatory, antimicrobial, antioxidant, anticancer, antimalarial, and hepato-protective properties of these compounds make them valuable in developing new treatments for various diseases and health conditions. Continued research on the isolation, characterization, and pharmacological evaluation of aliphatic compounds from *Bidens* species will further enhance our understanding of their therapeutic potential and help to develop novel drugs.

10.4 Therapeutic Applications of Aliphatic Compounds from *Bidens*

The genus *Bidens*, belonging to the Asteraceae family, is renowned for its rich phytochemical diversity and pharmacological potential. Among the various classes of compounds found in *Bidens* species, aliphatic compounds stand out due to their significant therapeutic applications. Aliphatic compounds, characterized by their straight or branched carbon chains, exhibit a range of biological activities that are beneficial in treating various ailments. These compounds include essential fatty acids, polyacetylenes, and terpenoids, which have been extensively studied for their pharmacological properties. For instance, *Biden pilosa,* a widely distributed species, contains aliphatic compounds with notable anti-inflammatory, antimicrobial, and antioxidant activities. The pharmacological potential of these compounds is attributed to their ability to modulate key biochemical pathways and interact with specific cellular targets, making them promising candidates for drug development (Xuan and Khanh 2016).

Biden pilosa, commonly known as the Spanish needle, is one of the most extensively studied species in the genus for its therapeutic applications. The aliphatic compounds in *B. pilosa* have demonstrated significant anti-inflammatory effects, primarily by inhibiting proinflammatory cytokines and enzymes such as COX-2 and iNOS (Rodriguez-Otero et al. 2023). Additionally, *B. pilosa* exhibits antimicrobial properties against a broad spectrum of pathogens, including bacteria, fungi, and viruses, making it a potential source of novel antimicrobial agents. The antioxidant activity of aliphatic compounds from *B. pilosa* helps in mitigating oxidative stress-related damage, in various chronic diseases such as cancer, cardiovascular disorders, and neurodegenerative conditions. These therapeutic effects highlight the multifaceted pharmacological potential of aliphatic compounds in *B. pilosa,* which could be harnessed for developing new treatments for inflammation, infections, and oxidative stress-related diseases (Abate et al. 2021).

The pharmacological potential of *Bidens bipinnata*, or Spanish needles, is also noteworthy. Aliphatic compounds from *B. bipinnata* have been studied for their anticancer properties, particularly in inhibiting cancer cells growth and inducing apoptosis. The anticancer activity is attributed to the ability of these compounds to modulate cell signaling pathways involved in cell proliferation, survival, and apoptosis. Furthermore, *B. bipinnata* exhibits antimicrobial properties, which are effective against a range of pathogenic bacteria and fungi, supporting its use as a natural antimicrobial agent. The anti-inflammatory properties of aliphatic compounds in *B. bipinnata* help in reducing inflammation and pain, which are beneficial in managing inflammatory conditions. Additionally, the antioxidant activity of these compounds helps in protecting cells from oxidative stress, contributing to their overall health benefits. The diverse pharmacological activities of *B. bipinnata* highlight its potential as a valuable medicinal plant for developing new treatments for cancer, infections, inflammation, and oxidative stress-related diseases (Zhang et al. 2020).

Another notable species, *Biden tripartita*, commonly known as three-lobe beggarticks, has been traditionally used in folk medicine for its health benefits. The aliphatic compounds isolated from *B. tripartita* have shown promising results in managing conditions such as diabetes, hypertension, and liver disorders. Studies have demonstrated that these compounds possess hypoglycemic properties, which are beneficial in controlling blood sugar levels in diabetic patients. The antihypertensive effects of aliphatic compounds in *B. tripartita* are attributed to their ability to induce vasodilation and improve vascular function, thereby reducing blood pressure. Moreover, the hepatoprotective properties of these compounds help protect the liver from damage induced by toxins and oxidative stress, supporting their use in liver disease management. The traditional and modern therapeutic applications of *B. tripartita* underscore its potential as a valuable medicinal plant with diverse health benefits (Rodin et al. 2024).

Biden frondosa, known as devil's beggarticks, also contains aliphatic compounds with notable pharmacological activities. These compounds have been found to exhibit significant antimalarial properties, making *B. frondosa* a potential source of new antimalarial drugs. The antimalarial activity is primarily attributed to the inhibition of *Plasmodium falciparum*, the parasite responsible for malaria, by disrupting its life cycle and preventing its replication. In addition to its antimalarial effects, *B. frondosa* exhibits anti-inflammatory and analgesic properties, which are beneficial in managing pain and inflammation associated with various conditions. The aliphatic compounds in *B. frondosa* also demonstrate antioxidant activity; this lowers the risk of chronic illnesses and shields cells from oxidative damage. The multifaceted pharmacological activities of *B. frondosa* highlight its potential as a source of therapeutic agents for treating malaria, inflammation, pain, and oxidative stress-related disorders (Khatua et al. 2024).

Biden parviflora, commonly known as small flower beggarticks, is another species with significant therapeutic potential. The aliphatic compounds isolated from *B. parviflora* have shown potent antifungal activity, making them effective against various fungal infections. The antifungal properties are attributed to the ability of these compounds to disrupt the cell membrane integrity of fungi, leading to their death. In addition to its antifungal effects, *B. parviflora* exhibits anti-inflammatory properties, which are beneficial in reducing inflammation and pain. The antioxidant activity of aliphatic compounds in *B. parviflora* helps protect cells from oxidative damage, contributing to their overall health benefits. Moreover, the antimicrobial properties of these compounds make *B. parviflora* a potential source of new antimicrobial agents for treating bacterial and fungal infections. The therapeutic applications of *B. parviflora* highlight its potential as a valuable medicinal plant with diverse health benefits (Atchan et al. 2023).

The pharmacological potential of aliphatic compounds from *Bidens* species is vast and diverse, encompassing a wide range of therapeutic applications. The anti-inflammatory, antimicrobial, antioxidant, anticancer, antimalarial, and hepatoprotective properties of these compounds make them valuable in developing new treatments for various diseases and health conditions. The therapeutic potential of *Bidens* species such as *B. pilosa*, *B. tripartita*, *B. frondosa*, *B. bipinnata*, and *B. parviflora* underscores their significance as valuable sources of medicinal compounds. Future research focusing on the isolation, characterization, and pharmacological evaluation of aliphatic compounds from *Bidens* species will further enhance our understanding of their therapeutic potential and facilitate the development of novel drugs (Xuan and Khanh 2016)

10.5 Conclusion

This chapter sets the stage for future investigations into the multifaceted aspects of Genus *Bidens*, underscoring its importance in both scientific and practical contexts. The chemistry of *Bidens* encompasses a diverse array of aliphatic natural products with intriguing structural diversity and pharmacological capability. From fatty acids and esters to polyketides and polyacetylenes, these compounds offer a glimpse into the biochemical richness of *Bidens* plants and their potential applications in medicine, agriculture, and industry. By unraveling the biosynthetic pathways and biological activities of aliphatic compounds from *Bidens* species, researchers can unlock new opportunities for drug discovery, crop improvement, and environmental sustainability, paving the way for a deeper understanding of the natural world and its myriad wonders.

References

Abate SM, Mantefardo B, Nega S, Chekole YA, Basu B, Ali SA et al (2021) Global burden of acute myocardial injury associated with COVID-19: a systematic review, meta-analysis, and meta-regression. Ann Med Surg 68:102594

Ahn D, Kim DK (2016) Antioxidant components of the aerial parts of Bidens frondosa L. Korean J Pharmacogn 47(2):110–116

Alamgir A (2017) Therapeutic use of medicinal plants and their extracts: volume 1. Springer, Cham

Albertsson A-C, Varma IK (2002) Aliphatic polyesters: synthesis, properties and applications. In: Degradable aliphatic polyesters, Advances in polymer science, vol 157. Springer, Berlin, pp 1–40

Atchan APN, Monthe OC, Tchamgoue AD, Singh Y, Shivashankara ST, Selvi MK et al (2023) Anti-inflammatory, antioxidant activities, and phytochemical characterization of edible plants exerting synergistic effects in human gastric epithelial cells. Antioxidants 12(3):591

Bartolome AP, Villaseñor IM, Yang W-C (2013) Bidens pilosa L. (Asteraceae): botanical properties, traditional uses, phytochemistry, and pharmacology. Evid Based Complement Alternat Med 2013(1):340215

Bo Y, Yuan L-P, Zhang J-J, Meng D-D, Jing H, Dai H-J et al (2012) Total flavonoids of Bidens bipinnata L. a traditional Chinese medicine inhibits the production of inflammatory cytokines of vessel endothelial cells stimulated by sera from Henoch–Schönlein purpura patients. J Pharm Pharmacol 64(6):882–887

Cai F-J, Wang L, Zhao W, Tian J-L, Kong D-G, Liu Q et al (2022) Phytochemical and chemotaxonomic investigations on the whole herbs of Bidens procera LC Xu ex XW Zheng. Biochem Syst Ecol 101:104395

Chalchat J-C, Petrovic S, Maksimovic Z, Gorunovic M (2009) Composition of essential oil of Bidens cernua L. Asteraceae from Serbia. Journal of Essential Oil Research 21(1): 41–42

Chen Y, Luo J, Zhang Q, Kong L (2016) Identification of active substances for dually modulating the renin–angiotensin system in Bidens pilosa by liquid chromatography–mass spectrometry-based chemometrics. J Funct Foods 21:201–211

Chiang Y-M, Chuang D-Y, Wang S-Y, Kuo Y-H, Tsai P-W, Shyur L-F (2004) Metabolite profiling and chemopreventive bioactivity of plant extracts from Bidens pilosa. J Ethnopharmacol 95(2–3):409–419

Crowe DR, Parker WH (1981) Hybridization and agamospermy of Bidens in northwestern Ontario. J Int Assoc Plant Taxon 30(4):749–760

Deba F, Xuan TD, Yasuda M, Tawata S (2008) Chemical composition and antioxidant, antibacterial and antifungal activities of the essential oils from Bidens pilosa Linn. var. Radiata. Food Control 19(4):346–352

Dembitsky VM, Savidov N, Poroikov VV, Gloriozova TA, Imbs AB (2018) Naturally occurring aromatic steroids and their biological activities. Appl Microbiol Biotechnol 102:4663–4674

Gagare S, Patil P, Jain A (2024) Natural product-inspired strategies towards the discovery of novel bioactive molecules. Future J Pharm Sci 10(1):55

Grombone-Guaratini MT, Solferini VN, Semir J (2004) Reproductive biology in species of Bidens L. (Asteraceae). Sci Agric 61:185–189

Hu H-M, Bai S-M, Chen L-J, Hu W-Y, Chen G (2018) Chemical constituents from Bidens bipinnata Linn. Biochem Syst Ecol 79:44–49

Idris AH, Aramendi Ecenarro E, Leroux B, Jaureguibeitia X, Yang BY, Shaver S et al (2023) Bag-valve-mask ventilation and survival from out-of-hospital cardiac arrest: a multicenter study. Circulation 148(23):1847–1856

Khatua S, Bose S, Cordeiro LM, Acharya K (2024) The emergence of mushrooms as novel resources of potential prebiotics: an updated view. CRC Press, Boca Raton, pp 271–314

Kumada M, Tamao K, Aliphatic organopolysilanes. (1968) Advances in organometallic chemistry, vol 6. Elsevier, New York, pp 19–117

Lima Silva F, Fischer DCH, Fechine Tavares J, Sobral Silva M, Filgueiras de Athayde-Filho P, Barbosa-Filho JM (2011) Compilation of secondary metabolites from Bidens pilosa L. Molecules 16(2):1070–1102

Mutuma S, Mwatha R, Kamere I, Barasa S, Musau S, Otiato J, et al. Women's participation in management and leadership positions in the public service in Kenya: closing the gender gap. 2022

Norton WT (1984) Isolation and characterization of myelin. Springer, Myelin, pp 161–199

Ravichandran A, Subramanian P, Manoharan V, Muthu T, Periyannan R, Thangapandi M et al (2018) Phyto-mediated synthesis of silver nanoparticles using fucoidan isolated from Spatoglossum asperum and assessment of antibacterial activities. J Photochem Photobiol B 185:117–125

Rodin MN, Baeva VM, Bokov DO, Voronov VM, Samylina IAJF (2024) Comparative study of morphological and anatomical features of Bidens L. species. Farmaciya (Pharmacy) 73(2):25–33

Rodriguez-Otero P, Ailawadhi S, Arnulf B, Patel K, Cavo M, Nooka AK et al (2023) Ide-cel or standard regimens in relapsed and refractory multiple myeloma. N Engl J Med 388(11):1002–1014

Rosenthal GA, Berenbaum MR (2012) Herbivores: their interactions with secondary plant metabolites: ecological and evolutionary processes. Academic press, San Diego

Scogin R, Zakar K (1976) Anthochlor pigments and floral UV patterns in the genus Bidens. Biochem Syst Ecol 4(3):165–167

Singh YD, Jena B, Ningthoujam R, Panda S, Priyadarsini P, Pattanayak S et al (2020) Potential bioactive molecules from natural products to combat against coronavirus. Adv Tradit Med 22:1–12

Tomczykowa M, Leszczyńska K, Tomczyk M, Tryniszewska E, Kalemba D (2011) Composition of the essential oil of Bidens tripartita L. roots and its antibacterial and antifungal activities. J Med Food 14(4):428–433

Tomczykowa M, Tomczyk M, Leszczyńska K, Kalemba D (2017) Flavonoids and essential oil of Bidens cernua of Polish origin and in vitro antimicrobial activity of the oil. Rec Nat Prod 11(5):468–473

Wahjudi E, Armanu A, Hadiwidjojo D, Solimun S (2024) Lecturer performance in focus: an extensive systematic literature review and analysis. SA J Human Res Manag 22:2477

Xuan TD, Khanh TD (2016) Chemistry and pharmacology of Bidens pilosa: an overview. J Pharm Investig 46:91–132

Zahara K, Bibi Y, Tabassum S, Bashir T, Haider S, Araa A et al (2015) A review on pharmacological properties of Bidens biternata: a potential nutraceutical. Asian Pac J Trop Dis 5(8):595–599

Zavala JA, Gog L, Giacometti R (2017) Anthropogenic increase in carbon dioxide modifies plant–insect interactions. Ann Appl Biol 170(1):68–77

Zhang J-J, Dong X, Cao Y-Y, Yuan Y-D, Yang Y-B, Yan Y-Q et al (2020) Clinical characteristics of 140 patients infected with SARS-CoV-2 in Wuhan, China. Allergy 75(7):1730–1741

Zhu YJ, Li CH, Yu SH, Jiang HQ, Tian ZH, Wang WQ et al (2021) Two new polyacetylene glucosides and a new caffeoyl derivative with angiogenic activity from Bidens parviflora Wild. Phytochem Lett 42:82–86

Flavonoids Isolated from *Bidens* Species 11

Tayyiba Afzal

Abstract

Genus *Bidens* contains numerous compounds including phenylpropanoids, alkaloids, triterpenoids, flavonoids, and organic acids, with flavonoids being one of the main effective compounds. Significant flavonoid content has been reported in various *Bidens* species, such as 37.67 mgQE/g in the methanol fraction from *B. engleri*. HPLC analysis of *B. tripartita* revealed total flavonoid contents being 1.230% in herbs and 0.649% in flowers. *B. pilosa* L. has a complex flavonoid profile comprising aurones, chalcones, flavanones, flavones, and flavonols. It has been widely used in Chinese ethnopharmacology for treating malaria, hepatitis, diabetes, inflammation, digestive disorders, and hypertension. Various studies have identified numerous flavonoids in *B. pilosa*, including astragalin with anticandidal properties and a variety of kaempferol and quercetin derivatives. *B. cernua* L. has also been studied for its flavonoid content, revealing compounds such as tilitanin, cynaroside, and maritimetin. *B. pilosa* var. radiata Sch Bip. contains flavonoid glycosides with significant antioxidant and anticancer activities. *B. parviflora* Willd. has been analyzed for its flavonoid profile, yielding compounds like luteolin 7-O-β-D-neohesperidoside and quercetin 3-O-β-D-glucoside. *B. humilis* is reported to contain new natural flavonoids such as luteolin 7-O-β-D-glucopyranosyl-(1 → 6)-β-D-galactopyranoside. *B. biternata* (Lour.) Merr and Sherff and *B. bipinnata* L. have also been recognized for their significant flavonoid content and associated health benefits. Advances in pharmacological studies have revealed that these compounds in *Bidens* species exhibit antibacterial, antioxidant, anti-inflammatory, hypolipidemic, analgesic, hepatoprotective, antitumor, and antidiabetic activities.

T. Afzal (✉)
Department of Plant Biology, Institute of Environmental Biology, Faculty of Biology and Animal Science, Wrocław University of Environmental and Life Sciences, Wrocław, Poland
e-mail: tayyiba.afzal@upwr.edu.pl

Y. Bibi et al. (eds.), *The Genus Bidens*,
https://doi.org/10.1007/978-981-96-4257-1_11

Keywords

Bidens · Flavanones · Quercetin · Flavanomarein · Astragalin · Antioxidant · HPLC

11.1 Introduction

Plants are known to contain hundreds of bioactive components. They utilize these active compounds to support their growth and protection against pathogens (Ali et al. 2022). Throughout the kingdom of plants, flavonoids are a common class of low-molecular-weight phenolic substances. In most families of angiosperms, flavonoids are easily recognized as floral pigments, but their presence is not limited to flowers; every part of plants has them.

Flavonoids are a class of naturally occurring compounds with different phenolic structures. They can be located in fruits, vegetables, flowers, root, stem, bark, grains, and beverages such as tea and wine (Ngameni et al. 2013). The health benefits of these organic compounds have been widely reported and efforts are being made to extract the active components known as flavonoids (Wolniak et al. 2007).

Within the genus *Bidens*, flavonoids have been identified in different species. Some common flavonoids associated with *Bidens* are Quercetin, Kaempferol, Apigenin, Luteolin, and Rutin. Advances in pharmacological studies revealed these compounds in *Bidens* species have antibacterial, antioxidant, anti-inflammatory, hypolipidemic, analgesic, hepatoprotective, antitumor, and antidiabetic activities (Rice-Evans et al. 1996). Similarly significant amount of flavonoids that was 37.67 mgQE/g was reported in the methanol fraction from *Bidens engleri* (Compaore et al. 2018). Certain species have shown a range of individual flavonoids that played a significant role in adjustment and resistance against both biotic as well as abiotic stressors in plants (Bayala et al. 2019).

The flavonoids are common in genus *Bidens* due to the plant's ability to adapt to environmental challenges. Because of this, *B. pilosa* and other species in the genus have developed defense mechanisms that include the production of flavonoids. Numerous environmental stresses have an impact on the production of flavonoids in *Bidens* (Bartolome et al. 2013). These consist of UV radiation exposure, herbivore attacks, infections by pathogens, and variations in light or temperature. The plant activates particular metabolic pathways that result in the synthesis of flavonoids when it encounters these difficulties. By acting as protective agents, these substances buffer the plant from oxidative stress-related damage and increase its general resilience.

11.2 Flavanoids Identified from Key Species of Genus *Bidens*

11.2.1 Flavonoids from *B. tripartita*

HPLC analysis of *B. tripartite* revealed that 7-O-glucosides of luteolin and isookanin are major flavonoid components present in its flower extracts. Cynaroside is

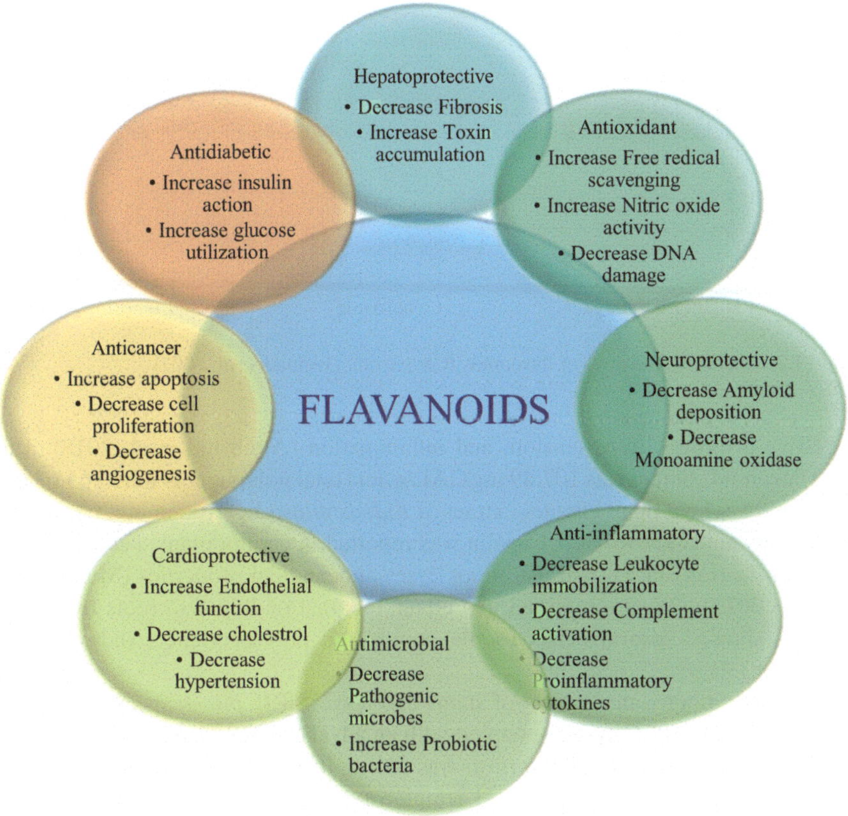

Fig. 11.1 Bioactivities of flavonoids

predominant in vegetative parts of plant, while flavanomarein is predominant in flowers. Flavanomarein is one of the rare flavanoids in the kingdom of Plantae (Fig. 11.1). Total flavonoid contents (in % of dry weight) were 1.230 and 0.649 (as quercetin equivalent) in the herbs and flowers, respectively (Wolniak et al. 2007) Methanol extract of *B. tripartita* revealed higher total phenolic (84.24 mgGAE/g extract) and flavonoid contents (49.00 mgRE/g extract) (Uysal et al. 2018).

11.2.2 Flavonoids from *B. pilosa* L.

With 58 distinct compounds identified, the complex flavonoid profile of *B. pilosa* comprises aurones, chalcones, flavanones, flavones, and flavonols with a wide range of O-methylation patterns and glycosylations (Lima Silva et al. 2011).

B. pilosa L. is an annual plant species from China, is a part of Chinese ethnopharmacology, as well as a famous constituent in herbal infusions. It has been extensively applied in the treatment of malaria, hepatitis, diabetes, inflammation,

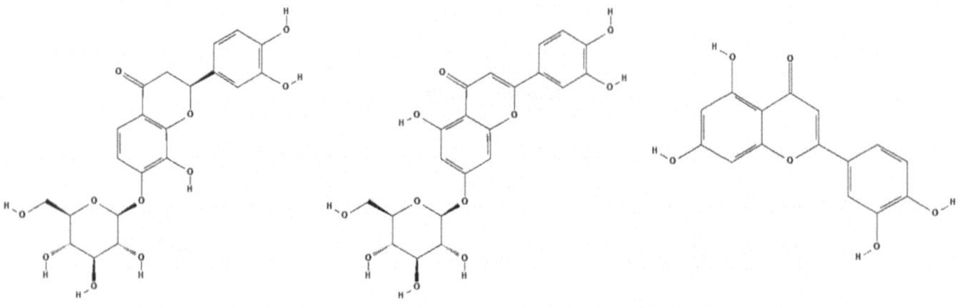

Flavanomarein Cynaroside Luteolin

Fig. 11.2 Chemical structures of flavonoids *B. tripartita*. (Wolniak et al. 2007)

digestive disorders, hypertension, and inflammation (Almosnid et al. 2018). The total flavonoid content was 107.49 mg GAE/g, and total polyphenols were 165.63 mg QE/g in the whole plant ethanol extract of *Bidens pilosa* L. (Kuo et al. 2021). One of the flavonoids known as astragalin was reported in certain plant species as well as *B. pilosa* (Fig. 11.2). Astragalin possesses anticandidal properties (Mtenga and Ripanda 2022). Oliveira reported that ethanol extract of *B. pilosa* roots by HPLC analysis showed several peaks for flavonoids and polyacetylene (Andrade-Neto et al. 2008).

Pressurized hot water extract of stems and leaves of *B. pilosa* species has been analyzed on a UHPLC system coupled with an advanced MS instrument. A total of 28 flavonoids in extracts of *B. pilosa* was found, listed as kaempferol-3-*O*-glycoside, kaempferol-3-*O*-glucuronide, kaempferol-3-*O*-glucoside isomers (1 and 2), kaempferol-3-acetyl-glycoside isomers (1–4), quercetin 3,7-diglucuronide, quercetin-3-rhamnosylhexoside isomers (1 and 2), quercetin-3-*O*-gluconoride isomer (1 and 2), quercetin-3-glycoside isomers (1 and 2), okanin triacetylglucoside (1–3), tetrahydroxyflavanone triacetylglucoside isomer (1–3), okanin diacetylglucoside isomer (1–5), and tetrahydroxyflavanone diacetylglucopyranoside isomer (1 and 2) (Gbashi et al. 2017). *B. pilosa* comprises flavonoids and their derivatives, including aglycones, aglycosides, aurones, and okanin glycosides, in most of its plant parts. It has yielded 20 flavonoid glycosides (Xuan and Khanh 2016).

11.2.3 Flavonoids from *Bidens cernua* L.

Tomczykowa conducted a study on the ethyl acetate, butanolic extracts, and diethyl ether extract of *B. cernua* L. Flavonoids were isolated by successive column chromatography on the polyamides with eluting solvent system as constant phase of high polarity. As an end result of series separation methods and purification on Sephadex (LH-20), flavonoids such as tilitanin, cynaroside, luteolin, cosmos in, maritimetin, and flavanomarein were identified (Tomczykowa et al. 2017).

11.2.4 Flavonoids from *Bidens pilosa* L. var. radiata. Sch Bip

B. pilosa is reported to contain six substances from entire plant ethanol, including flavonoid glycosides, along with 3,6-dimethoxyapigenin, quercetin, isoquercitrin, 5,6,7,4′-tetramethoxylflavone, and astragalin. The results of ABTS and DPPH-free radical scavenging pathway revealed that quercetin was most efficient with 10.7 and 17.5 μmol/L half-maximal inhibitory concentrations. 3,6-Dimethoxyapigenin (5,3′,4′-trihydroxy-3,7-dimethoxylflavone) and 5,6,7,4′-tetramethoxylflavone showed considerable growth preventive action on colon carcinoma cell line (RKO cells), and the IC50 value was found as 17.68 and 39.08 μmol/L, correspondingly (Yi et al. 2016).

11.2.5 Flavonoids from *Bidens parviflora* Willd

Fractionated whole plant extract of *B. parviflora* have been analyzed by HPLC, NMR, silica gel column chromatography, and ODS column chromatography to study complete chemical profile. Fifteen flavonoids were obtained for the first time from *B. parviflora*. Some of the compound were isolated as luteolin 7-*O*-*β*-D-neohesperidoside; quercetin 3-*O*-*β*-D-glucoside; acacetin 7-*O*-(*α*-D-apio-furanosyl) (1 → 6)-*β*-D-glucoside; 4-methoxyl-3,2′,3′-trihydroxy-chalcone: 4′-*O*-*β*-D-glucoside; quercetin 3-*O*-*β*-D-glucoside; quercetin; taxifolin; 3-*O*-*β*-D-rutinoside; acacetin 7-*O*-*β*-D-glucoside; luteolin; 2′-hydroxy-3,4,4′,6′-tetramethoxychalcone; kaempferol; 3,4,2′,4′-tetrahydroxychalcone and sulfuretin through spectroscopic techniques (^{1}H-NMR, ^{13}C-NMR, UV, IR, and MS) and similarities with earlier related studies (Li et al. 2008).

11.2.6 Flavonoids from *Bidens humilis*

B. humilis is reported to contain three new natural flavonoids: luteolin 7-*O*-*β*-D-glucopyranosyl-(1 → 6)-*β*-D-galactopyranoside, (2*S*)-isookanin 7-*O*-(2″-acetyl)-*α*-L-arabinopyranoside, and (2*S*)-isookanin 7-*O*-*α*-L-arabinopyranoside. These have been found in the *B. humilis* aerial parts through a bioassay-guided approach (Saltos et al. 2015).

11.2.7 Flavonoids from *Bidens biternata* (Lour.) Merr and Sherff

Bidens biternata (Lour.) Merr and Sherff is known to be a reliable source of important health-promoting entities (Zahara et al. 2015). It contains a significant amount of flavonoid (0.39 mg/g) that imparts antioxidant activity to plants (Zahara et al. 2019).

11.2.8 Flavonoid Composition of *Bidens bipinnata* L.

Shen et al. (2015) conducted the HPLC analysis of *B. bipinnata* L. It revealed the presence of catechin, hyperin, isoquercitrin, astragalin, and quercetin. Yang et al. (2021) reported 12 flavonoids from *B. bipinnata* L. such as rutin, hyperoside, isoquercitrin, quercitrin, and luteolin.

11.3 Conclusion or Summary

Genus *Bidens* contains various compounds such as phenylpropanoids, alkaloids, triterpenoids, flavonoids, and organic acids, with flavonoids being notably effective. For instance, *B. engleri* has a significant flavonoid content of 37.67 mgQE/g. *B. tripartita* features major flavonoids like luteolin 7-O-glucosides and isookanin, with flavanomarein in its flowers being particularly rare in the plant kingdom. *B. pilosa L.* has a complex flavonoid profile with 58 distinct compounds and is extensively used in traditional Chinese medicine for various treatments. Additionally, studies have identified numerous flavonoids in *B. cernua L., B. pilosa L. var. radiata, B. parviflora, B. humilis, B. biternata,* and *B. bipinnata L.*, highlighting their diverse and significant pharmacological potential (Fig. 11.3 and Tables 11.1 and 11.2).

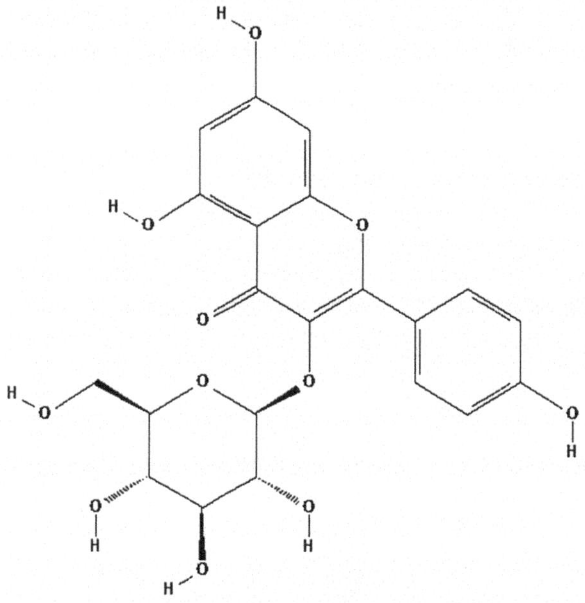

Fig. 11.3 Astragalin from *B. pilosa*. (Mtenga and Ripanda 2022)

Table 11.1 Flavonoid profile of *Bidens* species

Plant name	Plant part	Compound	References
Bidens tripartita L.	Flower	Flavanomarein, 7-O-glucosides of luteolin, 7-O-glucosides of isookanin	Wolniak et al. (2007)
	Leaves	Cynaroside	
Bidens pilosa L.	Whole plant	Astragalin	Mtenga and Ripanda (2022)
Bidens pilosa L.	Stem and leaves	Kaempferol-3-O-glycoside, kaempferol-3-O-glucuronide, kaempferol-3-O-glucoside isomers (1 and 2), kaempferol-3-acetyl-glycoside isomers (1–4), quercetin 3,7-diglucuronide, quercetin-3-rhamnosylhexoside isomers (1 and 2), quercetin-3-O-gluconoride isomer (1 and 2), quercetin-3-glycoside isomers (1 and 2), okanin triacetylglucoside (1–3), tetrahydroxyflavanone triacetylglucoside isomer (1–3), okanin diacetylglucoside isomer (1–5), tetrahydroxyflavanone diacetylglucopyranoside isomer (1 and 2)	Gbashi et al. (2016)
Bidens cernua L.	Whole plant	Tilitanin, cynaroside, luteolin, cosmosiin, maritimetin, flavanomarein	Tomczykowa et al. (2017)
Bidens pilosa L. var. radiata. Sch Bip. (BP)	Whole plant	3,6-Dimethoxyapigenin, quercetin, isoquercitrin, 5,6,7,4′-tetramethoxylflavone, astragalin	Yi et al. (2016)
Bidens parviflora Willd.	Whole plant	Luteolin 7-O-β-D-neohesperidoside, quercetin 3-O-β-D-glucoside, acacetin 7-O-(α-D-apio-furanosyl)(1 → 6)-β-D-glucoside, 4-methoxyl-3,2′,3′-trihydroxy-chalcone, 4′-O-β-D-glucoside, quercetin 3-O-β-D-glucoside, quercetin, taxifolin, 3-O-β-D-rutinoside, acacetin 7-O-β-D-glucoside, luteolin, (12) 2′-hydroxy-3,4,4′,6′-tetramethoxychalcone, kaempferol, 3,4,2′,4′-tetrahydroxychalcone, sulfuretin	Li et al. (2008)
Bidens humilis	Ariel parts	Luteolin 7-O-β-D-glucopyranosyl-(1 → 6)-β-D-galactopyranoside, (2S)-isookanin 7-O-(2″-acetyl)-α-L-arabinopyranoside, (2S)-isookanin 7-O-α-L-arabinopyranoside	Saltos et al. (2015)

Table 11.2 Flavonoid content in different *Bidens* species

Plant species	Flavonoid content	References
B. tripartita	49.00 mg RE/g	Uysal et al. (2018)
B. pilosa L.	107.49 mg GAE/g	Son et al. (2022)
B. tripartite ethyl acetate extract	297.62 ± 20.84 mg QE/g	Orhan et al. (2016)
B. pilosa leaves	4.95–5.91% dry weight (in terms of quercetin)	Bimenyindavyi et al. (2023)
Bidens cernua L.	1.62% hyperoside equivalent	Tomczykowa et al. (2017)
Bidens biternata	0.39 mg/g	Zahara et al. (2019)
B. pilosa L. leaves	250.6 mg/gm rutin	Abdulkader et al. (2022)
B. pilosa L. stem	294 mg.gm rutin	

References

Ali S, Alam M, Khatoon F, Fatima U, Elasbali AM, Adnan M et al (2022) Natural products can be used in therapeutic management of COVID-19: probable mechanistic insights. Biomed Pharmacother 147:112658

Almosnid NM, Zhou X, Jiang L, Ridings A, Knott D, Wang S et al (2018) Evaluation of extracts prepared from 16 plants used in Yao ethnomedicine as potential anticancer agents. J Ethnopharmacol 211:224–234

Andrade-Neto VF, Brandão MGL, Nogueira F, Rosário VE, Krettli AU (2008) Ampelozyziphus amazonicus Ducke (Rhamnaceae), a medicinal plant used to prevent malaria in the Amazon Region, hampers the development of Plasmodium berghei sporozoites. Int J Parasitol 38(13):1505–1511

Bartolome AP, Villaseñor IM, Yang W-C (2013) Bidens pilosa L. (Asteraceae): botanical properties, traditional uses, phytochemistry, and pharmacology. Evid Based Complement Alternat Med 2013(1):340215

Bayala B, Hilou A, Sombié PAD, Coulibaly AY, Nacoulma O, Simpore J (2019) Ethnomedicinal and pharmacognostical investigation on three medicinal asteraceae species from burkina faso: bidens engleri oe schulz, Acanthospermum hispidum DC. and Ageratum conyzoides L. Int J Pharm Biol Sci 9(4):329–342

Compaore M, Bakasso S, Meda RNT, Nacoulma OG (2018) Antioxidant and anti-inflammatory activities of fractions from Bidens engleri OE Schulz (Asteraceae) and Boerhavia erecta L. (Nyctaginaceae). Medicines (Basel) 5(2):53

Gbashi S, Njobeh P, Steenkamp P, Madala N (2017) Pressurized hot water extraction and chemometric fingerprinting of flavonoids from Bidens pilosa by UPLC-tandem mass spectrometry. CyTA J Food 15(2):171–180

Kuo TF, Yang G, Chen TY, Wu YC, Tran Nguyen Minh H, Chen LS et al (2021) Bidens pilosa: nutritional value and benefits for metabolic syndrome. Food Front 2(1):32–45

Li Y-L, Li J, Wang N-L, Yao X-S (2008) Flavonoids and a new polyacetylene from Bidens parviflora Willd. Molecules 13(8):1931–1941

Lima Silva F, Fischer DCH, Fechine Tavares J, Sobral Silva M, Filgueiras de Athayde-Filho P, Barbosa-Filho JM (2011) Compilation of secondary metabolites from Bidens pilosa L. Molecules 16(2):1070–1102

Mtenga DV, Ripanda AS (2022) A review on the potential of underutilized Blackjack (Biden Pilosa) naturally occurring in sub-Saharan Africa. Heliyon 8(6):e09586

Ngameni B, Fotso GW, Kamga J, Ambassa P, Abdou T, Fankam AG et al (2013) Flavonoids and related compounds from the medicinal plants of Africa. In: Medicinal plant research in Africa. Elsevier, London, pp 301–350

Rice-Evans CA, Miller NJ, Paganga G (1996) Structure-antioxidant activity relationships of flavonoids and phenolic acids. Free Radic Biol Med 20(7):933–956

Saltos MBV, Puente BFN, Milella L, De Tommasi N, Dal Piaz F, Braca A (2015) Antioxidant and free radical scavenging activity of phenolics from bidens humilis. Planta Med 81(12/13):1056–1064

Shen A-Z, Li X, Hu W, Chen F-H (2015) Total flavonoids of Bidens bipinnata L ameliorate experimental adjuvant-induced arthritis through induction of synovial apoptosis. BMC Complement Altern Med 15:1–9

Tomczykowa M, Tomczyk M, Leszczyńska K, Kalemba D (2017) Flavonoids and essential oil of Bidens cernua of Polish origin and in vitro antimicrobial activity of the oil. Rec Nat Prod 11(5):468–473

Uysal S, Ugurlu A, Zengin G, Baloglu MC, Altunoglu YC, Mollica A et al (2018) Novel in vitro and in silico insights of the multi-biological activities and chemical composition of Bidens tripartita L. Food Chem Toxicol 111:525–536

Wolniak M, Tomczykowa M, Tomczyk M, Gudej J, Wawer I (2007) Antioxidant activity of extracts and flavonoids from Bidens tripartita. Acta Pol Pharm 64(5):441–447

Xuan TD, Khanh TD (2016) Chemistry and pharmacology of Bidens pilosa: an overview. J Pharm Investig 46:91–132

Yang X, Bai Z-F, Zhang Y, Cui H, Zhou H-L (2021) Flavonoids-rich extract from Bidens bipinnata L protects pancreatic β-cells against oxidative stress-induced apoptosis through intrinsic and extrinsic pathways. J Ethnopharmacol 275:114097

Yi J, Wu J-G, Wu Y-B, Peng W (2016) Antioxidant and anti-proliferative activities of Flavonoids from Bidens pilosa L var radiata Sch Bip. Trop J Pharm Res 15(2):341–348

Zahara K, Bibi Y, Tabassum S, Bashir T, Haider S, Araa A et al (2015) A review on pharmacological properties of Bidens biternata: a potential nutraceutical. Asian Pac J Trop Dis 5(8):595–599

Zahara K, Bibi Y, Qayyum A, Nisa S (2019) Investigation of antimicrobial and antioxidant properties of Bidens Biternata. Iran J Sci Technol Trans A Sci 43:725–734

Son N H, Tuan N T, Tran T M (2022). Investigation of chemical composition and evaluation of antioxidant, antibacterial and antifungal activities of ethanol extract from Bidens pilosa L. Food Sci Technol 42, e22722.

Orhan N, İçöz Ü G, Altun L, Aslan M (2016). Anti-hyperglycaemic and antioxidant effects of Bidens tripartita and quantitative analysis on its active principles. Iran J basic med sci 19(10): 1114.

Abdulkader O M, Sharaf A E A, Fouda H M, Elhaw M H (2022). Phyto- Biochemical Constituents and GC-MS Analysis of Bidens pilosa Leaves and Stems. Egypt Acad J Biol Sci H Botany, 13(2): 1-9.

Bimenyindavyi E, Khusnetdinova L Z, Timofeeva O A (2023) Phenolic Compounds of Plants Bidens tripartita (L.) and Bidens pilosa (L.) from Different Locations. Russ J Plant Physiol, 70 (7): 160

Terpenoids Isolated from Genus *Bidens* 12

Sehrish Imran, Muhammad Hasnain, Abdul Manan Yousaf, Yamin Bibi, Abdul Qayyum, and Farooq Ahmad

Abstract

The genus *Bidens* belongs to the family Asteraceae, comprising numerous species distributed globally. It is recognized for its rich diversity of secondary metabolites, with terpenoids being particularly noteworthy, including monoterpenes, sesquiterpenes, and diterpenes, each contributing distinct aromatic and biological properties. Terpenoids regulate essential processes in plant development and growth, acting as signaling molecules and modulating responses to abiotic and biotic stresses. This chapter explores the diverse roles and biosynthesis of terpenoids in the genus *Bidens*. Terpenoids in *Bidens* species, including *Bidens pilosa, Biden tripartita,* and *Bidens bipinnata*, play crucial roles in plant biology and human health. The chapter covers their classification, biosynthesis pathways (MVA and MEP), evolutionary insights into terpene synthase genes, and functional roles in plant development and stress responses. Applications in medicine, industry (flavors, fragrances), and future perspectives in biotechnology are also discussed.

S. Imran (✉) · M. Hasnain · A. M. Yousaf
Department of Botany, PMAS Arid Agriculture University, Rawalpindi, Pakistan
e-mail: sehrishimran07@gmail.com; hasnainscholar99@gmail.com; manankashmiry21@gmail.com

Y. Bibi
Department of Botany, Rawalpindi Women University, Rawalpindi, Punjab, Pakistan
e-mail: yamin.bibi@f.rwu.edu.pk

A. Qayyum
Department of Agronomy, The University of Haripur, Haripur, Pakistan
e-mail: aqayyum@uoh.edu.pk

F. Ahmad
Sustainable Development Study Centre, Government College University, Lahore, Pakistan
e-mail: fagondal82@gmail.com

Y. Bibi et al. (eds.), *The Genus Bidens*,
https://doi.org/10.1007/978-981-96-4257-1_12

Keywords

Bidens · Asteraceae · Terpenoids · Diterpenes · *Pilosa*

12.1 Introduction

Terpenoids, also referred to as isoprenoids, are a vast group of naturally occurring organic compounds that originate from five-carbon isoprene units. and are known for their aromatic qualities and biological activities (Mabou and Yossa 2021). Terpenoids, which represent the most structurally diverse and abundant class of floral volatiles, encompass over 40,000 individual compounds (Muhlemann et al. 2014). These compounds are found in the majority of plant species and play a crucial role in plant development and growth. They are integral to membrane structures (sterols C30), serve as photosynthetic pigments (carotenoids C40), and include essential phytohormones such as abscisic acid (C15) and gibberellins (C20). Additionally, ubiquinones are involved in mitochondrial electron transport. Many terpenoids, including mono-, sesqui-, and diterpenes, act as secondary metabolites, playing crucial roles in plant-environment and plant-plant interactions (Dudareva et al. 2013). Among all terpenoids, monoterpenes and sesquiterpenes are the most commonly studied classes. This is due to their widespread distribution in the plant kingdom and their crucial roles in both plant functions and human society (Muhlemann et al. 2014). Terpenoids also provide numerous benefits to humans. For instance, limonene is commonly utilized as a fragrance ingredient in cosmetic products (Ravichandran et al. 2018). β-Selinene is a compound found in various species of *Bidens* and has potential medicinal properties (Ogunbinu et al. 2009).

Therefore, this chapter will focus more on these classes of terpenes within the plant genus *Bidens*. Ongoing research explores these species' ecological roles, medicinal benefits, and adaptability, underscoring the genus's diverse and significant impact. While previous studies have touched on terpenoid diversity and biosynthesis in various plant species, this chapter focuses exclusively on *Bidens*, highlighting its unique terpenoid profiles and their evolutionary underpinnings. By integrating recent genomic insights, the chapter unveils novel aspects of terpenoid biosynthesis pathways (MVA and MEP) and their functional roles in plant biology, stress responses, and ecological interactions. Furthermore, the discussion extends to the medicinal and industrial applications of *Bidens* terpenoids, offering insights into their potential therapeutic benefits and biotechnological applications. This approach not only enhances our understanding of *Bidens* as a terpenoid-rich genus but also underscores its relevance in both scientific research and practical applications in medicine and industry.

12.2 Diversity of Terpenoids in Genus *Bidens*

Terpenoids constitute a diverse class of organic compounds synthesized by a wide range of plants, including *Bidens* species such as *Bidens pilosa* and *Bidens bipinnata*. These compounds are derived from isoprene units and are known for their structural diversity and biological activities (Adefegha et al. 2024; Lorigooini et al. 2020). The number of isoprene units in terpenoids determines their classification, also their structural complexity, and properties. These are hemiterpenes (C5), monoterpenes (C10), sesquiterpenes (C15), diterpenes (C20), sesterterpenes (C25), triterpenes (C30), tetraterpenes (C40), and polyterpenes (C > 40) (Ashour et al. 2010). Four species of the genus *Bidens* have been chemically investigated, including *Bidens pilosa, Bidens tripartita, Bidens biternata,* and *Biden alba.*

12.2.1 Monoterpenoids

Overall, there are 28 monoterpenes that have been isolated from *Bidens pilosa*.. Among the monoterpenes identified, there are eight specific compounds. These include acyclic monoterpenes such as camphene, m-cymol, 3-carene, and cis-verbenol. Additionally, monocyclic monoterpenes such as limonene, terpinolene, and α-phellandrene have been identified. Furthermore, bicyclic compounds such as myrcene, β-myrcene, p-cymen-8-ol, and α-terpineol have also been noted (Xuan and Khanh 2016). Limonene and α-pinene are the most common. The essential oils extracted from the roots of *Bidens tripartita* have yielded 106 compounds. Among these, notable terpenoids include α-pinene, p-cymene, ocimene, linalool, and p-cymene-9-ol. These terpenoids are distinguished for their diverse aromatic properties and potential therapeutic benefits (Tomczykowa et al. 2011; Pozharitskaya et al. 2010). Another study reported that α-pinene is the main ingredient extracted from the essential oil of the aerial part of *Bidens alba* (Priestap et al. 2008).

12.2.2 Sesquiterpenoids

The majority of sesquiterpenes are either monocyclic, bicyclic, or tricyclic. However, there are exceptions such as caryophyllene oxide and elixene, which are linear sesquiterpenes isolated from *B. pilosa* (Priestap et al. 2008; Deba et al. 2008). There are 58 sesquiterpenes (C15) that have been isolated from *Bidens Pilosa* (Xuan and Khanh 2016). The majority of components, such as monoterpenes and sesquiterpenes, are present in the essential oils extracted from different sections of *Bidens pilosa* (Table 12.1). β-Selinene is found in various species of *Bidens* and has potential medicinal properties. Indeed, β-bisabolene, β-elemene, silphiperfol-6-ene, and α-phellandrene are cosmene, all sesquiterpenoids that have been isolated from the essential oil of the roots of *Bidens tripartita* (Tomczykowa et al. 2011; Pozharitskaya et al. 2010). β-Caryophyllene, germacrene D, and β-copaene have been isolated from the essential oils of *Bidens alba* and *Bidens pilosa*. These sesquiterpenoids contribute to the aromatic profile of these plants and may also exhibit a range of biological properties such as antioxidant, anti-inflammatory, and antibacterial (Priestap et al. 2008).

12.2.3 Diterpenoids

The entire *Bidens pilosa* plant yielded the acyclic diterpenes phytol, pimaradiene, and spathulenol. Furthermore, it has been determined that the essential oil extracted from the shoots contains two tricyclic diterpenes: phytyl heptanoate and precocene I (Priestap et al. 2008; Deba et al. 2008; Amvam Zollo et al. 1995).

12.2.4 Tri- and Tetraterpenes

Pentacyclic polyterpenes (triterpenes) compounds, sandaracopimara-8(14),15-diene, phytanic acid, 1-eicosene, lupeol, the one acyclic polyterpene compound, and friedelan-3β-ol have also been identified in entire plants (Geissberger and Séquin 1991; Chang et al. 2000). However, the only compound found in the leaf was tetraterpenes, lupeol acetate.

12.3 Biosynthesis of Terpenoids

Terpenoids, the biggest class of floral volatiles produced by plants, include 556 fragrance components. Terpenoids are made from two interconvertible C5 units: isopentenyl diphosphate (IPP) and its allelic isomer dimethylallyl diphosphate (DMAPP). These five-carbon units serve as precursors for the synthesis of

Table 12.1 Terpenoid composition of the plant of genus *Bidens*

Compound name	Plant parts	Plant spp.	Plant origin	References
Monoterpenes				
(e)-β-Ocimene	LF (EO)	*B. tripartita*	Cameroon	Amvam Zollo et al. (1995) and Pozharitskaya et al. (2010)
m-Cymol	LF, FL (EO)		Japan	Deba et al. (2008), Xuan and Khanh (2016), and Tomczykowa et al. (2011)
β-Ocimene	LF, FL (EO)		Japan	Deba et al. (2008) and Pozharitskaya et al. (2010)
p-Cymen-8-ol	LF, FL(EO)		Japan	Deba et al. (2008) and Tomczykowa et al. (2011)

(continued)

Table 12.1 (continued)

Compound name	Plant parts	Plant spp.	Plant origin	References
Limonene	Roots, R (EO)	*B. pilosa*	Argentina	Priestap et al. (2008)
Perillene	FL, S (EO)		Argentina	Priestap et al. (2008)
Sabinene	LF, S, R		Cameroon	Priestap et al. (2008)
Trans-pinocarveol	S (EO)		Nigeria	Ogunbinu et al. (2009)
Terpinolene	LF (EO)		Cameroon	Amvam Zollo et al. (1995)
(Z)-β-Ocimene	LF, FL (EO)		Cameroon, Argentina	Priestap et al. (2008) and Amvam Zollo et al. (1995)
γ-Terpinene	LF, FL (EO)		Cameroon, Japan	Deba et al. (2008) and Amvam Zollo et al. (1995)
3-Carene	FL (EO)		Japan	Deba et al. (2008)
(4Z,6e)-2,6-Dimethyl-2,4,6-octatriene	FL, LF (EO)		Argentina, Japan	Priestap et al. (2008) and Amvam Zollo et al. (1995)
Borneol	R (EO)		Argentina	Priestap et al. (2008)
Cis-Verbenol	LF, FL (EO)		Japan	Deba et al. (2008)
β-Pinene	LF, FL, S (EO)		Cameroon, Argentina	Priestap et al. (2008) and Amvam Zollo et al. (1995)
Camphene	LF, FL (EO)		Cameroon, Japan	Deba et al. (2008) and Amvam Zollo et al. (1995)
Myrcene, β-myrcene	LF		Cameroon	Xuan and Khanh (2016), Priestap et al. (2008), and Amvam Zollo et al. (1995)
Terpinen-4-ol	LF (EO)		Japan	Amvam Zollo et al. (1995)
α-Terpineol	LF, S, R (EO)		Cameroon, Argentina	Priestap et al. (2008)
1,8-Cineole	LF (EO)		Cameroon	Amvam Zollo et al. (1995)
4-Terpineol	LF, FL (EO)		Japan	Deba et al. (2008)
α-Phellandrene	LF, FL (EO)		Cameroon, Japan	Deba et al. (2008) and Amvam Zollo et al. (1995)
β-Phellandrene	LF, FL, S, R (EO)		Argentina	Priestap et al. (2008)
α-Pinene	LF, FL, S, R (EO)	*Bidens tripartite, B. pilosa, B. alba*	Argentina, Japan	Priestap et al. (2008), Amvam Zollo et al. (1995), and Tomczykowa et al. (2011)
Linalool, β-linalool	LF, S, FL (EO)	*B. tripartita* *B. pilosa*	Cameroon, Argentina	Deba et al. (2008), Amvam Zollo et al. (1995), and Priestap et al. (2008)

(continued)

Table 12.1 (continued)

Compound name	Plant parts	Plant spp.	Plant origin	References
Sesquiterpenes				
Acorenone B	LF (EO)	*B. pilosa*	Nigeria	Ogunbinu et al. (2009)
Allo-Aromadendrene	S (EO)		Nigeria	Ogunbinu et al. (2009)
Bicyclogermacrene	LF, S		Brazil	Ogunbinu et al. (2009)
Cedr-8(15)-en-9α-ol	LF (EO)		Nigeria	Ogunbinu et al. (2009)
Cis-Calamenen-10-ol	S (EO)		Nigeria	
Cyclosativene	LF, S (EO)		Nigeria	
Daucene	LF (EO)		Nigeria	
Epi-10-γ-Eudesmol	S (EO)		Nigeria	
Epi-Longipinanol	L (EO)		Nigeria	
Epoxy alloaromadendrene	S (EO)		Nigeria	
Elixene	LF, FL (EO)		Japan	Deba et al. (2008)
Farnesene, (E)-β-Farnesene	LF, FL (EO)		Cameroon, Japan	Deba et al. (2008) and Amvam Zollo et al. (1995)
Humulene oxide II	LF, S (EO)		Nigeria	Ogunbinu et al. (2009)
Intermedeol	S (EO)		Nigeria	
Isoledene	LF, FL (EO)		Japan	Deba et al. (2008)
Selina-3,11-dien-6α-ol	LF (EO)		Nigeria	Ogunbinu et al. (2009)
Selina-3,7(11)-diene	LF		Brazil	
Trans-Calamenen-10-ol	LF, S (EO)		Nigeria	
Trans-α-Bergamotene	LF (EO)		Nigeria	
Valencene	S (EO)		Nigeria	
Z-γ-Bisabolene	LF		Brazil	Grombone-Guaratini et al. (2004)
β-Cedrene	S (EO)		Nigeria	Ogunbinu et al. (2009)
β-Selinene	LF (EO)		Nigeria	
α-Cadinol	LF, FL, S (EO)		Cameroon, Argentina	Priestap et al. (2008)
α-Calacorene	LF (EO)		Nigeria	Ogunbinu et al. (2009)
α-Bergamotene	LF, FL (EO)		Japan	Deba et al. (2008)
α-Cubebene	LF, FL, R (EO)		Cameroon, Japan	Deba et al. (2008) and Priestap et al. (2008)
α-Gurjunene	LF (EO)		Nigeria	Ogunbinu et al. (2009)
α-Humulene	LF, FL, R (EO)		Cameroon, Argentina	Priestap et al. (2008) and Amvam Zollo et al. (1995)
α-Ylangene	LF, FL (EO)		Cameroon, Japan	Deba et al. (2008) and Amvam Zollo et al. (1995)
β-Bourbonene	LF, FL (EO)		Japan	Deba et al. (2008)
β-Bisabolene	LF, FL (EO)		Japan	Deba et al. (2008)

(continued)

Table 12.1 (continued)

Compound name	Plant parts	Plant spp.	Plant origin	References
β-Cubebene	LF, FL (EO)		Japan	Deba et al. (2008)
(−)-β-Cadiene	LF, FL (EO)		Japan	Deba et al. (2008)
β-Elemene	LF, S, R (EO)		Cameroon, Argentina	Amvam Zollo et al. (1995) and Priestap et al. (2008)
β-Gurjunene	LF, FL		Brazil	Grombone-Guaratini et al. (2004)
γ-Cadinene	LF (EO)		Cameroon	Amvam Zollo et al. (1995)
γ-Muurolene	S (EO)		Nigeria	Ogunbinu et al. (2009)
τ-Muurolene	LF, FL (EO)		Japan	Deba et al. (2008)
τ-Cadinol	LF (EO)		Nigeria	Ogunbinu et al. (2009)
τ-Cadinene	LF, FL (EO)		Japan	Deba et al. (2008)
δ-Elemene	WP (EO)		Argentina	Priestap et al. (2008)
δ-Cadinene	LF, FL, S (EO)		Cameroon, Argentina	Priestap et al. (2008) and Amvam Zollo et al. (1995)
1-epi-Cubenol	LF (EO)		Nigeria	Ogunbinu et al. (2009)
14-oxy-α-muurolene	LF (EO)		Nigeria	
14-Hydroxy-δ-cadinene	LF (EO)		Nigeria	
Caryophyllene oxide	LF, FL, S (EO)		Japan	Deba et al. (2008) and Priestap et al. (2008)
Epi-cedrol	R (EO)		Argentina	Priestap et al. (2008)
(E)-nerolidol; trans-Nerolidol	LF, FL (EO)		Argentina	Priestap et al. (2008)
Precocene 1	LF (EO)		Cameroon	Amvam Zollo et al. (1995)
Spathulenol	LF, S, R (EO)		Cameroon, Argentina	Priestap et al. (2008) and Amvam Zollo et al. (1995)
T-Muurolol	LF (EO)		Cameroon	Amvam Zollo et al. (1995)
α-Caryophyllene	LF, FL (EO)		Japan	Deba et al. (2008)
α-Muurolene	LF		Brazil	Grombone-Guaratini et al. (2004)
Cosmene	R(EO)	*B. tripartita*	Hawaii	Pozharitskaya et al. (2010)
β-Bisabolene,	R			Tomczykowa et al. (2011)
β-Elemene,				
Silphiperfol-6-ene				
α-Fellandren				
Germacrene A	LF, S (EO)	*B. alba*	Nigeria	Ogunbinu et al. (2009)
Germacrene-D	LF, FL, S (EO)		Cameroon	Amvam Zollo et al. (1995)
α-Copaene	LF, FL, S, R (EO)		Cameroon, Argentina	Priestap et al. (2008) and Amvam Zollo et al. (1995)
β-Caryophyllene	LF, FL, S, R (EO)			

(continued)

Table 12.1 (continued)

Compound name	Plant parts	Plant spp.	Plant origin	References
Diterpenes				
Pimaradiene	S (EO)	*B. pilosa*	Argentina	Priestap et al. (2008)
Phytyl heptanoate	LF		Not stated	Iqbal and Zuleika (1995)
Phytol	WP		China	Chang et al. (2000)
Sandaracopimara-8(14),15-diene	S (EO)		Nigeria	Ogunbinu et al. (2009)
1-Eicosene	LF, S (EO)		Nigeria	
Triterpenes				
Friedelin	LF, S, FL, SD	*B. pilosa*	China, Tanzania	Geissberger and Séquin (1991) and Chang et al. (2000)
Friedelan-3 β-ol	LF, S, FL, SD		China, Tanzania	Chang et al. (2000) and Sarg et al. (1991)
Lupeol	PP		Egypt	Sarg et al. (1991)
Lupeol acetate	PP		Egypt	
Squalene	WP		China	Chang et al. (2000)
β-Amyrin	PP		Egypt	Sarg et al. (1991)
Tetraterpene				
B-carotene	Leaf	*B. alba*	Zimbabwe	Benhura and Chitsiku (1997)

FL flower, *LF* leaf, *S* shoot, *R* root, *EO* essential oil, (–) not detected

terpenoids (Muhlemann et al. 2014). Terpenes are structurally formed of single or multiple five-carbon units, which can be classified as sesquiterpenes, monoterpenes, sesquiterpenes, diterpenes, triterpenes, tetraterpenes, or high polyterpenes based on the number of five-carbon units. Five carbon isoprene building blocks are formed by two distinct pathways: cytosolic MVA and the plastidial MEP pathways (Chen et al. 2016). The classical independent MVA route is distributed in the cytosol, peroxisomes, and endoplasmic reticulum (Vranová et al. 2013). The MVA pathway consists of six enzymatic steps that are initiated by sequentially condensation of three acetyl CoA molecules. These molecules are then reduced to the MVA pathway, which is then followed by two consecutive stages of phosphorylation and decarboxylation, which yields the final product, IPP (Tholl 2015).

The methylerythritol phosphate (MEP) pathway occurs in the plastids, including some dissemination in the endoplasmic reticulum (ER) and peroxisomes. The condensation of pyruvate and glyceraldehyde-3-phosphate is the first of seven enzymatic stages in the MEP pathway (Muhlemann et al. 2014). DMAPP, which is produced when IPP isomerizes, can serve as a substrate for the biosynthesis of hemiterpenes. Alternatively, geranyl diphosphate (GPP) is produced by the combination of DMAPP with one IPP unit, which is mediated by GPP synthases (GPS). The combination of one IPP unit with one GPP molecule produces farnesyl pyrophosphate (FPP), with FPP synthases (FPS) catalyzing this process. GPP and FPP serve as precursors for the biosynthesis of monoterpenes and sesquiterpenes,

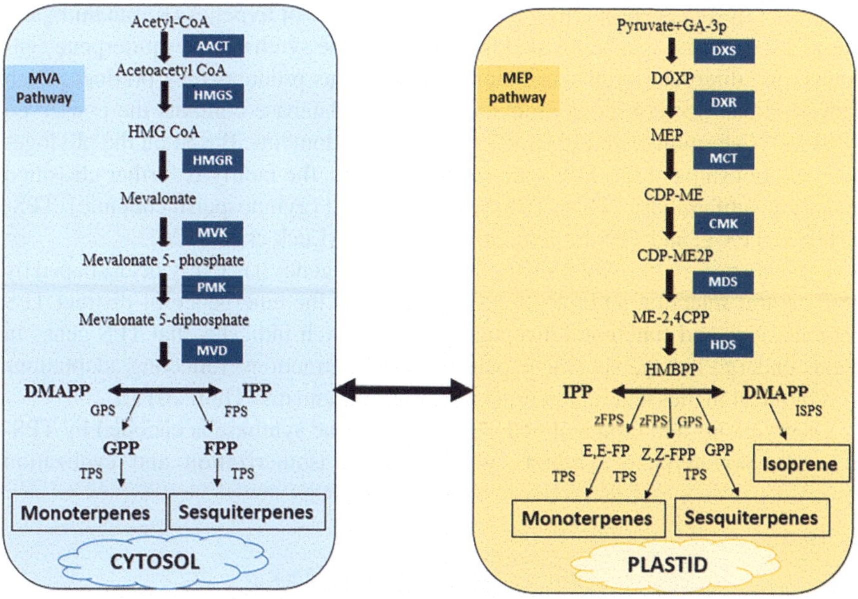

Fig. 12.1 Pathways and enzymatic reactions involved in terpenoids synthesis

respectively (Vranová et al. 2013). Moreover, both pathways cooperate with one another for a period of time and supply the IPP precursor needed for the production of terpenoids (Fig. 12.1).

In plants, sesquiterpenes biosynthesis varies between different parts of the plant, with roots predominantly involving the mevalonate (MVA) pathway and leaves primarily utilizing the methylerythritol phosphate (MEP) pathway. Recent studies indicate that the MEP pathway extends beyond the plastids, being present in the endoplasmic reticulum (ER) and peroxisomes as well. This broader localization suggests a more intricate regulatory network for isoprenoid biosynthesis than previously understood, offering greater flexibility and control over isoprenoid production. Such a distribution is crucial for the synthesis of specific secondary metabolites in response to environmental and developmental signals, highlighting the complexity and adaptability of plant metabolic processes (Vranová et al. 2013; Kuzuyama and Seto 2012).

12.4 Functional Evolution of Terpene Synthesis

The functional evolution of terpene synthases (TPSs) in the plant refers to how these enzymes have diversified and adapted to produce a variety of terpenoids with specific functions. Terpene synthases are crucial enzymes that catalyze the biosynthesis of terpenoids, which play essential roles in plant defense, communication, and stress adaptation. TPS is the main enzyme that produces terpene molecules, and because of its diverse properties, it increases the range of terpenes (Karunanithi and Zerbe 2019). These can be divided into monoterpene synthases, sesquiterpene synthases, and diterpene synthases based on the various products they produce. Each TPS gene has two conserved domains; the Pfam database contains the conserved PF03936 (C-terminus) and PF01397 (N-terminus) domains. Based on the phylogenetic relationships of the TPS gene family in plants, the family is further classified into eight subfamilies: TPS-a, TPS-b, TPS-c, TPS-d (gymnosperm endemic), TPS-e, TPS-f, TPS-g, and TPS-h (*Selaginella* endemic) (Luck et al. 2020).

In plant species, the evolutionary history of TPS genes has likely been shaped by both natural selection and genetic drift, leading to the emergence of distinct TPS gene families and functional diversification. Research indicates that TPS genes in plants undergo lineage-specific expansions and contractions, reflecting adaptations to ecological niches and interactions with other organisms (Tholl 2015).

Dicotyledons and monocotyledons: Sesquiterpene synthase is encoded by TPS-a, while monoterpene synthase, which aids in isomerization and cyclization

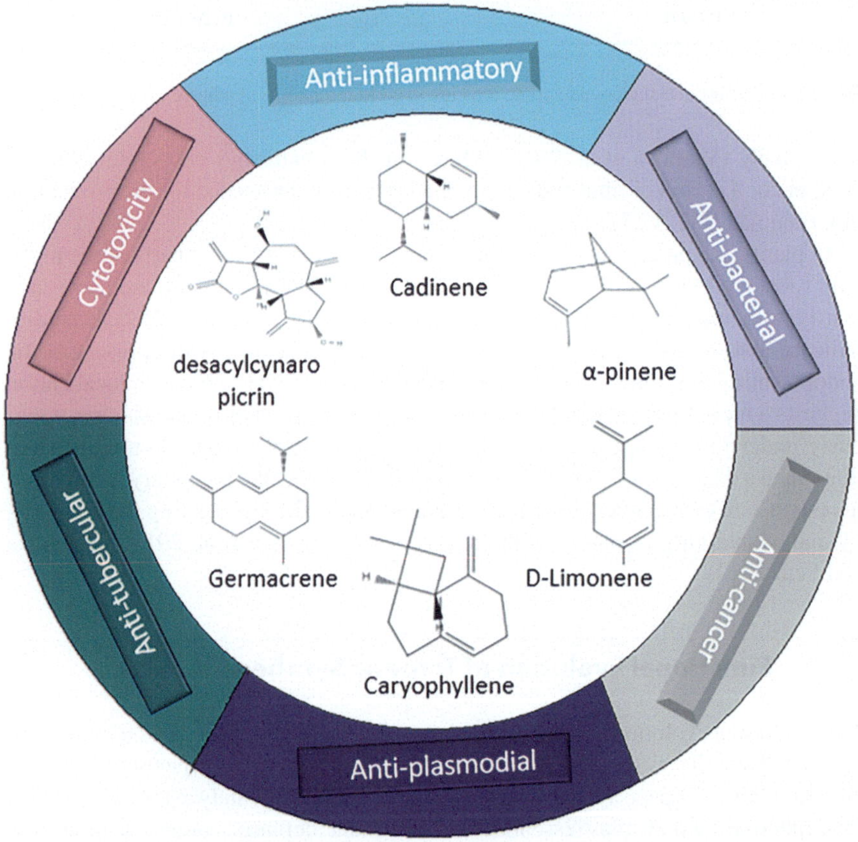

Fig. 12.2 Reported and traditional therapeutic application of terpenes and terpenoids

processes, is produced by TPS-b, which is exclusive to angiosperms. Cobaltyl pyrophosphate synthase is catalyzed by TPS-c, which belongs to a long-extinct evolutionary branch. The enzymes cobalt pyrophosphate/kaurene synthase, which are essential for the synthesis of gibberellic acid, are encoded by TPS-e/f; TPS-g is unique to angiosperms, but it also makes a monoterpene synthase (Chen et al. 2024). Although advancements in genome sequencing technology have been made, high-quality genomes of the genus *Bidens* have not yet been identified. Furthermore, comprehensive research on the TPS (terpene synthase) genes in *Bidens* plants remains to be conducted.

Scientists conducted Chloroplast P genome-wide comparative analysis among the *Bidens pilosa* Linn., *Bidens bipinnata* Linn., and *Bidens alba* var. radiata., with others published in *Bidens*. According to phylogenetic research, these species clustered together and eventually formed a monophyletic clade. This suggests that *Bidens* plants may be subject to radiation adaptive selection in response to the changing global environment. Furthermore, intergenic areas of ndhD-ccsA, ndhI-ndhG, ndhF-rpl32, trnL_UAG-rpl32, ndhE-psaC, matK-rps16, rps2-atpI, cemA-petA, and petN-psbM were shown to be candidate markers of molecular identification for *Bidens* plants, according to mutation hotspot analysis and in silico PCR analysis.

12.5 Terpenoids and Their Applications

12.5.1 Terpenoids in Medicine: Therapeutic Potentials

Terpenoids exhibit a broad range of medicinal properties (Fig. 12.2) and are ubiquitous in nature, having been used in traditional medicine for centuries. Many compounds are commercially available, primarily as dietary supplements; however, some are also registered as pharmaceutical drugs. In folk medicine, *Bidens* species have been utilized as herbal remedies to manage and treat a variety of illnesses. For example, *B. pilosa* for its antiseptic, hypoglycemic, blood-pressure-lowering, and liver-protective properties (Deba et al. 2008); *B. bipinnata* for acute nephritis, malaria, and chronic diarrhea; and *B. humilis* for fever and headaches (Zahara et al. 2022a). Another study provides valuable insights into the antimicrobial properties of extracts from the aerial parts of *Bidens cernua* var. *radiata*, *Bidens frondosa*, and *Bidens tripartita*. These findings suggest a possible rationale for the traditional use of *Bidens* species in folk medicine to treat various skin diseases, injuries, and colds, given their observed antimicrobial activities. Further research is needed to fully understand and validate the therapeutic potential of *Bidens* species in modern medicinal contexts.

Bidens pilosa is recognized for its rich content of sesquiterpenoids and diterpenoids. Among them, phenylheptatriyne stands out for its potent anti-inflammatory and antimicrobial activities. Research has shown that phenylheptatriyne possesses properties that can help alleviate inflammation and combat microbial infections. Additionally, acetyl-phenylheptadiyne, another compound found in *Bidens pilosa*, has been studied for its anti-inflammatory effects (Mtenga and Ripanda 2022).

Likewise, caryophyllene oxide and β-sitosterol extracted from *Bidens pilosa* exhibit anti-inflammatory, antioxidant, and anticancer properties. Recent studies underscore the cytotoxic potential of 8-desacyl-cynaropicrin, positioning *Bidens tripartita* as a promising candidate for further exploration in cancer treatment (Wagatsuma et al. 2021). Cadinene is a member of the sesquiterpenes isolated from genus *Bidens* and possesses anti-inflammatory properties. Germacrene D from *Bidens pilosa* exhibits significant antibacterial activity against various pathogens. Certain sesquiterpene lactones have been found to exhibit cytotoxic effects on cancer cells (Chen et al. 2024). Research has elucidated the specific mechanisms by which carotenoids act as antioxidants in *Bidens* species. They contribute to cellular defense against oxidative stress by neutralizing reactive oxygen species (ROS) and increasing the antioxidant defense system (Lima Silva et al. 2011). The roots ethanol extract of *Bidens biternata* is reported to have significant anticancer potential (Zahara et al. 2022b).

The presence of these terpenoids in plant of genus *Bidens* underscores its potential as a source of natural compounds with medicinal applications. Further exploration into their specific mechanisms of action and therapeutic potential could pave the way for new pharmaceutical developments.

12.5.2 Bidens Terpenoids in Industry: Flavors, Fragrances, and Beyond

These volatile terpenoids have numerous significant applications in human civilization, particularly in the food and pharmaceutical industries. Their effectiveness and potential are harnessed as flavor enhancers. In the flavor and fragrance industries, terpenoids such as those found in *Bidens pilosa* and *Bidens alba* contribute to unique aromatic profiles. In the food industry, terpenoids are widely used as natural flavor enhancers. Compounds such as limonene, pinene, and linalool are used to enhance the sensory qualities of food, beverages, perfumes, and cosmetics. Their aromatic properties contribute to the flavors of various foods and beverages, improving the sensory experience for consumers. β-Linalool and α-Phellandrene are terpenoids isolated from the essential oil of *Biden tripartita* and are often used in perfumery and aromatherapy due to their calming effects (Tomczykowa et al. 2005). In the pharmaceutical industry, terpenoids are valued for their medicinal properties. Their multifarious biological properties, such as their anti-inflammatory, antibacterial, and anticancer properties, render them advantageous in the creation of medicinal compounds. For instance, β-caryophyllene is known for its anti-inflammatory effects (Mtenga and Ripanda 2022). Beyond flavor and fragrance, terpenoids are also used in agriculture as natural pesticides and herbicides. Their bioactive properties help in managing pests and diseases without relying on synthetic chemicals, promoting sustainable agricultural practices. Examples include *Bidens pilosa* oil, rich in terpenoids, which is widely used as biopesticides (Xuan and Khanh 2016). Since terpenes can slow down the development of bacterial resistance to antibiotics, they can be added to animals in place of conventional antibiotics. Terpene-containing plant oils have shown encouraging in vivo bactericidal action because of their lipid-organizing qualities. Terpenoids are widely employed in antibacterial soaps, household goods, and cosmetics. These compounds enhance the antimicrobial efficacy and stability of these products (Fokunang et al. 2020).

12.6 Advances in Genomic Studies of *Bidens* Terpenoids

Recent genomic studies of *Bidens* species have significantly advanced our understanding of terpenoids biosynthesis. By identifying and characterizing the key genes and pathways involved, researchers have laid the groundwork for innovative biotechnological approaches and synthetic biology applications to optimize and expand the use of *Bidens* terpenoids.

12.6.1 Biotechnological Approaches

Biotechnological methods are being utilized to enhance the production and diversity of terpenoids in plants. By genetic modification and metabolic engineering manipulating the biosynthetic pathways, scientists can increase the yield of specific terpenoids. For instance, overexpressing key enzymes or introducing regulatory genes can significantly boost the production of desired terpenoids (Zhang et al. 2020). Advanced tissue culture methods enable the mass propagation of *Bidens* plants with superior terpenoids profiles. This ensures a consistent supply of plant material for industrial applications, mitigating the limitations of natural variability and environmental influences.

12.6.2 Synthetic Biology

Synthetic biology offers transformative potential for the production of *Bidens* terpenoids. Key developments include engineering microbial hosts like bacteria and yeast to harbor synthetic terpenoids biosynthesis pathways, so it is possible to produce these compounds on an industrial scale. This method reduces production costs and allows for the creation of novel terpenoids that are not naturally found in *Bidens* species (Yuzawa et al. 2018). Synthetic biology tools, including CRISPR-Cas9, enable precise editing of genetic elements that control terpenoids biosynthesis. This fine-tuning can enhance pathway efficiency and terpenoids yield, making the production process more effective. The integration of genomic studies with biotechnological and synthetic biology approaches promises a bright future for the production and application of *Bidens* terpenoids. Continued research and development are expected to lead to the discovery of new terpenoids with unique properties and applications, benefiting industries such as pharmaceuticals, agriculture, and biofuels. The scalability and sustainability of terpenoids production will also be significantly improved, addressing both economic and environmental concerns.

12.7 Conclusion

The genus *Bidens*, commonly known as beggarticks or bur marigolds, encompasses a diverse group of flowering plants distributed across various ecosystems worldwide. One of the significant aspects of *Bidens* species is their rich source of terpenoids, which have garnered attention for their diverse biological activities and potential applications in medicine, agriculture, and industry. In this chapter, we aim to provide a comprehensive overview of the terpenoids isolated from the genus *Bidens,* highlighting their chemical structures, biosynthesis pathways, pharmacological properties, and therapeutic potentials. *Bidens* terpenoids exhibit remarkable structural diversity, comprising mono-, sesqui-, di-, and triterpenoids, as well as their derivatives. These compounds are primarily biosynthesized through the mevalonate or methylerythritol phosphate pathways within *Bidens* plants, leading to a wide array of chemical structures and functional groups. The pharmacological properties of *Bidens* terpenoids hold significant promise for the development of novel therapeutic agents and nutraceuticals to address various human health conditions. For instance, the antimicrobial activity of *Bidens* terpenoids makes them potential candidates for combating drug-resistant pathogens, while their antioxidant and anti-inflammatory properties may offer preventive and therapeutic strategies for oxidative stress-related disorders and inflammatory diseases. Furthermore, *Bidens* terpenoids show potential as adjuvants in cancer therapy, either as chemopreventive agents or sensitizers for enhancing the efficacy of conventional chemotherapeutic drugs. Additionally, the hepatoprotective and antidiabetic effects of *Bidens* terpenoids suggest their utility in managing liver disorders and diabetes mellitus. Continued research efforts aimed at exploring the chemical diversity, biological activities, and applications of *Bidens* terpenoids hold promise for the development of novel drugs, nutraceuticals, and agricultural products to address various human health challenges and societal needs.

References

Adefegha SA, Mostardeiro V, Morsch VM, Morel AF, Da Cruz IBM, Somacal S et al (2024) Protective effect and antioxidative efficacy of *Moringa stenopetala* leaf extract in BV-2 microglial cell. Food Chem Adv 5:100767

Amvam Zollo P, Kuiaté J, Menut C, Lamaty G, Bessiere J, Chalchat J et al (1995) Aromatic plants of tropical Central Africa. Part XX. The occurrence of 1-phenylhepta-1, 3, 5-triyne in the essential oil of *Bidens pilosa* L. from Cameroon. Flavour Fragr J 10(2):97–100

Benhura MA, Chitsiku IC (1997) The extractable β-carotene content of Guku (*Bidens pilosa*) leaves after cooking, drying and storage. Int J Food Sci Technol 32(6):495–500

Chang C, Zhang Y, Li T (2000) Interannual and interdecadal variations of the East Asian summer monsoon and tropical Pacific SSTs. Part I: roles of the subtropical ridge. J Clim 13(24):4310–4325

Chen Y, Luo J, Zhang Q, Kong L (2016) Identification of active substances for dually modulating the renin–angiotensin system in *Bidens pilosa* by liquid chromatography–mass spectrometry-based chemometrics. J Funct Foods 21:201–211

Chen Z, Wu Z, Zamfir E, Zhang K, Zhang Y, Timofte R, et al (2024) Ntire 2024 challenge on image super-resolution (x4): methods and results. Proceedings of the IEEE/CVF conference on computer vision and pattern recognition

Deba F, Xuan TD, Yasuda M, Tawata S (2008) Chemical composition and antioxidant, antibacterial and antifungal activities of the essential oils from *Bidens pilosa* Linn. var. Radiata. Food Control 19(4):346–352

Dudareva N, Klempien A, Muhlemann JK, Kaplan I (2013) Biosynthesis, function and metabolic engineering of plant volatile organic compounds. New Phytol 198(1):16–32

Fokunang ET, Fonmboh DJ, Mballa RN, Nyuyki AB, Fokunang LB, Kaba N et al (2020) Pharmacovigilance of natural herbal medicines research for efficacy, safety and quality assurance of phytomedicine products. J Complement Altern Med Res 12:21–37

Geissberger P, Séquin U (1991) Constituents of *Bidens pilosa* L.: do the components found so far explain the use of this plant in traditional medicine? Acta Trop 48(4):251–261

Grombone-Guaratini MT, Solferini VN, Semir J (2004) Reproductive biology in species of Bidens L. (Asteraceae). Sci Agric 61:185–189

Iqbal S, Zuleika M (1995) Eighty-seven days of orotracheal intubation. Anaesthesia 50(4):343–344

Karunanithi PS, Zerbe P (2019) Terpene synthases as metabolic gatekeepers in the evolution of plant terpenoid chemical diversity. Front Plant Sci 10:1166

Kuzuyama T, Seto H (2012) Two distinct pathways for essential metabolic precursors for isoprenoid biosynthesis. Proc Jpn Acad Ser B Phys Biol Sci 88(3):41–52

Lima Silva F, Fischer DCH, Fechine Tavares J, Sobral Silva M, Filgueiras de Athayde-Filho P, Barbosa-Filho JM (2011) Compilation of secondary metabolites from Bidens pilosa L. Molecules 16(2):1070–1102

Lorigooini Z, Sadeghi Dehsahraei K, Bijad E, Habibian Dehkordi S, Amini-Khoei H (2020) Trigonelline through the attenuation of oxidative stress exerts antidepressant-and anxiolytic-like effects in a mouse model of maternal separation stress. Pharmacology 105(5–6):289–299

Luck K, Kim D-K, Lambourne L, Spirohn K, Begg BE, Bian W et al (2020) A reference map of the human binary protein interactome. Nature 580(7803):402–408

Mabou FD, Yossa IBN (2021) TERPENES: structural classification and biological activities. IOSR J Pharm Biol Sci 16:25–40

Mtenga DV, Ripanda AS (2022) A review on the potential of underutilized Blackjack (*Biden Pilosa*) naturally occurring in sub-Saharan Africa. Heliyon 8(6):e09586

Muhlemann JK, Klempien A, Dudareva NJ (2014) Floral volatiles: from biosynthesis to function. Plant Cell Environ 37(8):1936–1949

Ogunbinu AO, Okeniyi SO, Flamini G, Cioni PL, Ogunwande IA, Olayinka ET (2009) Essential oil-bearing plants from Nigeria: studies on *Vernonia perrottetii* (leaf and stem bark), young leaves from *Eucalyptus decaisneana* and immature leaves of *Hyptis suaveolens*. J Essent Oil Res 21(2):154–158

Pozharitskaya O, Shikov A, Makarova M, Kosman V, Faustova N, Tesakova S et al (2010) Anti-inflammatory activity of a HPLC-fingerprinted aqueous infusion of aerial part of *Bidens tripartita* L. Phytomedicine 17(6):463–468

Priestap HA, Bennett BC, Quirke JME (2008) Investigation of the essential oils of *Bidens pilosa* var. minor, *Bidens alba* and *Flaveria linearis*. J Essent Oil Res 20(5):396–402

Ravichandran A, Subramanian P, Manoharan V, Muthu T, Periyannan R, Thangapandi M et al (2018) Phyto-mediated synthesis of silver nanoparticles using fucoidan isolated from Spatoglossum asperum and assessment of antibacterial activities. J Photochem Photobiol B 185:117–125

Sarg T, Salama O, El-Domiaty M, Bishr M, Mansour SES, Weight E (1991) Iridoid glucosides from Gentiana kurroo Royle. Alexandria J Pharm Sci 5:84–86

Tholl D (2015) Biosynthesis and biological functions of terpenoids in plants. Adv Biochem Eng Biotechnol 148:63–106

Tomczykowa M, Gudej J, Majda T, Gora J (2005) Essential oils of *Bidens tripartita* L. J Essent Oil Res 17(6):632–635

Tomczykowa M, Leszczyńska K, Tomczyk M, Tryniszewska E, Kalemba D (2011) Composition of the essential oil of *Bidens tripartita* L. roots and its antibacterial and antifungal activities. J Med Food 14(4):428–433

Vranová E, Coman D, Gruissem W (2013) Network analysis of the MVA and MEP pathways for isoprenoid synthesis. Annu Rev Plant Biol 64(1):665–700

Wagatsuma T, Tanaka K, Iino Y, Satou T, Khan MSH, Tawaraya K (2021) Burr marigold (*Bidens tripartita* L.) roots directly and immediately scavenge rhizosphere methane with highly exuded hydrogen peroxide via a rhizosphere Fenton reaction. Plant Soil 459:289–313

Xuan TD, Khanh TD (2016) Chemistry and pharmacology of *Bidens pilosa*: an overview. J Pharm Investig 46:91–132

Yuzawa S, Mirsiaghi M, Jocic R, Fujii T, Masson F, Benites VT et al (2018) Short-chain ketone production by engineered polyketide synthases in *Streptomyces albus*. Nat Commun 9(1):4569

Zahara K, Bibi Y, Arshad M, Kaukab G, Al Ayoubi S, Qayyum A (2022a) In-vitro examination and isolation of antidiarrheal compounds using five bacterial strains from invasive species *Bidens bipinnata* L. Saudi J Biol Sci 29(1):472–479

Zahara K, Bibi Y, Masood S, Nisa S, Sher A, Ali N et al (2022b) Isolation and identification of bioactive compounds from bidens spp. using HPLC-DAD and GCMS analysis and their biological activity as anticancer molecules. Molecules 27(6):1927

Zhang J-j, Dong X, Cao Y-y, Yuan Y-d, Yang Y-b, Yan Y-q et al (2020) Clinical characteristics of 140 patients infected with SARS-CoV-2 in Wuhan, China. Allergy 75(7):1730–1741

Part IV

Phytochemical-Based Evidence of Pharmacological Properties

Antimicrobial Activities of the *Bidens* Species

13

Huma Mehreen Sadaf

Abstract

There is a great history of using plants and their derivatives to treat microbial illnesses with fewer side effects. The chapter underscores the importance of exploring *Bidens* species for antimicrobial potential and traditional medicinal uses and highlighting important bioactive compounds responsible for their antimicrobial activity. The search for alternative methods provides an avenue to develop an affordable, stable, sustainable, and efficient system for treatment and allopathic delivery systems. This dual perspective emphasizes the value of integrating traditional knowledge with scientific research to uncover novel therapeutic agents from natural sources. The comprehensive evidence positions *Bidens* species as valuable candidates for further pharmacological investigations and the development of new therapeutic agents. This chapter serves as a bridge to the gap between conventional knowledge and modern scientific techniques and underscores the importance of the *Bidens* genus in the wide perspective of medicinal plant research.

Keywords

Bidens · Remedies · Antimicrobial · *Pilosa* · *Antibacterial*

13.1 Introduction

The belief that some plants had healing potentials was widely accepted before man discovered that microorganisms exist. Natural remedies were commonly utilized until the first part of the twentieth century when people started using synthetic

H. M. Sadaf (✉)
Department of Botany, The University of Chakwal, Chakwal, Punjab, Pakistan

Y. Bibi et al. (eds.), *The Genus Bidens*,
https://doi.org/10.1007/978-981-96-4257-1_13

medications instead as they were more profitable, more effective and could be patented (Varro 1999; Pichersky and Gang 2000). A vast variety of phytochemicals found in plants have been used for ages in folk or ethnical medical remedies (Greathead 2003). In traditional medicine, it is used to treat neoplasms, diathesis, wounds, metabolic disorders, scrofulous, seborrhea, psoriasis, and skin conditions. It is also used as a sudorific and diuretic agent (Rybalchenko et al. 2010). Gram-positive, Gram-negative, and fungal microorganisms were used to test the essential oils' antimicrobial properties. In particular, the oil demonstrated potent antibacterial and antifungal properties against strains of Gram-positive bacteria (Tomczykowa et al. 2017).

The genus *Bidens* L., which belongs to the Asteraceae family, comprises approximately 280 species distributed across various regions worldwide (Connelly et al. 2009). Most of these species are found in America, while only four to five species are known in Europe. In Turkey, the *Bidens* genus is represented by four species and five taxa (Korkmaz et al. 2011). Phytochemical research on *Bidens* species has identified a variety of bioactive compounds such as polyacetylenes (Chang et al. 2001), flavonoids (Wang and Weisberg 2000), essential oils, coumarins, polysaccharides, carotenoids, amines, mineral elements (Tomczykowa et al. 2008), and chalcones (Redl et al. 1994). Several species within this genus are employed in traditional medicine to treat various conditions, including inflammation, rheumatism (arthritis), diabetes, enteritis, angina (pharyngitis), malaria, metabolic disorders, skin diseases, seborrhea, psoriasis, wounds, and tumors (Wang et al. 2003).

In traditional medicine across various countries, different parts of *Bidens pilosa*, in different forms, such as juice, powder, or decoction, are used orally to treat a variety of diseases and disorders (Lima Silva et al. 2011). Moreover, medicinal plants have been accounted to possess various pharmaceutical and biological properties (Bartolome et al. 2013). *Bidens frondosa* L. is an annual species that flourishes on nutrient-rich soil along the coastline of lakes and rivers. This plant originally is native to North America and has been widespread across central Europe since the nineteenth century. The seed infusion of the plant made as syrup with honey had significant medicinal value for several ailments including cold, laryngeal infections, acute bronchial and whooping cough (Crowe et al. 1990).

Bidens cernua and *Bidens tripartita* are two prominent representatives of the genus among the six species recorded in Poland (Stępień and Rosadziński 2020). Bidens rhodina is recognized as an important plant used against multi-infectious microorganism that causes significant damage to crops, particularly affecting tropical fruits . *Bidens cernua L.* has been traditionally used in folk medicine for the treatment of various conditions, including metabolic disturbances, diathesis, skin diseases, scrofula, seborrhea, psoriasis, neurodermatitis, wounds, and neoplasms. Additionally, it has been utilized as a diuretic and sudorific agent (Yang et al. 2012). *B. biternata* has been found to possess a wide range of nutritional components, including carbohydrates, proteins, reducing sugars, amino acids, free fatty acids, crude fiber, lipids, and vitamins. Additionally, it contains low concentrations of anti-nutritional factors such as phytic acid, total phenols, and tannic acid (Sukumaran and Parthiban 2014). *Bidens tripartita* herb has been found to contain flavonoids

(Korcelli-Olejniczak 2004), polysaccharides, carotenoids, amines, lactones, mineral elements, coumarins (Tomczykowa et al. 2005), and volatile oils (Tomczykowa et al. 2005).

The antibacterial properties of specific flavonoids such as flavanomarein, cynaroside, and luteolin isolated from bur-marigold, along with various extracts from the herb and flowers, were assessed using the electron paramagnetic resonance method. Among the tested extracts, all except the butanol extracts showed antibacterial activity, while none displayed inhibitory effects against the tested fungi. Conversely, both essential oils exhibited strong antifungal activity but had minimal bacteriostatic effects. Furthermore, aqueous extracts of *B. tripartita* demonstrated anti-inflammatory activity against carrageenan-induced acute paw edema in rats (Thompson and Wolniak 2008). Bioactive exopolysaccharides extracted from fruits of Bidens rhodinai shows antimutagenic and immunodulatory effects. (Martins et al. 2008). Additionally, macrophorins E, F, and G isolated from *Bidens berengeriana* exhibited potent antibacterial and antifungal activity against phytopathogenic fungi, along with cytotoxic effects.

13.2 Antibacterial Potential of the Genus *Bidens*

Bidens species have demonstrated significant antibacterial activity in studies using different solvent extracts. Ethanol and methanol extracts from multiple *Bidens* plants exhibited potent antibacterial effects against both Gram-positive and Gram-negative bacteria.

These crude extracts efficiently suppress bacterial growth which highlights the significance of *Bidens* species as a potent source of antimicrobial agents. Aqueous extract of *Bidens* species presented remarkable antibacterial potential signifying broad spectrum antibacterial potential against various pathogens. These findings emphasize the significance of *Bidens* species in the investigation of natural antimicrobial bioactive compounds best for the pharmaceutical and therapeutic potential.

13.3 Antifungal Potential of the Genus *Bidens*

Many studies have reported the significant antifungal properties of *Bidens* species against different fungal strains with different solvents extracted from these species. Methanolic and ethanolic fractions of the multiple *Bidens* species showed effectiveness against several fungal strains. The minimum inhibitory concentration (MIC) value highlights the efficiency of these plants as a potent source of antifungal agents. These findings underscore the importance of *Bidens* species in the search for natural antifungal compounds, which could be beneficial for pharmaceutical and agricultural applications.

13.3.1 Mechanism of Action

One of the biggest obstacles to the effective treatment of microbial infections is the emergence of drug-resistant bacteria (Prabuseenivasan et al. 2006). The screening of plant-based medicinal compounds has attracted a lot of attention in the last 50 years. The acetone, methanol, and water extracts from the roots of the *Bidens* genus have minimum inhibitory concentrations (MICs) that are efficient against microorganisms (Chang et al. 2000). Some mechanisms through which *Bidens* exhibit antimicrobial role are discussed as follows.

13.3.1.1 Anticandidal Mechanism

Bidens genus contains astragalin, a flavonoid that has anticandidal activity and may be used as an anticandidal agent. In addition to its potential antibacterial activities, astragalin can prevent the growth of fungal biological films, which is essential to their pathogenicity, without disrupting the yeast hyphae transition. Before using astragalin to create further treatment possibilities, it is crucial to study its anticandidal processes both in vivo and in vitro. Human fibroblast cells do not react toxically to astragalin, suggesting that it is safe to use alone or in combination with existing commercial antifungals to treat *Candida albicans* infections. Consequently, there is a chance to offer a strong antibacterial action at a lower dose. Examining the proliferation of fungal cells with and without astragalin, the findings showed a slight decrease in the percentage of hyphal cells after astragalin was treated for 4 h at MIC (Ivanov et al. 2022). A related study showed that kaempferol containing honey flavonoid extract might lower the percentage of cells that transitioned from yeast to hyphae (Candiracci et al. 2012).

13.3.1.2 Loss of Cell Membrane Integrity in Pathogens

Astragalin's potent antifungal action is attributed in part to its potent influence on the fungus membrane's integrity (Ivanov et al. 2022). It has been shown that astragalin can change the shape of the cell membrane in the parasite *Trypanosoma cruzi*. *Bidens* genus has antitrypanosomal characteristics, especially when it comes to the recognition of tryptophan's butyl and propyl derivatives. It has been discovered that certain substances can successfully block the cell cycle of *Trypanosoma brucei* (Nardella et al. 2020). Astragalin has the potential to interfere with sterol synthesis.

13.3.1.3 Immunomodulatory Mechanism

The ability of plants to reduce inflammation is linked with types of bioactive chemical constituents such as phenolics and alkaloids with a wide range of structural diversity. Pathophysiology of autoimmune diseases is linked with a boosted immune response that could be managed to reduce pain and inflammation by a therapeutic field of bioactive potent components such as quercetin and terpenes. Bioactive polyacetylene compounds including cytopiloyne (CP) act as anti-inflammatory agents and manage immune response. CP reduces proinflammatory Th1 response while boosting the regulatory Th2 reaction in type 1 diabetes mellitus. In infectious diseases, CP promotes the phagocytic potential, phagolysosomal combination, and enzymatic lysosomal activity of the macrophages that induce antimicrobial response. Therefore, CP may signify a diverse approach to managing infectious problems particularly those exhibiting resistance to the standard treatments.

Although the action and mechanism of CP as an immune modulatory response is unclear, the results signify potential for anticancer therapy and antimicrobial agents. For example, CP can increase macrophage phagocytic activity and T-cell production of interferon gamma (IFN-γ) in *Listeria monocytogenes*, *Candida parapsilosis*, and *Eimeria tenella* infection models (Chung et al. 2016). It is yet unknown how CP causes cytotoxicity, triggers immunogenic cancer cell death, and affects immune system cells implicated in the antitumor response activation (Rodriguez-Otero et al. 2023). The hydroalcoholic extract of the *Bidens* genus was shown to include 13 compounds (targeting AMY2A) by Liquid Chromatography(LC)-Mass spect+rophotometery (MS) analysis. These compounds demonstrated strong suppression of alpha-amylase (AMY2A) in vitro experiments, which supports the in silico findings regarding their antidiabetic action. These results support the inhibitory activity of pure substances and concentrated fractions that may be used in future drug discovery procedures (Galagali et al. 2024).

13.3.2 Comparative Analysis

An annual plant, *Bidens frondosa* L., grows along the banks of rivers and lakes in moist, nutrient-rich mud soils, or muddy sand soils. Also referred to as devil's beggarticks, this plant is consumed by humans. Additionally, it helps with acute bronchial and laryngeal attacks brought on by cold exposure. Herbal remedies for irritation, inflammation, bleeding in the respiratory system, and uterine hemorrhage employ this plant's infusions (Icoz 2016). *Bidens frondosa* has been shown to have antimalarial and antidiarrheal properties (Zhao and Lee 1999).

Commonly known as bur-marigold, *Bidens tripartita* L. is used as a sudorific, diuretic, and anti-inflammatory in folk medicine. The World Health Organization describes the antimalarial, choleretic, anti-ulcer, and photoprotective properties of *B. tripartita* L. (Herba Bidentis). In the study of Tomczykowa et al. (2011), the antibacterial and antifungal properties of the essential oil were used to test against 11 Gram-negative and 8 Gram-positive bacterial species, and also against 10 fungal strains. The oil exhibited a strong antifungal activity. In 2020, the results of the Mendel et al. study showed that *B. tripartita* may be beneficial for a variety of functional gastrointestinal problems, such as the constipation-predominant phases of irritable bowel syndrome and dyspeptic symptoms, for which gastroprokinetics is typically the recommended course of treatment.

The annual herb *Bidens pilosa* is a global native of tropical and Central America. Native Americans and peoples of Africa utilize it widely as a traditional remedy to cure respiratory infections dysentery, diarrhea and colic, and a wide range of illnesses (Arthur et al. 2012). Boiling water extract of *B. pilosa* aerial parts has anti-allergic and anti-inflammatory effects in Japan (Horiuchi and Seyama 2006). According to Rabe and Van Staden (1997), the Zulu tribe in South Africa uses an aqueous extract of the leaves to cure colic, diarrhea, and dysentery. The fact that *B. pilosa* is widely used in Africa and other parts of the world suggests that this plant could provide useful medications to treat a range of illnesses, including malaria. Despite being viewed as an invasive species in many nations, the benefits could exceed the environmental hazards posed by the weed. Strong cytotoxic and antiproliferative properties, as well as antibacterial and antifungal activity, have all been shown for ethanolic extracts. Numerous isolated polyacetylenes and flavonoids have demonstrated antibacterial, antihelmintic, and protozoocidal properties (Arthur et al. 2012).

Brandão et al. (2006) studied the antimalarial activity of nine *Bidens* species, namely *B. frondosa*, *B. biternata*, *B. tripartitus*, *B. ferulaefolia*, *B. maximovicziana*, *B. bipinnata*, *B. campylotheca*, *B. pilosa*, and *B. parviflora*. *B. pilosa* L. was collected around Pampulha Lake, Brazil, and other species of *Bidens* were collected outside Brazil. The findings showed that *Plasmodium falciparum* development was inhibited in vitro up to 90% by crude ethanol (whole plant, leaves, and roots), butanol, and chloroform extracts of *B. pilosa* L. All species of *Bidens*, which had aliphatic acetylenes were also highly active, whereas extracts of *B. parviflora* and *B. biternata* with none of the three acetylenes, respectively as documented in the literature, were inactive or had borderline activity in vitro. The fractions partially reduced the *Plasmodium berghei parasitemia* in mice in vivo.

Icoz et al. (2016) studied the effects of extracts of aqueous and ethanolic aerial parts of *Bidens cernua L. var. radiata* DC., *B. frondosa L.*, and *B. tripartita L.* for their antimicrobial activities against Gram-positive and Gram-negative bacteria. Results of the study demonstrated that ethanolic extracts of the examined species had superior antibacterial efficacy compared to aqueous extracts. No antibacterial activity was observed for all the aqueous extracts. Only aqueous extracts of *B. frondosa* and *B. tripartita* showed antifungal activity against *Candida albicans*. It is conceivable that the use of *Bidens* species for the treatment of various skin diseases and injuries, and also cold in folk medicine is because of its antimicrobial activity.

13.4 Clinical Implications

Bidens species have great potential for applications in modern medicine due to their wide range of antimicrobial qualities. Microbial biofilm production is inhibited by *Bidens* species extracts confirming antimicrobial aptitude against many infectious bacterial pathogens such as *Staphylococcus aureus*, which showed its clinical implications (Boakye et al. 2023).

13.4.1 Potential Uses in Modern Medicine

Considering the significant antimicrobial potential of the *Bidens* genus, bioactive components isolated from the plants could be incorporated in several modern products such as gels, lotion, and ointments, as topical formulations. Different bacterial and fungal infections can be managed by the extracts of plants and encapsulated forms of the active constituents that might be used as an oral supplement to boost and strengthen the different immune responses against microbial pathogenic strains (Boakye et al. 2023). *Bidens* extracts are considered as an alternative natural antimicrobial source of novel bioactive agents in light of the present challenges faced by antibiotic resistance. These work against the pathogenic strains of bacteria which showed resistance to the antibiotics due to their distinct mechanism and mode of action which includes rupturing of microbial cell membrane and inhibition of vital active metabolic enzymes (Ngameni et al. 2013). The synergistic effect of *Bidens* extracts and antibiotics might be adjuvant to traditional antibiotics which may enhance healing outcome and efficiency. According to studies, this approach significantly improves the effectiveness of allopathic modern medicine such as broad-spectrum ciprofloxacin and amoxicillin by minimizing side effects and lowering dosage requirements.

13.5 Future Research Directions

The significant scientific evidence is a compelling argument for further investigations and isolation of bioactive antimicrobial constituents from *Bidens*. The knowledge achieved is a fundamental support in the design of novel bioactive phytochemical constituents focused on the treatment of microbial infections that imbalance the immune responses. Future research should be performed to concentrate on the clinical trials for the confirmation of effects in the patients. Furthermore, the investigations should be based on dosage, formulations, and standardization of the bioactive components of *Bidens* for therapeutic properties. Preliminary studies have shown the low toxicity of *Bidens* extract; however, comprehensive clinical and toxicological investigations are required to confirm the findings (Ngameni et al. 2013).

13.6 Summary

As shown in this chapter, an extensive investigation underscores the potential of *Bidens* as a significant source of broad-spectrum potent bioactive compounds. Several plant parts such as stem, leaves, roots, and flowers of *Bidens* species were thoroughly examined against pathogenic microbes employing a range of solvent extracts including methanolic, ethanolic, and aqueous extracts. *Bidens pilosa* showed the most significant antimicrobial potential against various pathogenic microorganisms. This research highlights the crucial role of potent bioactive antimicrobial components isolated from *Bidens* species. The key components contributing to distinct bioactivities include polysaccharides, flavonoids, and essential oils that can be isolated. For example, essential oils exhibit significant potential against certain fungal pathogens extracted from *Bidens frondosa*. These investigations underscore the diverse nature and complexity of the bioactive phytocompounds within *Bidens* species which suggest promising avenues for extracting solvents for applications in the therapeutic and pharmaceutical industry. This perspective focuses on the value of integrating traditional knowledge with the current scientific investigations to uncover the pharmacological potential and development of novel bioactive therapeutic agents from the genus *Bidens*.

References

Arthur G, Naidoo K, Coopoosamy R (2012) *Bidens pilosa* L.: agricultural and pharmaceutical importance. J Med Plants Res 6(17):3282–3287

Bartolome AP, Villaseñor IM, Yang W-C (2013) *Bidens pilosa* L. (Asteraceae): botanical properties, traditional uses, phytochemistry, and pharmacology. Evid Based Complement Alternat Med 2013(1):340215

Boakye H, Atabila A, Hinneh T, Ackah M, Ojo-Benys F, Bello AI (2023) The prevalence and determinants of non-communicable diseases among Ghanaian adults: a survey at a secondary healthcare level. PLoS One 18(2):e0281310

Brandão MGL, Cosenza GP, Moreira RA, Monte-Mor RLM (2006) Medicinal plants and other botanical products from the Brazilian Official Pharmacopoeia. Rev Bras 16:408–420

Candiracci M, Piatti E, Dominguez-Barragan M, Garcia-Antras D, Morgado B, Ruano D et al (2012) Anti-inflammatory activity of a honey flavonoid extract on lipopolysaccharide-activated N13 microglial cells. J Agric Food Chem 60(50):12304–12311

Chang C, Zhang Y, Li T (2000) Interannual and interdecadal variations of the East Asian summer monsoon and tropical Pacific SSTs. Part I: roles of the subtropical ridge. J Clim 13(24):4310–4325

Chang J-S, Chiang L-C, Chen C-C, Liu L-T, Wang K-C, Lin C-C (2001) Antileukemic activity of *Bidens pilosa* L. var. minor (Blume) Sherff and Houttuynia cordata Thunb. Am J Chin Med 29(02):303–312

Chung BC, Mashalidis EH, Tanino T, Kim M, Matsuda A, Hong J et al (2016) Structural insights into inhibition of lipid I production in bacterial cell wall synthesis. Nature 533(7604):557–560

Connelly W, Shenton F, Lethbridge N, Leurs R, Waldvogel H, Faull R et al (2009) The histamine H4 receptor is functionally expressed on neurons in the mammalian CNS. Br J Pharmacol 157(1):55–63

Crowe JH, Carpenter JF, Crowe LM, Anchordoguy TJ (1990) Are freezing and dehydration similar stress vectors? A comparison of modes of interaction of stabilizing solutes with biomolecules. Cryobiology 27(3):219–231

Galagali A, Patil VS, Hiremath K, Sampat GH, Patil R, Virge R et al (2024) Investigation of alpha amylase inhibitors from *Bidens pilosa* L. by in silico and in vitro studies. In Silico Pharmacol 12(1):9

Greathead H (2003) Plants and plant extracts for improving animal productivity. Proc Nutr Soc 62(2):279–290

Horiuchi M, Seyama Y (2006) Antiinflammatory and antiallergic activity of *Bidens pilosa* L. var. radiata SCHERFF. J Health Sci 52(6):711–717

Icoz G (2016) The interconnectedness of the past, the present and the future: where Turkey–EU relations have been, and where they are heading. J Contemp Eur Stud 24(4):494–508

Ivanov M, Ćirić A, Stojković D (2022) Emerging antifungal targets and strategies. Int J Mol Sci 23(5):2756

Korcelli-Olejniczak E (2004) Funkcje metropolitalne Berlina i Warszawy w latach 1990–2002: współzależność pozycji w systemie miast Europy. IGiPZ PAN, Środkowej

Korkmaz H, Mumcu Ü, Alkan S, Conservation. (2011) A new alien species record for the flora of Turkey: *Bidens bipinnata* (Asteraceae). Biol Divers Conserv 4(2):49–54

Lima Silva F, Fischer DCH, Fechine Tavares J, Sobral Silva M, Filgueiras de Athayde-Filho P, Barbosa-Filho JM (2011) Compilation of secondary metabolites from *Bidens pilosa* L. Molecules 16(2):1070–1102

Martins RF, Ramos MF, Herfindal L, Sousa JA, Skærven K, Vasconcelos VM (2008) Antimicrobial and cytotoxic assessment of marine cyanobacteria-Synechocystis and Synechococcus. Mar Drugs 6(1):1–11

Nardella G, Brammer S, Surdu I (2020) Shame on who? The effects of corporate irresponsibility and social performance on organizational reputation. Br J Manag 31(1):5–23

Ngameni B, Fotso GW, Kamga J, Ambassa P, Abdou T, Fankam AG et al (2013) Flavonoids and related compounds from the medicinal plants of Africa. In: Medicinal plant research in Africa. Elsevier, London, pp 301–350

Pichersky E, Gang DR (2000) Genetics and biochemistry of secondary metabolites in plants: an evolutionary perspective. Trends Plant Sci 5(10):439–445

Prabuseenivasan S, Jayakumar M, Ignacimuthu S (2006) In vitro antibacterial activity of some plant essential oils. BMC Complement Altern Med 6:1–8

Rabe T, Van Staden J (1997) Antibacterial activity of South African plants used for medicinal purposes. J Ethnopharmacol 56(1):81–87

Redl K, Breu W, Davis B, Bauer RJ (1994) Anti-inflammatory active polyacetylenes from *Bidens campylotheca*. Planta Med 60(01):58–62

Rodriguez-Otero P, Ailawadhi S, Arnulf B, Patel K, Cavo M, Nooka AK et al (2023) Ide-cel or standard regimens in relapsed and refractory multiple myeloma. N Engl J Med 388(11):1002–1014

Rybalchenko N, Prykhodko V, Nagorna S, Volynets N, Ostapchuk A, Klochko V et al (2010) In vitro antifungal activity of phenylheptatriyne from *Bidens cernua* L. against yeasts. Fitoterapia 81(5):336–338

Stępień E, Rosadziński S (2020) Communities of the Bidentetea class of small coastal river valleys of the Western Pomerania (Poland). Ecologica Montenegrina 28:8–19

Sukumaran S, Parthiban B (2014) Vascular plant diversity of Udayagiri fort, Kanyakumari district, Tamilnadu, India. Biosci Dis 5(2):204–217

Thompson MV, Wolniak SM (2008) A plasma membrane-anchored fluorescent protein fusion illuminates sieve element plasma membranes in Arabidopsis and tobacco. Plant Physiol 146(4):1599–1610

Tomczykowa M, Gudej J, Majda T, Góra J (2005) Essential oils of *Bidens tripartita* L. J Essent Oil Res 17(6):632–635

Tomczykowa M, Tomczyk M, Jakoniuk P, Tryniszewska E (2008) Antimicrobial and antifungal activities of the extracts and essential oils of *Bidens tripartita*. Folia Histochem Cytobiol 46(3):389–393

Tomczykowa M, Leszczyńska K, Tomczyk M, Tryniszewska E, Kalemba D (2011) Composition of the essential oil of *Bidens tripartita* L. roots and its antibacterial and antifungal activities. J Med Food 14(4):428–433

Tomczykowa M, Tomczyk M, Leszczyńska K, Kalemba D (2017) Flavonoids and essential oil of *Bidens cernua* of Polish origin and in vitro antimicrobial activity of the oil. Rec Nat Prod 11(5):468–473

Varro ET (1999) Phytomedicines: back to the future. J Nat Prod 62(11):1589–1592

Wang C, Weisberg RH (2000) The 1997–98 El Niño evolution relative to previous El Niño events. J Clim 13(2):488–501

Wang M, Tran JH, Jacoby GA, Zhang Y, Wang F, Hooper DC et al (2003) Plasmid-mediated quinolone resistance in clinical isolates of *Escherichia coli* from Shanghai, China. Antimicrob Agents Chemother 47(7):2242–2248

Yang X-W, Huang M-Z, Jin Y-S, Sun L-N, Song Y, Chen H-S (2012) Phenolics from *Bidens bipinnata* and their amylase inhibitory properties. Fitoterapia 83(7):1169–1175

Zhao Q, Lee FS (1999) Mitogen-activated protein kinase/ERK kinase kinases 2 and 3 activate nuclear factor-κB through IκB kinase-α and IκB kinase-β. J Biol Chem 274(13):8355–8358

Anticancer and Antioxidant Activities of *Bidens* Species

14

Khair-un-Nisa and Saboon

Abstract

The Asteraceae family includes over 200 species of annual or perennial herbs in the genus *Bidens*. Its species are found all over the world. Numerous indigenous medicinal systems make use of certain species of *Bidens*. For instance, *Bidens pilosa, B. humilis, B. biternata, B. segetum, B. cernua, B. frondosa*, and *B. tripartita* are used for their antioxidant properties; *B. pilosa, B. alba, B. biternata*, and *B. bippinata* are used for their anticancer properties. This chapter provides current information on the phytochemical and pharmacological uses of the genus *Bidens* regarding their antioxidant and anticancer properties from various research works. It also offers preliminary data and recommendations for future fundamental and clinical research in this area. The majority of the scientific data examined for this analysis gives credence to the possibility of using the genus *Bidens* to cure aging-related conditions such as cancer, heart disease, stroke, and other conditions.

Keywords

Anticancer · Asteraceae · Antioxidant · *Bidens* · Cytotoxic

14.1 Introduction

About 200 species of annual or perennial herbs in the genus *Bidens* that is part of the Asteraceae family are found throughout the world. The *Bidens* species are found in North and South America, Polynesia, Europe, Africa, and Asia. These Plants are prickly and bristly, that why they commonly known as Spanish Needles. they also

Khair-un-Nisa · Saboon (✉)
Department of Botany, Women University Mardan, Mardan, Pakistan
e-mail: drsabaoon@wumardan.edu.pk

Y. Bibi et al. (eds.), *The Genus Bidens*,
https://doi.org/10.1007/978-981-96-4257-1_14

have many other different vernacular names like, stickseeds, tickseed Sunflowers, cobbler's pegs, beggarticks, stickseeds, blackjack, Marigold etc. The species are characterized by branched stems, opposing leaves, and composite flower heads that can resemble a single flower, though certain species may have white, pink, or red flowers. These flower heads are typically yellow or orange in color. They yield fruits with the characteristic barbs and hooks after they have flowered. Numerous native medical systems use plants in this genus for anticancer, antidiabetic, antimalarial, antioxidant, antihypertensive, anti-inflammatory, anti-allergic, antibacterial, and antiviral treatments (Baiwa et al. 2018). Flavonoids, polyacetylenes, and phenol glycosides are the essential nutrients found in *Bidens* species (Tommasi 1998).

14.2 Anticancer and Antioxidant Activity of the Key Species of Genus *Bidens*

Some bioactive substances found in *Bidens* species have been found to contain anti-cancer activities including flavonoids, polyacetylenes, phenolic acids, terpenoids, and saponins. Thus, these species are being investigated for their potential in cancer treatment. *Bidens* species have a variety of medicinal uses in the treatment of cancer due to their cytotoxic effects, induction of apoptosis, anti-angiogenic actions, immune system regulation, and chemoprevention.

Bidens species are also rich in antioxidants, including flavonoids, triterpenes, and polyphenols which support the defense of cells against free radical-induced oxidative damage. These species exhibit free radical scavenging, quenching, and reducing effects on reactive oxygen species (ROS), as well as antioxidant activity against membrane lipid peroxidation (Tsuda et al. 1994). Antioxidants play a critical role in maintaining human health and prevention of various diseases (Hamid et al. 2010).

14.2.1 *Biden pilosa* (Black Jack or Bitter Leaf)

B. pilosa is an annual plant and found in America, Africa, China, Japan, and other tropical and subtropical regions; however, it is native to South America (Tan et al. 2000). The whole plant including its roots and aerial parts has been found to have various biological properties.

According to Sundararajan et al. (2006), the KB IS A (human nasopharyngeal epidermal carcinoma cell line) cancer cell lines and human HeLa cell lines are cytotoxically affected by the EtOAc fraction and coarse methanolic extracts from the whole *B. pilosa* plants.

Kviecinski et al. (2011) tested the trichloromethane, ethyl acetate, and hydroalcoholic crude extracts, and the methyl alcohol fractions from this plant for antitumor activity. Brine shrimp, neutral red uptake, (MTT) 3-(4,5-dimethylthiazol-2-yl)-2, 5-diphenyltetrazolium bromide assay, and hemolytic tests were used to evaluate the toxicity of the extracts. Ehrlich ascites carcinoma was used in vivo experiments on isogenic BALB/c mice. The most dangerous of all was the trichloromethane portion.

Kviecinski et al. (2011) further assessed the anticancer properties of two extracts, namely SCF and hydroethanolic crude extract (HCE). The extracted HCE and other fractions such as methanol, chloroform, and (EtOAc) ethyl acetate from the aerial section of *B. pilosa* showed antioxidant and hepatoprotective properties. The ability to prevent lipid peroxidation and the in vitro free radical scavenging activity were assessed using the 1,1-diphenyl-2-picrylhydraz (DPPH) radical scavenger method. The HCE and the EtOAc fraction was also demonstrated to lower serum levels of enzymes linked to liver damage including alanine aminotransferase (ALT), lactate dehydrogenase (LDH), and aspartate aminotransferase (AST), as well as to stop the CCl4-induced depletion of reduced glutathione. In mouse hepatocytes, the EtOAc fraction stopped DNA fragmentation. On human breast cancer cells (MCF-7), both HCE and the EtOAc extracts exhibited cytotoxic activity that was concentration dependent.

Kumari (2008) reported that the leaves of *B. pilosa* show anticancer properties. They discovered phenyl-1,3,5-heptatriene against human breast, colon, liver and oral cancer cell lines based on a cytotoxicity-oriented fractional process. Phenyl-1,3,5-heptatriene, on the other hand, demonstrated lower activity against breast cancer.

Wu et al. (2013) examined the cytotoxic effects of *B. pilosa* Linn. *Var. Radiata Sch. Bip* against human tumor cell lines. UV spectrophotometry was used to determine the extracts' total phenolic and flavonoid contents. Using the MTT (thiazolyl blue tetrazolium bromide) method, the extracts' antiproliferative effects on four human tumor cells, namely HepG2, MGC 803, RKO, and MCF-7 were examined. When it came to the tumor cells, the EE-BP showed the most impressive antiproliferative impact.

Yue et al. (2019) examined in vitro cytotoxic effect of a petroleum ether extract of *B. pilosa* on the tumor cell lines, namely A549 (adenocarcinomic human alveolar basal epithelial cells), B16 (murine melanoma), CNE-2 (nasopharyngeal carcinoma), and HepG2 (hepatocellular carcinoma). In all four cell lines, A549 was the most susceptible cell line of the four human cancer cell lines to cytotoxic effects of the petroleum ether extracts. Using the A549-xenograft murine form, the *B. pilosa* extracts' in vivo anticancer activity was investigated.

Free radicals can harm cellular components through a variety of chemical processes (Muchuweti 2007), which can result in the development and advancement of neurological illnesses, cancer, cardiovascular disease, and aging (Chiang et al. 2010). Macrophages can create super-oxide anions, free radicals, and nitric acid to destroy bacteria. On the other hand, an immoderate production of free radicals in pathological settings is linked to a variety of diseases. It is well known that these plants contain a lot of antioxidant phytochemicals.

Chiang et al. (2004) evaluated the antioxidant activity of *B. pilosa* crude extracts, fractions, and compounds to scavenge free radicals by employing hypoxanthine/xanthine oxidase and 1,1-diphenyl-2-picrylhydrazyl (DPPH) tests. The results indicate that the crude extracts such as butanol, ethyl acetate, and water fractions had free radical scavenging activity using the DPPH and hypoxanthine/xanthine oxidase assays.

Deba et al. (2008) investigated the use of GC-MS (Gas chromatography/mass spectrometry) for the analysis of the essential oils extracted from *B. pilosa* flowers and fresh leaves. Out of the 44 components that were found, the primary chemicals found in leaves and flowers were b-caryophyllene and s-cadinene, respectively. The potential antioxidant properties of aqueous extracts and oils from the flowers and leaves were investigated using the b-carotene bleaching techniques and 2, 20 -diphenyl-1-picrylhydrazyl (DPPH). The results showed that the essential oils from flowers and leaves were superior to all aqueous extracts in the DPPH assay, while all extracts and essential oils inhibited the oxidation of linoleic acid in the β-carotene bleaching assay.

14.2.2 *Bidens alba* (Shepherd's Needles)

B. alba, commonly referred to as shepherd's needles, are distributed in Africa, Asia, and tropical and subtropical parts of North and South America (Knope et al. 2020). They are found in gardens, roadside ditches, agricultural fields, and disturbed habitats (Tsuruta et al. 2023). In the United States, *B. alba* is regarded as a grass, since it is an annual or short perennial (Jhala et al. 2013). The *B. alba* leaves are delicious and have been used in traditional medicine (Uludag et al. 2017). Due to the presence of bioactive constituents such as flavonoids, phenolic acids, and saponins, *B. alba* has been studied for its potential as a cancer treatment.

Ong et al. (2008) conducted a study to determine whether the protein extracts from *B. alba* could have anticancer properties against colorectal cancer in humans. When *B. alba* protein extracts were applied to human colorectal cancer SW480 cells, it resulted in notable alterations to DNA and cellular morphologies associated with apoptosis.

14.2.3 *Bidens biternata*

B. biternata (Lour.) Merr. and Sheriff is a common weed in cultivated regions. It is a member of the Asteraceae family. It is used as a medicinal drug to cure various illnesses, such as asthma, colds, cough, diarrhea, and hepatitis. Among the many compounds found in *B. biternata* that are utilized medicinally are terpenoids, phenolic acids, and flavonoids.

Pradeesh (2018) examined the in vitro anticancer activities of crude methanol extract and isolated bioactive compounds of *B. biternata* leaves. To assess the anticancer activity, many cell lines including L929, H9C2, and A375 were used. Zahara et al. (2019) studied the antioxidant and phytochemical properties of *B. biternata* extracts were used. Following a powdered whole plant extraction using methanol and chloroform, both qualitative and quantitative phytochemical screening was performed. Phosphomolybdenum assay, DPPH (2,2-diphenyl-1-picrylhydrazyl), and reducing power were used for antioxidant analysis. This species has considerable antioxidant activity with a known concentration of nutritious components, such as flavonoids and alkaloids, and a negligible concentration of antinutritional components, including phenols and tannic acid.

14.2.4 *Bidens bipinnata* (Spanish Needles Beggartick)

B. bipinnata, also known as Spanish needles or beggartick, is an annual herb that likely originated in eastern Asia and parts of North America. This plant can reach a height of 1 m and produces flowers from September to October, followed by seed ripening in October. The plant is hermaphroditic, meaning it has both male and female organs, and is pollinated by insects. Traditionally, the root and seeds are widely used as a remedy for bronchial and laryngeal disorders, while tea made from the leaves has been used as a vermifuge. The leaves have also been chewed to relieve sore throats, Ear drops made from the styptic plant juice have been utilized (Duke 2008). The plant's extract possesses bactericidal qualities.

Yang et al. (2013) investigated the in vitro antitumor potential from *B. bipinnata* L. extracts. The MTT test was employed to examine the potential inhibitory impact of varying doses of the extracts on human cervical carcinoma (HeLa) and hepatocellular carcinoma (HepG2) cell lines. The extract from *B. bipinnata* L. had a good inhibitory impact on HeLa and HepG2 human cell lines.

14.2.5 *Bidens humilis*

The plant *B. humilis* Kunth (also known as *B. triplinervia* Kunth) is used to treat fever, headaches, and liver ailments. An extract of the flower is used to relieve abdominal and stomach pain.

Saltos et al. (2015) isolated 11 compounds from the aerial parts of *B. humilis* including three new natural flavonoids using a bioassay-oriented approach. Two-dimensional nuclear magnetic resonance and other spectroscopic investigations were used to determine their structures. The antioxidant activity of all these compounds was evaluated using three different assays. In this research, the antioxidant capacity data obtained using different chemical approaches is integrated to apply the Relative Antioxidant Capacity Index (RACI) from a statistical standpoint.

14.2.6 *Bidens segetum*

B. segetum is an annual plant found in the Cerrado biome of some South American states. This plant has been studied for its antioxidant activity.

Franco et al. (2021) investigated the ethanolic extract from *B. segetum* (EEBs) that showed antifungal activity against *Cladosporium sphareospermum* and *C. cladosporiodes* as well as antioxidant capacity when examined using the free radical 2,2-diphenyl-1-picrylhydrazyl (DPPH) assay. Potential antioxidants and 5-O caffeoylquinic acid (5-CQA) were found in the PFFR3.3 subfraction from EEBs. Using the dihydroethidium assay (DHE), it was found that PFFR3.3 did not reduce superoxide anion in metastatic melanoma cells. The polyacetylene combination, known as the PP4 subfraction, exhibited antioxidant and antifungal properties against *Cladosporium*, as evidenced by the decreased levels of superoxide anion in melanoma cells.

14.2.7 *Bidens cernua* and *Bidens frondosa*

In folk medicine, *B. cernua L.* has been used to treat neoplasms, wounds, scrofulous, seborrhea, psoriasis, metabolic disorders, diathesis, and skin ailments. It has also been used as a drug and medicine. In addition to being used as food, *B. frondosa* fresh leaves and stems can help treat colds and upper respiratory infections. The extract is used to treat uterine hemorrhage, respiratory organ bleeding, inflammation, and irritation (Ezuruike and Prieto 2014).

Icoz et al. (2017) evaluated the standardized ethyl alcohol extracts from *B. frondosa L.* and *B. cernua L. var. Radiata* for use in assessing the antidiabetic and antioxidant profile, both in vivo and in vitro. Rats that were streptozotocin-induced, glucose-loaded, and normoglycemic were used to test the extracts' acute hypoglycemia potential. The two extracts demonstrated a significant and promising hypoglycemic impact across all in vivo models examined. The extracts' inhibitory properties against a-glucosidase and a-amylase were also determined. Further, measurements were made of the extracts' total flavonoid content and in vitro antioxidant activity. RP-HPLC-DAD was used to standardize the extracts using markers such as cynaroside and chlorogenic acid.

14.2.8 *Bidens tripartita*

B. tripartita L. or bur-marigold is widely distributed in the humid region of the Northern Hemisphere. The aerial portions of *B. tripartita* are used orally and externally in the treatment of skin diseases, acne, eczema, and psoriasis, as well as in the treatment of bladder, colds, diarrhea, enteritis, fevers, kidney complaints, and inflammation (Al-Snafi 2016; Orhan 2016). Phytochemical studies have revealed that *B. tripartita* contains secondary metabolites such as essential oil, flavonoids, hydroxy coumarins, polyacetylenes, phenolic acids, and sterols (Uysal et al. 2018).

Wolniak et al. (2007) examined the antioxidant activity of the herb and flowers of *B. tripartita* L. (Asteraceae) with the extracted solvents of varying polarity. Pure flavonoids that had been extracted from this plant, including luteolin, cynaroside, and flavanometrane, were evaluated for their antioxidant properties. Using a stable 1,1-diphenyl-2-picrylhydrazyl (DPPH) radical, electron paramagnetic resonance (EPR) spectroscopy was used to assess the radical-scavenging activity. Although the amount of flavonoids in flower heads is half that of the herb, the antioxidant activity was nearly two times higher in the flower extract. The n-BuOH component of several extracts exhibited persistent radical-scavenging activity, and the EPR spectra were captured in a manner that allowed the reaction kinetics to be tracked.

14.3 Conclusion

The genus *Bidens* is an important genus of the Asteraceae family. The traditional applications of this genus in medicine are supported by modern-day science as its species have a wide range of bioactivities. The significant antioxidant and anticancer properties of its species are due to its rich content of phenolic acids, flavonoids, and other bioactive compounds. The presence of compounds like paclitaxel in this genus further underlines its potential as a valuable source for developing antioxidant and anticancer treatments.

References

Al-Snafi AE (2016) Medicinal plants with antimicrobial activities (part 2): plant based review. Sch Acad J Pharm 5(6):208–239

Baiwa FI, Yakubu A, Kanya GU (2018) Nutritional food composition analysis of some traditional foods in Jigawa, Nigeria. Int J Sci Res (IJSR) 7(11):676–680

Chiang Y-M, Chuang D-Y, Wang S-Y, Kuo Y-H, Tsai P-W, Shyur L-F (2004) Metabolite profiling and chemopreventive bioactivity of plant extracts from *Bidens pilosa*. J Ethnopharmacol 95(2–3):409–419

Chiang G, Insel P, Tosun D, Schuff N, Truran-Sacrey D, Raptentsetsang S et al (2010) Hippocampal atrophy rates and CSF biomarkers in elderly APOE 2 normal subjects. Neurology 75(22):1976–1981

Deba F, Xuan TD, Yasuda M, Tawata SJFc. (2008) Chemical composition and antioxidant, antibacterial and antifungal activities of the essential oils from *Bidens pilosa* Linn. var. Radiat Ther 19(4):346–352

Decker-Franco C, Taibo CB, Di Rienzo JA, Alfonso V, Arneodo JD (2021) Comparative pathogenesis of generalist AcMNPV and specific RanuNPV in larvae of Rachiplusia nu (Lepidoptera: Noctuidae) following single and mixed inoculations. J Econ Entomol 114(2):1009–1014

Duke JA (2008) Duke's handbook of medicinal plants of Latin America. CRC Press, Boca Raton

Ezuruike UF, Prieto JM (2014) The use of plants in the traditional management of diabetes in Nigeria: Pharmacological and toxicological considerations. J Ethnopharmacol 155(2):857–924

Hamid A, Aiyelaagbe O, Usman L, Ameen O, Lawal A (2010) Antioxidants: its medicinal and pharmacological applications. Afr J Pure Appl Chem 4(8):142–151

Icoz UG, Orhan N, Altun L, Aslan M (2017) In vitro and in vivo antioxidant and antidiabetic activity studies on standardized extracts of two *Bidens* species. J Food Biochem 41(6):e12429

Jhala AJ, Ramirez AH, Singh M (2013) Tank mixing saflufenacil, glufosinate, and indaziflam improved burndown and residual weed control. Weed Technol 27(2):422–429

Knope ML, Bellinger MR, Datlof EM, Gallaher TJ, Johnson MA (2020) Insights into the evolutionary history of the Hawaiian *Bidens* (Asteraceae) adaptive radiation revealed through phylogenomics. J Hered 111(1):119–137

Kumari A, Kakkar P (2008) Screening of antioxidant potential of selected barks of Indian medicinal plants by multiple in vitro assays. Biomed Environ Sci 21(1):24–29

Kviecinski M, Benelli P, Felipe K, Correia J, Pich C, Ferreira S et al (2011) SFE from *Bidens pilosa* Linné to obtain extracts rich in cytotoxic polyacetylenes with antitumor activity. J Supercrit Fluid 56(3):243–248

Muchuweti M (2007) Phenolic composition and antioxidant properties of some spices. Am J Food Technol 2(5):414–420

Ong P-L, Weng B-C, Lu F-J, Lin M-L, Chang T-T, Hung R-P et al (2008) The anticancer effect of protein-extract from *Bidens alba* in human colorectal carcinoma SW480 cells via the reactive oxidative species-and glutathione depletion-dependent apoptosis. Food Chem Toxicol 46(5):1535–1547

Orhan F (2016) Alleviation of salt stress by halotolerant and halophilic plant growth-promoting bacteria in wheat (*Triticum aestivum*). Braz J Microbiol 47(3):621–627

Pradeesh S (2018) In vitro anticancer activity of *Bidens biternata* (Lour.) Merr. & Sheriff–An ethno medicinal plant of Kerala. Open Access Int J Sci Eng 3(1):15–20

Saltos MBV, Puente BFN, Milella L, De Tommasi N, Dal Piaz F, Braca A (2015) Antioxidant and free radical scavenging activity of phenolics from *Bidens humilis*. Planta Med 81(12/13):1056–1064

Sundararajan P, Dey A, Smith A, Doss AG, Rajappan M, Natarajan S (2006) Studies of anticancer and antipyretic activity of *Bidens pilosa* whole plant. Afr Health Sci 6(1):27–30

Tan PV, Dimo T, Dongo E (2000) Effects of methanol, cyclohexane and methylene chloride extracts of *Bidens pilosa* on various gastric ulcer models in rats. J Ethnopharmacol 73(3):415–421

Tommasi A (1998) Forward modeling of the development of seismic anisotropy in the upper mantle. Earth Planet Sci Lett 160(1–2):1–13

Tsuda T, Osawa T, Ohshima K, Kawakishi S (1994) Antioxidative pigments isolated from the seeds of *Phaseolus vulgaris* L. J Agric Food Chem 42(2):248–251

Tsuruta K, Shidara T, Miyagishi H, Nango H, Nakatani Y, Suzuki N et al (2023) Anti-inflammatory effects of Miyako *Bidens pilosa* in a mouse model of amyotrophic lateral sclerosis and lipopolysaccharide-stimulated BV-2 microglia. Int J Mol Sci 24(18):13698

Uludag A, Aksoy N, Yazlık A, Arslan ZF, Yazmış E, Uremis I et al (2017) Alien flora of Turkey: checklist, taxonomic composition and ecological attributes. NeoBiota 35:61–85

Uysal S, Ugurlu A, Zengin G, Baloglu MC, Altunoglu YC, Mollica A et al (2018) Novel in vitro and in silico insights of the multi-biological activities and chemical composition of *Bidens tripartita* L. Food Chem Toxicol 111:525–536

Wolniak M, Tomczykowa M, Tomczyk M, Gudej J, Wawer I (2007) Antioxidant activity of extracts and flavonoids from *Bidens tripartita*. Acta Pol Pharm 64(5):441–447

Wu J, Wan Z, Yi J, Wu Y, Peng W, Wu J (2013) Investigation of the extracts from *Bidens pilosa* Linn. var. radiata Sch. Bip for antioxidant activities and cytotoxicity against human tumor cells. J Nat Med 67:17–26

Yang Q-H, Yang J, Liu G-Z, Wang L, Zhu T-C, Gao H-L et al (2013) Study onin vitro anti-tumor activity of *Bidens bipinnata* L. extract. Afr J Tradit Complement Altern Med 10(3):543–549

Yue M, Shen H, Li W, Chen J, Ye W, Tian X et al (2019) Waterlogging tolerance of *Bidens pilosa* translates to increased competitiveness compared to native Bidens biternata. Plant Soil 437:301–311

Zahara K, Bibi Y, Qayyum A, Nisa S (2019) Investigation of antimicrobial and antioxidant properties of *Bidens biternata*. Iran J Sci Technol Trans A Sci 43:725–734

Anti-inflammatory and Analgesic Activity of *Bidens* Species

15

Shumaila Ijaz, Banzeer Ahsan Abbasi, Javed Iqbal, Zakir Ullah, Tabassum Yaseen, Mamoona Munir, Sobia Kanwal, Tariq Mahmood, Sayed Afzal Shah, Imran Khan, and Shobana Sampath

---

S. Ijaz
Department of Burn and Plastic Surgery, Shenzhen Institute of Translational Medicine, The First Affiliated Hospital of Shenzhen University, School of Biomedical Engineering, Shenzhen University Medical School, Shenzhen University, Shenzhen, China

B. A. Abbasi (✉) · M. Munir
Department of Botany, Rawalpindi Women University, Rawalpindi, Pakistan
e-mail: banzeer.abbasi@f.rwu.edu.pk; mamoona.munir@f.rwu.edu.pk

J. Iqbal (✉) · T. Yaseen
Department of Botany, Bacha Khan University, Charsadda, Khyber Pakhtunkhwa, Pakistan
e-mail: tabassumyaseen@bkuc.edu.pk

Z. Ullah · T. Mahmood
Department of Plant Sciences, Faculty of Biological Sciences, Quaid-i-Azam University Islamabad, Islamabad, Pakistan
e-mail: zakirullah@bs.qau.edu.pk; tmahmood@qau.edu.pk

S. Kanwal
Department of Biology and Environmental Sciences, Allama Iqbal Open University, Islamabad, Pakistan
e-mail: sobia.kanwal@aiou.edu.pk

S. A. Shah
Department of Biological Sciences (PWD Campus), National University of Medical Sciences (NUMS), Rawalpindi, Pakistan
e-mail: afzal.shah@numspak.edu.pk

I. Khan
Department of Food Science and Technology, The University of Haripur, Haripur, Pakistan
e-mail: imrankhan572@uoh.edu.pk

S. Sampath
Department of Biotechnology, Vel Tech Rangarajan Dr. Sagunthala R&D Institute of Science and Technology, Chennai, Tamil Nadu, India

Y. Bibi et al. (eds.), *The Genus Bidens*,
https://doi.org/10.1007/978-981-96-4257-1_15

Abstract

The genus *Bidens* belongs to the *Asteraceae* family of flowering plants comprising approximately 230 species that are found across the globe. These plants are known by various common names such as beggarticks, tickseeds, blackjack, cobbler's pegs, stickseeds, burr marigolds, Spanish needles, and tickseed sunflowers, which are based on the characteristics of their fruits, many of which are bristly and barbed. *Bidens* species have gained significant attention in recent years due to their potential therapeutic properties, particularly in the fields of anti-inflammatory and analgesic effects. This chapter presents a comprehensive analysis of the phytochemical composition, anti-inflammatory and analgesic properties, mechanisms of action, preclinical studies, clinical trials, therapeutic applications, safety considerations, and future research directions of *Bidens* species. The anti-inflammatory effects of *Bidens* species are manifested through the modulation of key inflammatory mediators and pathways, such as cytokines, prostaglandins, and nitric oxide. These plants have been found to exhibit analgesic properties by inhibiting pain signaling pathways and reducing pain perception. Mechanistic studies have shed light on the intricate pathways through which *Bidens* species exert their anti-inflammatory and analgesic effects, including the modulation of inflammatory enzymes, receptors, and signaling molecules. These mechanisms provide valuable insights into potential therapeutic targets for the identification of novel anti-inflammatory and analgesic agents derived from *Bidens* species. Preclinical studies have further confirmed the anti-inflammatory and analgesic activity of *Bidens* species in various animal models, paving the way for clinical trials to assess their therapeutic potential in humans. The potential therapeutic applications of *Bidens* species extend beyond inflammation and pain, encompassing a wide range of health conditions such as arthritis, diabetes, and cancer. However, it is crucial to carefully evaluate safety and toxicity considerations to ensure the safe utilization of *Bidens* species as medicinal agents. Future research directions for *Bidens* species involve exploring novel extraction methods, identifying specific bioactive compounds responsible for their therapeutic effects, elucidating additional mechanisms of action, and expanding their therapeutic applications in various disease models. *Bidens* species hold great promise as natural anti-inflammatory and analgesic agents, thus justifying the need for continued research in this domain.

Keywords

Bidens species · Anti-inflammatory · Analgesic · Therapeutic · Phytochemical

15.1 Authors Shumaila Ijaz and Banzeer Ahsan Abbasi have equally contributed to this chapter.Introduction

In the past, plants have been essential to the advancement of medicine because of their capacity to produce secondary metabolites that have the potential to have large biological effects (Achika et al. 2014). Traditional medicine employs a variety of plant-based therapies to address a wide range of conditions. The World Health Organization (WHO) reports that more than 80% of people on the planet still use folk and traditional medicine, the majority of which is derived from plant-based cures. Medications derived from plants that are used in traditional medicine are normally less expensive, easier to get, and possess fewer side effects than synthetic drugs. Recently, several traditionally used medicinal plants have been examined using more advanced techniques, which have produced the identification of numerous intriguing chemicals. These substances originating from plants can be utilized to create whole new medications or to modify ones that already exist (Eruygur et al. 2019).

The main source of medications is herbal medicine. Phytochemicals found in medicinal plants play a significant role in the management of certain inflammatory illnesses (Ayertey et al. 2024). For instance, alkaloid compounds produced from *Datura stramonium* and *Berberis vulgaris*, respectively, that have licenses to be marketed as potential anti-inflammatory medicines are hyoscine and berberine (Heinrich et al. 2021). When treating inflammatory agents, it is frequently suggested to employ medicinal plant extracts as an alternative to standard treatment procedures. Many studies have shown that, despite the lack of information about the secondary metabolites of medicinal plants, the majority of people in developing countries treat inflammatory illnesses primarily with medical herbs (Heinrich et al. 2021). It has been noted that a range of inflammatory diseases can be treated by secondary metabolites that have been separated from various parts of medicinal plants (Gonfa et al. 2021). Additionally, a lot of research is being done on the health advantages of natural items derived from food, herbs, and plants, as well as the chemical components that have anti-inflammatory properties in clinical trials and lab settings (Taleb et al. 2016).

Numerous investigations into the anti-inflammatory properties of *Bidens* species have shown encouraging findings. Extracts from the *Bidens* species have been shown by Jiang et al. to have strong anti-inflammatory properties. These effects may be attributed to the regulation of proinflammatory cytokines and pathways involved in inflammation (Jiang et al. 2023). These results were corroborated by Compaore et al., who showed that certain chemicals isolated from the *Bidens* species may reduce inflammatory responses in vitro, indicating that they may have been used as therapeutic agents for inflammatory illnesses (Compaore et al. 2018). Furthermore, several *Bidens* extracts have been shown to have antioxidant qualities by Xin et al. (Xin et al. 2021). This might help explain the extracts' anti-inflammatory benefits by lowering oxidative stress, which is a known cause of inflammation. Even if these results are encouraging, Fotso et al. pointed out that more investigation is required to completely understand the mechanisms of action and evaluate the effectiveness of these extracts in clinical settings (Fotso Fotso et al. 2014). Furthermore, by disrupting inflammatory pathways including the NF-κB signaling pathway, flavonoids and other phenolic compounds found in *Bidens* species contribute to their anti-inflammatory properties. These substances support a balanced immune response by both upregulating the synthesis of anti-inflammatory cytokines and suppressing the expression of inflammatory mediators (Pozharitskaya et al. 2010). Even if the data points to these processes, more investigation is required to completely understand the precise pathways at play and determine whether these results have any practical applications in human health.

Several research has investigated the analgesic effects of several species of *Bidens*, with encouraging findings. For example, research (Fotso Fotso et al. 2014) showed that *Bidens pilosa* has strong analgesic effects, suggesting that it might be used as a natural pain reliever via both central and peripheral pathways. According to another study, extracts from the *Biden*s species showed dose-dependent analgesic effects, indicating that they were effective in treating pain (Ezeonwumelu et al. 2018). Furthermore, bioactive substances including terpenoids and flavonoids which are well-known for their analgesic and anti-inflammatory properties were discovered in these plants according to a phytochemical investigation (Zapata et al. 2017). Though these results are promising, it is important to keep in mind that there is still variation in the potency of various *Biden*s species and that more clinical trials are needed to determine safety and effectiveness in humans (Fig. 15.1; Gutiérrez et al. 1998). Overall, the data points to the possibility of *Bidens* species serving as strong analgesics, which calls for more research into their possible medicinal uses.

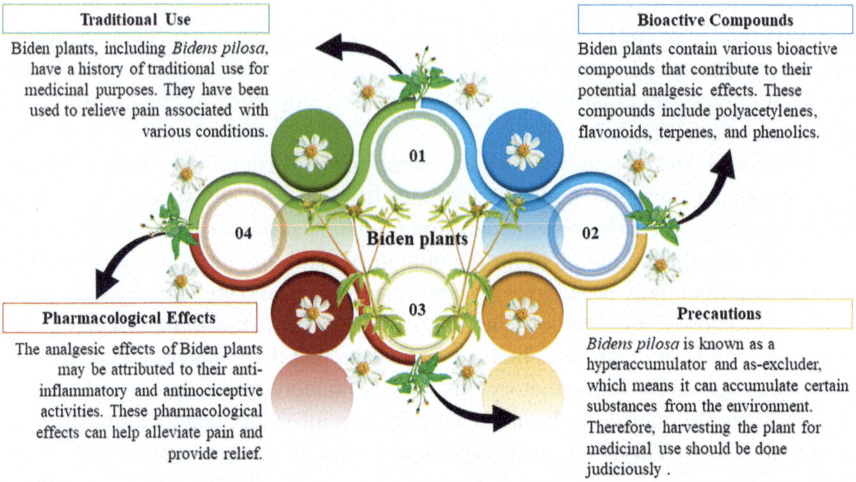

Fig. 15.1 Overview of *Bidens* species (traditional use, bioactive compounds, pharmacological effects, and precautions)

15.2 Anti-inflammatory Properties of *Bidens* Species

Numerous chronic illnesses and disorders, such as diabetes, Parkinson's, Alzheimer's, heart disease, respiratory conditions, kidney disease, liver disease, and cancer, have an intricate underlying mechanism of inflammation. Inflammation is linked to oxidative stress and excessive reactive oxygen species (ROS) generation, which triggers the production and release of proinflammatory cytokines (Lan et al. 2020). There are several natural products on the market right now that have anti-inflammatory properties. Chitosan and its derivatives have a variety of biological qualities, one of which is an anti-inflammatory action; the processes behind this effect are still partially known. Phytochemicals derived from plants have anti-inflammatory qualities. Inflammatory illnesses are treated in part by flavonoids, terpenoids, saponins, tannins, alkaloids, anthraquinones, and essential oils (Gonfa et al. 2023).

Throughout the inflammatory process (Fig. 15.2), host cells produce high levels of reactive oxygen species (ROS), reactive nitrogen species (RNS), cyclooxygenase-1 (COX-1), tumor necrosis factor-α (TNF-α), interleukins (ILs), and cyclooxygenase-2 (COX-2) as inflammatory bio-indicators. Although several models are utilized to evaluate the anti-inflammatory potential of phytochemicals, the two most often used tests for in vitro research were characterized as protein denaturation and erythrocyte membrane integrity (Otunola and Afolayan 2018). An intricate set of signaling pathways controls the amount of inflammation in the body. Drug development has therefore concentrated on the main targets that inhibited, neutralized, or stopped specific products, such as enzymes, by mimicking possible phytochemicals

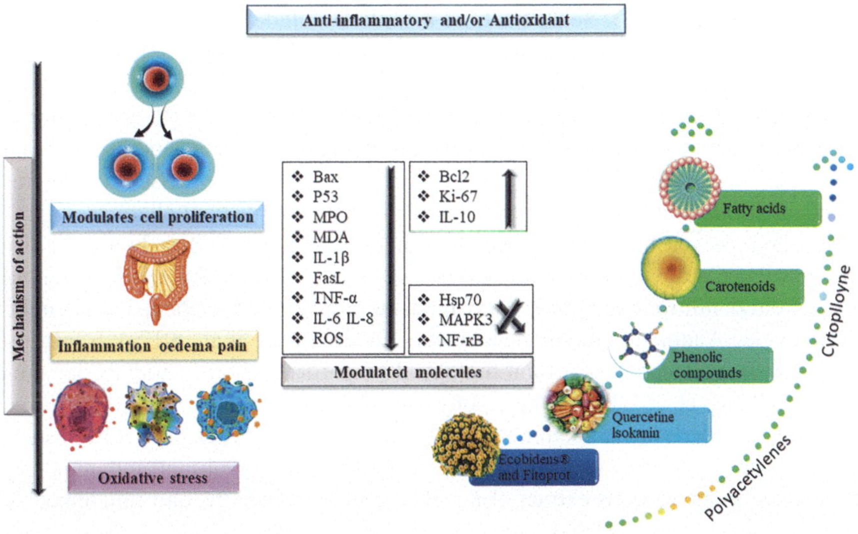

Fig. 15.2 Anti-inflammatory and/or antioxidant mechanism of action along with modulated molecules present in *Bidens* species

that are employed as anti-inflammatory medications (Grando-Lemaire et al. 1999). Nonsteroidal anti-inflammatory drugs (NSAIDs) are medications that are used to treat inflammation. However the primary reason to switch from synthetic medications to low-risk plant-based medications is the growing list of adverse effects associated with these pharmaceuticals, including heart attacks and strokes (Nemudzivhadi et al. 2014).

Worldwide, reports of the effectiveness of medicinal plant extracts as folk remedies for anti-inflammatory drugs have been published (de Cássia da Silveira e Sá et al. 2013). They provide an acceptable foundation for the development of novel anti-inflammatory medications and are abundant sources of phytochemicals used in herbal remedies. There have also been reports of the potent anti-inflammatory properties of isolated substances derived from medicinal plant extracts (Yang et al. 2021). To find possible anti-inflammatory actions of natural compounds of medicinal plants, further ethno pharmacological data is needed. As a result, numerous studies have demonstrated that the type of living cells involved and the distribution method employed both affect the anti-inflammatory properties of phytochemicals and their derivatives. Certain in vitro investigations have confirmed the impact of different doses of medicinal plant extracts and their nanoparticles on tumor cell lines and primary mammalian cells' anti-inflammatory properties (Heleno et al. 2015). For example, due to their structural characteristics, phytochemicals have been shown in vitro to have a suppressive effect when treating inflammatory illnesses. According to Singh et al. (Singh et al. 2018), green M/MONPs have been shown to have anti-inflammatory properties that have a major impact on inflammatory illnesses.

Globally dispersed, the genus *Bidens* (*Asteraceae*) has long been utilized in traditional Asian medicine. *Bidens* is used to treat 40 ailments with all of its parts (Bartolome et al. 2013). Some of these disorders include gastrointestinal hemorrhage, eczema, and antipyretic and analgesic effects. Numerous biological activities, including antioxidative (Wu et al. 2013), anticancer (Kviecinski et al. 2011), antidiabetic (Yang et al. 2013), antimicrobial (Tobinaga et al. 2009), and immunomodulatory activity, have been demonstrated by pharmacological studies involving *Bidens pilosa* L. extract (*B. pilosa* extract) (Chang 2004). The precise active ingredients in *B. pilosa* extract that are in charge of each of its distinct pharmacological activities, however, have not yet been thoroughly investigated. Based on its ability to suppress inflammation, we extracted isookanin, a phytochemical of the flavonoid class, from *B. pilosa* in this study using bioassay-guided fractionation. The phenolic flavonoid isookanin is found in the extract of *Bidens* (Yang et al. 2012a). Isookanin has been shown to possess certain biological properties, such as antidiabetic, antioxidant (Wang et al. 2017), and α-amylase inhibitory (Yang et al. 2012b). The mechanism underlying isookanin's anti-inflammatory effects has not yet been studied. As a result, in this chapter, we identified shokunin as an anti-inflammatory component in an aqueous extract of *B. pilosa* and examined the anti-inflammatory characteristics and related mechanisms of shokunin in LPS-activated macrophages.

Xin et al. (2021) investigated the molecular mechanism of isookanin in the inflammatory response. Their results showed that isookanin reduced the activation of phospho-p38 MAPK and phospho-JNK. These results are in line with a research on the anti-inflammatory effectiveness of genkwanin that was previously published by Gao et al. (2014): AP-1 expression was greatly decreased by genkwanin pretreatment, whereas NF-κB expression was not significantly altered. Additionally, it was demonstrated that genkwanin reduced the levels of LPS-induced phospho-p38 MAPK and phospho-JNK in the MAPK signaling pathway; phospho-ERK1/2, on the other hand, suggested that genkwanin had no effect.

According to a previously published study, these results are likely associated with the increased expression of the stress-responsive protein mitogen-activated protein kinase phosphatase-1 (MKP-1). The MKP-1 protein is located in the nucleus through its N terminus and preferentially dephosphorylates active p38 MAPK and JNK over ERK1/2 (Chi et al., 2006). It is an essential negative regulator of macrophage responsiveness to inflammatory stimuli and inhibits the production of proinflammatory cytokines (Chen et al., 2002). As a result, the varied ways that isookanin regulates MAPK phosphorylation imply that MKP-1 could be the upstream regulator. In further research, we intend to investigate mechanistic studies on the link with MKP-1. As an active anti-inflammatory component, isookanin was extracted from the *B. pilosa* aqueous extract using bioassay-guided fractionation. Our findings demonstrate that in RAW 264.7 cells exposed to LPS, isookanin inhibits the generation of NO and PGE2. Isookanin's inhibitory action was achieved by reducing AP-1 activation, which in turn suppressed the production of COX-2 and iNOS. Isookanin decreased the phosphorylation of p38 JNK and MAPK, which was connected to an anti-inflammatory action through the MAPK signaling pathway (Fig. 15.3).

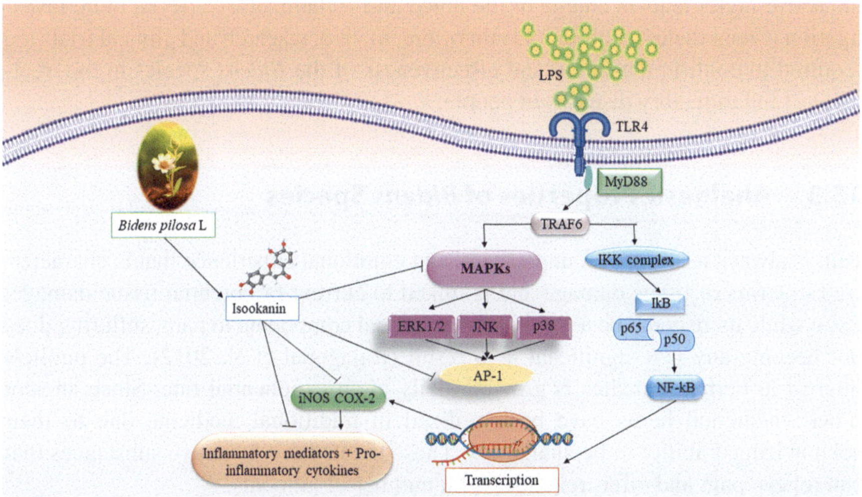

Fig. 15.3 Isookanin, which was derived from *Bidens pilosa* L., has an anti-inflammatory mechanism that involves inhibiting the TLR4-mediated AP-1 and MAPK signaling pathways that are triggered by LPS

Furthermore, the isookanin inhibited the synthesis of TNF-α, IL-6, IL-1β, and IL-8 in THP-1 cells stimulated by LPS. These results imply that isookanin may prove to be a useful therapeutic agent for inflammatory diseases. Nevertheless, in order to confirm isookanin's anti-inflammatory potential, it needs to be tested in vivo using reliable animal models of inflammation (Xin et al. 2021).

The biden species have variety of pharmacological benefits; B *Bidens tripartitus* have recently been discovered to exhibit considerable anti-inflammatory character stics. Furthermore, in a variety of in vitro and in vivo settings, substances originating from the *Bidens* species have demonstrated encouraging anti-inflammatory action, suggesting their possible therapeutic application.

To sum up, the investigation into the anti-inflammatory characteristics of the Genus *Bidens* has yielded significant knowledge about the possible medical advantages of these plants. Several *Bidens* species have been shown to have strong anti-inflammatory qualities after thorough testing and analysis. According to the research findings, flavonoids, phenolic acids, and essential oils which are bioactive substances found in *Bidens* may play a role in the plant's anti-inflammatory properties. These substances have inhibitory effects on important enzymes and mediators of inflammation, which lowers inflammation and its accompanying symptoms. Furthermore, the study emphasizes how diverse the anti-inflammatory activity of various *Bidens* species is. Different species demonstrated varying degrees of anti-inflammatory action, highlighting the significance of species selection in the context of prospective medicinal uses. *Bidens* species have anti-inflammatory qualities that could be useful in the creation of pharmaceutical and natural treatments for inflammatory diseases. Their efficacy, in addition to their low toxicity and negligible side effects, may make them a desirable substitute for traditional anti-inflammatory medications. To clarify the precise mechanisms of action and pinpoint the active ingredients in charge of the noted anti-inflammatory effects, more investigation is nonetheless required. Furthermore, in vivo, research and clinical trials are required to confirm the safety and effectiveness of the *Bidens* species in the treatment of inflammatory diseases in people.

15.3 Analgesic Properties of *Bidens* Species

Pain is always a subjective, unpleasant, and emotional experience that is characterized in terms of tissue damage and is linked to current or potential tissue damage. Even while there could be a significant emotional component to pain, suffering does not become any less significant as a result (Rajagopal et al. 2012). The public's interest in herbal remedies is growing daily at an exponential rate. Since ancient times, medicinal herbs have been utilized in traditional medicine due to their acknowledged ability to be analgesics. These plants have bioactive substances that can relieve pain and offer treatment for a number of ailments.

Due to its analgesic properties, medicinal plants have been utilized in traditional medicine across the globe. They have been used to treat a variety of illnesses associated pain, such as headaches, arthritis, menstrual cramps, and aches in the muscles.

Numerous bioactive substances found in medicinal plants contribute to their analgesic effects. Alkaloids, flavonoids, terpenoids, phenolic acids, and essential oils are some of these chemicals. Numerous therapeutic herbs have been investigated for their analgesic properties. For example, the bark of willow trees is known to have analgesic effects and has been utilized as a natural supply of salicylates, which are the building blocks of aspirin. The analgesic qualities of medicinal plants are attributed to a variety of pharmacological actions. Anti-inflammatory, antioxidant, and antinociceptive properties are among these benefits (Rauf et al. 2022).

Medicinal herbs have a variety of analgesic mechanisms, including interactions with pain receptors, suppression of inflammatory mediators, and neurotransmitter regulation. Additional research is required to determine the clinical efficacy and safety of medicinal plants as analgesics, even though preclinical studies and traditional use offer insightful information. To ascertain the proper dosages and assess their efficacy in humans, clinical trials are required. Traditional medicine has employed medicinal herbs for pain management due to their established analgesic qualities. These plants' analgesic properties are partly attributed to the bioactive chemicals they contain. Preclinical and ethnobotanical research has produced encouraging results, but more clinical trials are required to confirm their safety and effectiveness as human analgesics (Bouyahya et al. 2022).

Bidens species include phytochemicals such as isookanin, quercetin, and other flavonoids that are essential for their anti-inflammatory and analgesic properties. Isookanin, which was extracted from *Bidens pilosa*, has the ability to prevent the generation of cytokines and proinflammatory mediators, suggesting that it may be used to treat inflammatory conditions (Ruiz-Reyes et al. 2024). The plant's analgesic and anti-inflammatory benefits are further enhanced by the presence of quercetin and isookanin in the ethyl acetate fraction of *Bidens pilosa*, which are recognized for their anti-inflammatory and analgesic qualities (Xin et al. 2021). Furthermore, *Bidens* species have a wealth of phenolic compounds and flavonoids that support their potential as anti-inflammatory and antioxidant agents by modulating immune cells, regulating oxidative stress, and controlling the generation of cytokines (Fotso Fotso et al. 2014). Numerous research has demonstrated the phytochemical richness of *Bidens* species, which highlights their medicinal potential in the management of pain and inflammatory disorders.

Numerous research has demonstrated the considerable analgesic qualities of the *Bidens* species. Globally distributed, the genus *Bidens* has long been employed in conventional Asian medicine. *Bidens* is used to treat 40 ailments with all of its parts (Bartolome et al. 2013). Some of these disorders include gastrointestinal hemorrhage, eczema, and antipyretic and analgesic effects. An *Asteraceae* plant called *Bidens pilosa* has long been used to relieve pain. Polyacetylenes, flavonoids, and phenolics are among the substances found in *B. pilosa* that contribute to its analgesic effects. It has been discovered that some substances contain analgesic properties, which can lessen discomfort and pain. Furthermore, the herb has long been utilized in many traditional medical systems for its analgesic properties.

B. pilosa's analgesic qualities can help control pain and offer relief from a number of ailments. It has been demonstrated that the analgesic properties of the

B. pilosa plant lessen mouth pain in HIV/AIDS patients. This analgesic action raises the possibility of using it in the future to treat patients with oral lesions or ulcers. When utilized in oral health products, the analgesic qualities of *Bidens pilosa* L. can aid in lessening oral pain and suffering. Because *B. pilosa* contains natural substances that provide pain relief, people who are experiencing oral pain or discomfort may benefit from its analgesic capability. Because of this plant's analgesic properties, mouthwash formulations may want to include it to help reduce oral discomfort in addition to fighting oral bacteria (Wahjudi et al. 2024). In animal experiments, the ethyl acetate fraction of the methylene chloride/methanol (1:1) extract of *Bidens pilosa* leaves had notable analgesic effects. A number of pain models, such as the acetic acid writhing, hot plate, capsaicin, and formalin-induced pain models, showed analgesic effects. In each of the four nociception models, the extract significantly decreased pain, suggesting that it may have analgesic properties (Fotso Fotso et al. 2014). Even though analgesic drug development has advanced, the long-term use of these medications is hindered by side effects including tolerance or stomach lesions, which are observed with NSAIDs and opiate analgesics, respectively. It is crucial to look for strong analgesics derived from plants that have fewer side effects. The plant *Bidens pilosa* is one such potential. These results support the traditional usage of *Bidens* species in pain relief and offer scientific evidence for their analgesic effects, making them a viable natural pain reliever.

While the analgesic qualities of *Bidens pilosa* are well-known (Ruiz-Reyes et al. 2024), other *Bidens* have comparable effects. Such *Bidens pilosa*, *Bidens frondosa*, also called devil's beggarticks, has been shown to contain bioactive chemicals such as polyacetylenes and flavonoids that have analgesic characteristics. Furthermore, shepherd's needles, or *Bidens alba*, have a long history of use in pain management, suggesting that they may also have analgesic qualities. These *Bidens* species, such as *Bidens pilosa*, are excellent candidates for investigating natural analgesic substitutes for pharmaceutical drugs because they contain a variety of chemicals with biological activities that contribute to their therapeutic qualities (Ezeonwumelu et al. 2018).

According to Al-Snafi (Al-Snafi 2016), *Bidens*, more especially *Bidens tripartitus*, have a variety of pharmacological effects, including analgesic qualities. Furthermore, plants such as *Solanum* spp. have long been utilized for their analgesic and anti-inflammatory properties; flavonol glycosides and hydroxycinnamates are two of the chemicals that contribute to these pain-relieving properties.

The study's findings about the analgesic properties of *Bidens tripartita* are associated with its alcoholic extract's capacity to lengthen rats' latency times in response to thermally painful stimulation in the tail. Given that a higher latency time implies a decreased sensitivity to pain, this supports a pain-relieving effect. According to the study, the BT-alcoholic extract has analgesic effects because long-term use of it increased the latency time response in a statistically significant way when compared to the control group. It's important to note that ketoprofen, a well-known analgesic medication, was discovered to have stronger effects than the BT-alcoholic extract (Lupuşoru et al. 2017).

Studies done in Mexico have demonstrated the analgesic effects of the ethanolic extract of *Bidens odorata* leaves (BOE). The plant generally referred to as "aceitilla" has long been utilized in traditional medicine to treat a variety of ailments, including pain. Prior studies have indicated that BOE has modest levels of toxicity and mild anti-inflammatory properties. Furthermore, it has been discovered to block ATP-sensitive K+ channels in mice, hence suppressing nociception. In a recent study, the neurological activity of BOE was examined in more detail, and the antinociceptive effects of the compound were examined in relation to two widely used analgesics, naproxen, and paracetamol. These results add to the increasing corpus of research on the possible analgesic effects of Mexican medicinal herbs, especially *Bidens odorata.* It is advised to be cautious while using or ingesting *Bidens* species for medical purposes. Although medical uses for *Bidens*, like *Bidens pilosa*, date back centuries, it is important to see a specialist before ingesting any of these plants. Various factors, including individual health conditions, possible drug interactions, and the particular preparation of the plant, may influence the safety and optimum dosage for intake. For specific circumstances, species of *Bidens* can also be applied topically. To guarantee appropriate application techniques and ascertain whether there are any possible hazards or contraindications, it is advisable to speak with medical professionals or herbalists.

Any possible allergies or sensitivity to *Bidens* or related species should be considered serious. Furthermore, accurate plant identification is essential to prevent confusion with closely related but potentially dangerous or deadly species. It is crucial to make sure that *Bidens* species are free of pollutants and pesticides and come from reliable sources before using them. By doing so, possible hazards can be reduced and the plant material's quality can be guaranteed. It is critical to understand that different people react differently to medicinal herbs. One person's definition of safe and effective may differ from another's. Consequently, it is best to begin with modest dosages and keep an eye out for any possible negative effects (Bartolome et al. 2013). It is important to remember that while there is some data supporting the analgesic qualities of some species of *Bidens* plant, more studies are required to confirm their effectiveness and safety as analgesics for people. To assess their efficacy and establish the proper dosages and usage, clinical trials are required.

15.4 Clinical Trials and Evidence-Based Research on Anti-inflammatory and Analgesic of *Bidens* Species

Numerous plant species have been the subject of evidence-based medicine and clinical trials to investigate their potential as anti-inflammatory and analgesics. According to studies, extracts from *Bidens pilosa* and *Cymbopogon citratus* can help control high blood pressure by lowering oxidative stress, enhancing renal function, and preventing increases in heart rate, blood pressure, and lipid profiles (Tcheutchoua et al. 2022). *Bidens odorata* has shown neuropharmacological effects, including synergistic antinociceptive interactions with paracetamol and naproxen, as well as anticonvulsant, antidepressant-like, anxiolytic, and sedative activities

(Alonso-Castro et al. 2020). Extracts from *Bidens pilosa* have demonstrated chemoprotective properties against rats' induction of breast cancer, lowering the development of adenocarcinomas and oxidative stress indicators (Arroyo et al. 2010). The chemical composition and acute toxicity of *Bidens tripartita* have been investigated; the results indicate low toxicity levels and the possibility of using this plant as an alternative therapy. Furthermore, positive immunological responses have been shown by biocompatibility assessments of *Bidens tripartita* extracts in rats, indicating its safe in vivo use (Sandu et al. 2015). Potential adverse effects that should be taken into consideration have been disclosed by the experiments utilizing *Bidens*, notably *Bidens pilosa* and *Bidens tripartita*. Extracts from *Bidens pilosa* have demonstrated acute toxicity in rodents, including adverse inotropic and chronotropic effects on isolated rat hearts, and cardio-inhibition in rabbits. These findings indicate that this herb should be used with caution (Kouakou et al. 2013). However, *Bidens tripartita* has been reported to be mildly toxic, exhibiting symptoms in mice such as anorexia and respiratory distress, even if no histological lesions were seen in the organs that were studied (Sandu et al. 2015). Furthermore, a study on a combination of extracts from *Bidens pilosa* and *Cymbopogon citratus* in rats revealed preventive effects on a number of parameters impacted by high blood pressure, suggesting that when using these plant extracts for medicinal purposes, caution should be taken regarding dosage and potential side effects (Tcheutchoua et al. 2022).

The use of methanolic and ethanolic extracts of the *Bidens* plant, specifically *Bidens pilosa*, has been shown in long-term trials to have significant chemoprotective effects against induced breast cancer in rats, as evidenced by the reduction in mammary adenocarcinoma development and oxidative stress markers (Singh et al. 2017). Furthermore, long-term plant diversity trials have shown that variables like site fertility, richness, and functional group composition can affect plant community C:N ratios; high diversity plots result in a slowdown in temporal variability and a convergence of C:N ratios over time. These results demonstrate the value of long-term experiments with the *Bidens* for studying the dynamics of plant elemental composition in a variety of ecological situations, as well as for cancer research.

15.5 Conclusion

This chapter examines the anti-inflammatory and analgesic potentials of several species of the plant *Bidens*. The chapter summarizes the most recent findings about the impact of *Bidens* species on pain and inflammation, emphasizing their potential as herbal treatments for pain and inflammation. The *Bidens* species is a valuable candidate for therapeutic uses due to its exceptional analgesic and anti-inflammatory characteristics. By clarifying the mechanisms of action of *Bidens*, scientists have been able to understand how these plants reduce pain and inflammation. The anti-inflammatory and analgesic properties of *Bidens* species have been well-supported by preclinical research, opening the door for additional investigation in clinical trials. Growing hope surrounds the promise of *Bidens* species as potent natural treatments for a range of painful and inflammatory illnesses as evidence-based research and clinical trials on these species unfold. Beyond just treating pain and inflammation, *Bidens* have a wide range of potential medical uses, including the treatment of cancer, diabetes, and arthritis. Although there is hope for the therapeutic benefits of *Bidens* species, safety and toxicity concerns must be taken into account in order to guarantee their use as medicinal agents. Thorough evaluation of safety profiles is essential to determine the ideal dosage and reduce any possible side effects, protecting the health of people looking for other forms of treatment. Research on the anti-inflammatory and analgesic effects of *Bidens* species is promising for the field of natural medicine. With a focus on safety and efficacy, *Bidens* show enormous promise to improve healthcare through their numerous medicinal applications, provided that clinical investigations and a thorough knowledge of their mechanisms of action are conducted. Prospective avenues for investigation encompass investigating innovative techniques for extraction, recognizing bioactive substances, comprehending supplementary mechanisms of action, and broadening the scope of therapeutic uses. *Bidens* species have the potential to be effective natural analgesics and anti-inflammatory medicines, and hence, more research in this area is necessary.

References

Achika JI, Arthur DE, Gerald I, Adedayo A (2014) A review on the phytoconstituents and related medicinal properties of plants in the Asteraceae family. IOSR J Appl Chem 7(8):1–8

Alonso-Castro S, Pardo E, Charca L, Pino M, Fernández S, Alperi M et al (2020) Performance of the ASAS health index for the evaluation of spondyloarthritis in daily practice. J Rheumatol 47(10):1483–1489

Al-Snafi AE (2016) Medicinal plants with antimicrobial activities (part 2): Plant based review. Sch Acad J Pharm 5(6):208–239

Arroyo R, Martín V, Maldonado A, Jiménez E, Fernández L, Rodríguez JM (2010) Treatment of infectious mastitis during lactation: antibiotics versus oral administration of Lactobacilli isolated from breast milk. Clin Infect Dis 50(12):1551–1558

Ayertey W, Sharifi A, Yoshida Y (2024) The impact of increase in block pricing on electricity demand responsiveness: Evidence from Ghana. Energy 288:129858

Bartolome AP, Villaseñor IM, Yang W-C (2013) Bidens pilosa L. (Asteraceae): botanical properties, traditional uses, phytochemistry, and pharmacology. Evid Based Complement Alternat Med 2013(1):340215

Bouyahya A, El Hachlafi N, Aanniz T, Bourais I, Mechchate H, Benali T et al (2022) Natural bioactive compounds targeting histone deacetylases in human cancers: recent updates. Molecules 27(8):2568

Chang L (2004) Epidemiology and quality of life in functional gastrointestinal disorders. Aliment Pharmacol Ther 20:31–39

Compaore TR, Soubeiga ST, Ouattara AK, Tchelougou D, Bisseye C, Bakouan DR et al (2018) APOBEC3G expression and HIV-1 infection in Burkina Faso. J Public Health Afr 9(3):907

de Cássia da Silveira e Sá R, Andrade LN, de Sousa DP (2013) A review on anti-inflammatory activity of monoterpenes. Molecules 18(1):1227–1254

Eruygur N, Koçyiğit U, Taslimi P, Ataş M, Tekin M, Gülçin İ (2019) Screening the in vitro antioxidant, antimicrobial, anticholinesterase, antidiabetic activities of endemic Achillea cucullata (Asteraceae) ethanol extract. South Afr J Bot 120:141–145

Ezeonwumelu I, Bártolo I, Martin F, Abecasis A, Campos T, Romero-Severson EO et al (2018) Accidental father-to-son HIV-1 transmission during the seroconversion period. AIDS Res Hum Retroviruses 34(10):857–862

Fotso Fotso A, Mediannikov O, Diatta G, Almeras L, Flaudrops C, Parola P et al (2014) MALDI-TOF mass spectrometry detection of pathogens in vectors: the Borrelia crocidurae/Ornithodoros sonrai paradigm. PLoS Negl Trop Dis 8(7):e2984

Gao J, Sun S-P, Zhu W-P, Chung T-S (2014) Chelating polymer modified P84 nanofiltration (NF) hollow fiber membranes for high efficient heavy metal removal. Water Res 63:252–261

Gonfa FT, Lemu YK, Koricha ZB (2021) Predictors of women's awareness of common non-communicable diseases screening during preconception period in Manna District, Southwest Ethiopia: implication for wellness check-up. BMC Health Serv Res 21:1–10

Gonfa YH, Gelagle AA, Hailegnaw B, Kabeto SA, Workeneh GA, Tessema FB et al (2023) Optimization, characterization, and biological applications of silver nanoparticles synthesized using essential oil of aerial part of Laggera tomentosa. Sustainability 15(1):797

Grando-Lemaire V, Guettier C, Chevret S, Beaugrand M, Trinchet J-C (1999) Hepatocellular carcinoma without cirrhosis in the West: epidemiological factors and histopathology of the non-tumorous liver. J Hepatol 31(3):508–513

Gutiérrez B, Arranz Ma J, Collier DA, Vallès V, Guillamat R, Bertranpetit J et al (1998) Serotonin transporter gene and risk for bipolar affective disorder: an association study in a Spanish population. Biol Psychiatry 43(11):843–847

Heinrich L, Feickert M, Stark G, Cranmer K (2021) pyhf: pure-Python implementation of HistFactory statistical models. J Open Source Softw 6(58):2823

Heleno SA, Barros L, Martins A, Morales P, Fernandez-Ruiz V, Glamoclija J et al (2015) Nutritional value, bioactive compounds, antimicrobial activity and bioaccessibility studies with wild edible mushrooms. LWT Food Sci Technol 63(2):799–806

Jiang Y, Wei C, Jiao Q, Li G, Alyemeni MN, Ahmad P et al (2023) Interactive effect of silicon and zinc on cadmium toxicity alleviation in wheat plants. J Hazard Mater 458:131933

Kouakou AC, N'Guessan FK, Dadie AT, Montet D, Djè MK (2013) Production et commercialisation de l'adjuevan, poisson fermenté de Côte d'Ivoire. Cahiers Agric 22(6):559–567. (1)

Kviecinski M, Benelli P, Felipe K, Correia J, Pich C, Ferreira S et al (2011) SFE from Bidens pilosa Linné to obtain extracts rich in cytotoxic polyacetylenes with antitumor activity. J Supercrit Fluids 56(3):243–248

Lan J, Ge J, Yu J, Shan S, Zhou H, Fan S et al (2020) Structure of the SARS-CoV-2 spike receptor-binding domain bound to the ACE2 receptor. Nature 581(7807):215–220

Lupuşoru CE, Popa EG, Sandu RB, Buca BR, Mititelu-Tarţău L, Lupuşoru RV (2017) The influence of Bidens tripartita extracts on psychomotor abilities and cognitive functions in rats. Farmacia 65(2):284–288

Nemudzivhadi V, Masoko P, Medicine A (2014) In vitro assessment of cytotoxicity, antioxidant, and anti-inflammatory activities of Ricinus communis (euphorbiaceae) leaf extracts. Evid Based Complement Alternat Med 2014(1):625961

Otunola GA, Afolayan A (2018) In vitro antibacterial, antioxidant and toxicity profile of silver nanoparticles green-synthesized and characterized from aqueous extract of a spice blend formulation. Biotechnol Biotechnol Equip 32(3):724–733

Pozharitskaya O, Shikov A, Makarova M, Kosman V, Faustova N, Tesakova S et al (2010) Anti-inflammatory activity of a HPLC-fingerprinted aqueous infusion of aerial part of Bidens tripartita L. Phytomedicine 17(6):463–468

Rajagopal K, Maheswari S, Kathiravan G (2012) Diversity of endophytic fungi in some tropical medicinal plants–a report. Afr J Microbiol Res 6(12):2822–2827

Rauf A, Khalil AA, Khan M, Anwar S, Alamri A, Alqarni AM et al (2022) Can be marine bioactive peptides (MBAs) lead the future of foodomics for human health? Crit Rev Food Sci Nutr 62(25):7072–7116

Ruiz-Reyes J, Heredia-Bobadilla RL, Ávila-Akerberg V, Tejocote-Perez M, Gómez-Ortiz Y, Domínguez-Vega H et al (2024) Assessing functional connectivity and anthropogenic impacts on Ambystoma altamirani populations in Bosque De Agua, Central Mexico. Eur J Wildlife Res 70(5):85

Sandu RE, Buga AM, Uzoni A, Petcu EB, Popa-Wagner A (2015) Neuroinflammation and comorbidities are frequently ignored factors in CNS pathology. Neural Regen Res 10(9):1349–1355

Singh G, Passsari AK, Singh P, Leo VV, Subbarayan S, Kumar B et al (2017) Pharmacological potential of Bidens pilosa L and determination of bioactive compounds using UHPLC-QqQ LIT-MS/MS and GC/MS. BMC Complement Altern Med 17:1–16

Singh R, Upadhyay A, Chandra P, Singh D (2018) Sodium chloride incites reactive oxygen species in green algae Chlorococcum humicola and Chlorella vulgaris: implication on lipid synthesis, mineral nutrients and antioxidant system. Bioresour Technol 270:489–497

Taleb H, Maddocks SE, Morris RK, Kanekanian AD (2016) Chemical characterisation and the anti-inflammatory, anti-angiogenic and antibacterial properties of date fruit (Phoenix dactylifera L). J Ethnopharmacol 194:457–468

Tcheutchoua YC, Bilanda DC, Mengue Ngadena YS, Djomeni Dzeufiet PD, Owona PE, Goufani RBÁ et al (2022) Acute and subchronic toxicity studies on the aqueous extract of the plant mixture (Bidens pilosa and Cymbopogon citratus aerial parts) in rat model. J Toxicol 2022(1):1998433

Tobinaga S, Sharma MK, Aalbersberg WG, Watanabe K, Iguchi K, Narui K et al (2009) Isolation and identification of a potent antimalarial and antibacterial polyacetylene from Bidens pilosa. Planta Med 75(06):624–628

Wahjudi E, Armanu A, Hadiwidjojo D, Solimun S (2024) Lecturer performance in focus: An extensive systematic literature review and analysis. SA J Human Resour Manag 22:2477

Wang X, Shen Y, Wang S, Li S, Zhang W, Liu X et al (2017) PharmMapper 2017 update: a web server for potential drug target identification with a comprehensive target pharmacophore database. Nucleic Acids Res 45(W1):W356–WW60

Wu J, Wan Z, Yi J, Wu Y, Peng W, Wu J (2013) Investigation of the extracts from Bidens pilosa Linn. var. radiata Sch. Bip for antioxidant activities and cytotoxicity against human tumor cells. J Nat Med 67:17–26

Xin S, Liu G, Ma X, Gong J, Ma B, Yan Q et al (2021) High efficiency heterogeneous Fenton-like catalyst biochar modified CuFeO2 for the degradation of tetracycline: economical synthesis, catalytic performance and mechanism. Appl Catal B Environ 280:119386

Yang Y, Hu C, Abu-Omar MM (2012a) Conversion of glucose into furans in the presence of AlCl3 in an ethanol–water solvent system. Bioresour Technol 116:190–194

Yang X-W, Huang M-Z, Jin Y-S, Sun L-N, Song Y, Chen H-S (2012b) Phenolics from Bidens bipinnata and their amylase inhibitory properties. Fitoterapia 83(7):1169–1175

Yang Q-H, Yang J, Liu G-Z, Wang L, Zhu T-C, Gao H-L et al (2013) Study onin vitro anti-tumor activity of Bidens bipinnata L. extract. Afr J Tradit Complement Altern Med 10(3):543–549

Yang X, Bai Z-F, Zhang Y, Cui H, Zhou H-L (2021) Flavonoids-rich extract from Bidens bipinnata L protects pancreatic β-cells against oxidative stress-induced apoptosis through intrinsic and extrinsic pathways. J Ethnopharmacol 275:114097

Zapata D, Salazar-Gutierrez M, Chaves B, Keller M, Hoogenboom G (2017) Predicting key phenological stages for 17 grapevine cultivars (Vitis vinifera L). Am J Enol Vitic 68(1):60–72

Other Bio Activities of Genus *Bidens* Species

16

Tomi Lois Adetunji, Olalekan Bukunmi Ogunro, Olamide Omolafe Ogunremi, and Olumayowa Rebecca Andrew

Abstract

Genus *Bidens* is an important genus of the Asteraceae family and is widely distributed in tropical and subtropical regions. The members of this genus have been used in traditional systems of medicine, i.e., Ayurveda, African, and Chinese medicine. Their plants are known to heal wounds, cure snake bites, treat the common cold, fever, etc. Upon evaluation, their extracts have demonstrated antihyperglycemic effects, potentially aiding in managing diabetes. They also possess antihypertensive activity, antimalarial activity, immunomodulatory activity, uterotonic effects, wound healing activity, and antiulcer activity. The future of the *Bidens* species lies in rigorous scientific exploration, collaboration, and responsible utilization. Its diverse bioactivities offer exciting possibilities for healthcare and well-being. However, further research is essential to fully understand its mechanisms and therapeutic potential.

T. L. Adetunji (✉)
Unit for Environmental Sciences and Management, Faculty of Natural and Agricultural Sciences, North-West University, Potchefstroom, South Africa

O. B. Ogunro
Drug Discovery and Pharmacology Laboratory, Department of Biological Sciences, KolaDaisi University, Ibadan, Nigeria

O. O. Ogunremi
Institute for Sustainable Development (I4SD), Abiola Ajimobi Technical University, Ibadan, Oyo, Nigeria

O. R. Andrew
Research and Development Department, Kellogg Tolaram Nigeria Limited, Surulere, Lagos, Nigeria

Y. Bibi et al. (eds.), *The Genus Bidens*,
https://doi.org/10.1007/978-981-96-4257-1_16

Keywords

Bidens · Antimalarial · Antihypertensive · Wound · Uterotonic

16.1 Antidiabetic Activity

A total of 1200 plant species of the *Bidens* genera have been examined for their ability to treat diabetes (Lin et al. 2020). Multiple studies have indicated that *B. pilosa* has the potential to effectively cure both type 2 and type 1 diabetes in animals. Type 1 Diabetes (T1D) is distinguished by the immune system's attack on pancreatic β-cells, leading to high blood sugar levels (Baiwa et al. 2018). As of now, there is no known remedy for T1D. The differentiation of T helper (Th) cells into distinct phenotypes is crucial in the development of T1D. In this context, the reduction of Th1 cells and the activation of Th2 cells have a positive effect in mitigating the disease (Chang et al. 2001). Studies on *B. pilosa* have discovered potential chemicals for the treatment of T1D.

Studies have demonstrated that the butanol component of *B. pilosa* can hinder the growth of T-cells and reduce the presence of Th1 cells and cytokines, while simultaneously promoting the presence of Th2 cells and cytokines (Chang 2004). Three specific polyynes, namely cytopiloyne, 3-β-D-Glucopyranosyl-1-hydroxy-6(E)-tetradecene-8,10,12-triyne, and 2-β-D-Glucopyranosyloxy-1-hydroxy-5(E)-tridecene-7,9,11-triyne (Chang et al. 2007), which were extracted from *B. pilosa*, have demonstrated noteworthy efficacy in preventing the development of T1D in animal models. Furthermore, cytopiloyne from *B. pilosa* exhibited strong anti-T1D activity by suppressing the development of naïve $CD4^+$ T cells into Th1 cells and encouraging their differentiation into Th2 cells. The impact of cytopiloyne therapy on the development of T1D was confirmed through in vivo investigations, which demonstrated a considerable delay and decrease in the occurrence of the disease in NOD mice. Cytopiloyne, through the Fas ligand route, mechanistically triggers the production of GATA-3, a key regulator of Th2 cell development, and induces the death of invading $CD4^+$ T cells in pancreatic islets (Chang et al. 2001).

In addition, the administration of cytopiloyne from *B. pilosa* has been reported to reduce the levels of serum IFN-γ and IgG2c, which are linked with the Th1 immune response, and an increase in the levels of serum IL-4 and IgE, which are associated with the Th2 immune response. Significantly, cytopiloyne specifically reduced the number of $CD4^+$ T cells in NOD mice, while having no impact on $CD8^+$ T cells. This emphasizes the distinct immunomodulatory effects of cytopiloyne in T1D. To summarize, *B. pilosa* and its active component cytopiloyne have great potential as alternatives for treating T1D. They achieve this by influencing the development of Th cells and reducing autoimmune reactions against pancreatic β-cells. These findings indicate a possible treatment approach for controlling T1D and justify the need for additional investigation in clinical environments (Chiang et al. 2007).

*Bidens pilos*a, when examined for its potential in treating both T1D and Type 2 Diabetes (T2D), demonstrated notable therapeutic results in animal models. The polyyne compound cytopiloyne is a strong inhibitor of T-cell proliferation in individuals with T1D when activated by IL-2 with Con A or anti-CD3 antibody. Cytopiloyne inhibits the differentiation of Th1 cells, enhances the differentiation of Th2 cells, and partially reduces the number of $CD4^+$ T cells, therefore delaying the beginning of T1D in NOD mice. These effects were accomplished by utilizing processes that involve the induction of death in invading $CD4^+$ T lymphocytes in pancreatic islets through the Fas ligand (Chang et al. 2007). Conversely, research conducted on T2D in db/db mice has shown notable antidiabetic properties of *B. pilosa* extracts and polyynes, particularly compounds 3-*β*-*D*-Glucopyranosyl-1-hydroxy-6(*E*)-tetradecene-8,10,12-triyne and 2-β-D-Glucopyranosyloxy-1-hydroxy-5(E)-tridecene-7,9,11-triyne. These substances decreased blood glucose levels, diminished meal consumption, and enhanced insulin secretion. Specifically, when a precise ratio of compounds was combined, it caused major reduction of glucose levels and inhibition of hunger in mice that were fed with the mixture. In addition, aqueous extracts of *B. pilosa* increased insulin production, boosted glucose tolerance, and decreased levels of glycosylated hemoglobin (HbA1c) in db/db mice (Ubillas et al. 2000).

In addition, different varieties of *B. pilosa* showed a decrease in postprandial glucose levels and a rise in serum insulin levels in db/db mice, with the effect being depending on the dosage. Out of these different versions, the BPR extract that had higher concentrations of cytopiloyne was very successful in reducing blood glucose levels and enhancing insulin levels (Chien et al. 2009a). Cytopiloyne was discovered to boost insulin secretion and expression in pancreatic β-cells through the calcium/DAG/PKCα cascade mechanism. Nevertheless, it failed to reduce blood glucose levels in mice treated with streptozotocin (STZ), suggesting that its effectiveness relies on the presence of intact β-cells (Chang et al. 2015). These studies demonstrate the potential of *B. pilosa* and its polyynes, particularly cytopiloyne, as attractive candidates for treating both T1D and T2D. They provide insights into the mechanisms of action and therapeutic effects of these substances in managing diabetes.

Furthermore, the sub-extracts of *B. tripartita* that contained n-butanol and ethyl acetate showed strong antidiabetic properties according to in vivo bioactivity-guided fractionation. Additionally, these sub-extracts exhibited potent in vitro enzyme inhibitory properties. Total phenol concentration, flavonoid content, and radical scavenging activity were all highest in the ethyl acetate sub-extract. The primary components identified by LC-MS studies were cynaroside, luteolin, and chlorogenic acid. These results imply that *B. tripartita* has a strong potential to prevent diabetes and that the active ingredients in the plant may help treat diabetes and its after effects (Orhan et al. 2016).

Another study evaluated the aqueous leaf extract from *B. gardneri* as an antidiabetic agent. The extract inhibited disaccharidases and advanced glycation end

Table 16.1 Chemical composition of the genus *Bidens* and their associated biological properties

Name	Classification	Molecular formula	Biological activities	References
(3E,5E,11E)-tridecatriene-7,9-diyne-1,2,13-triol-2-O-β-D-glucopyranoside	Polyacetylene glucoside	–	Anti-inflammatory	Li et al. (2008)
(R)-1,2-dihydroxytrideca-3,5,7,9,11-pentayne	Polyyne	$C_{13}H_8O_2$	Antimalarial and antibacterial	Tobinaga et al. (2009)
1,2-Dihydroxytrideca-5,7,9,11-tetrayne	Polyyne	$C_{13}H_{12}O_2$	Anti-angiogeneic	Wu et al. (2013)
1,2-Dihyroxy-5(E)-tridecene-7,9,11-triyne	Polyyne	$C_{13}H_{14}O_2$	Anti-angiogeneic	Tadesse et al. (1995)
			Antiproliferative	Tadesse et al. (1995)
1,3-Dihyroxy-6(E)-tetradecene-8,10,12-triyne	Polyyne	$C_{14}H_{16}O_2$	Anti-angiogeneic	Wu et al. (2013)
1-Phenyl-1,3-diyn-5-en-7-ol-acetate	Polyyne	$C_{15}H_{12}O_2$	Antimalarial	Pereira et al. (1999)
1-Phenylhepta-1,3,5-triyne	Polyyne	$C_{13}H_8$	Antimicrobial	Wang et al. (2023)
			Antimalarial	Karis and Ryding (1994)
			Cytotoxic	
			Antifungal	Rybalchenko et al. (2010)
2′-butoxyethylconiferin	Phenylpropanoid glucoside	–	Anti-inflammatory	Le et al. (2015)
2-O-β-Glucosyltrideca-11(E)-en-3,5,7,9-tetrayn-1,2-diol	Polyyne	$C_{19}H_{20}O_7$	Immunosuppressive and anti-inflammatory	Tan et al. (2000)
2-β-D-Glucopyranosyloxy-1-hydroxy-5(E)-tridecene-7,9,11-triyne	Polyyne	$C_{19}H_{24}O_7$	Antidiabetic	Dimo et al. (2003)
			Anti-inflammatory	Nguelefack et al. (2006)
			Antimalarial and antibacterial	Tobinaga et al. (2009)
2-β-D-Glucopyranosyloxy-1-hydroxytrideca-5,7,9,11-tetrayne	Polyyne	$C_{19}H_{22}O_7$	Antidiabetic	Champagnat (1951)
			Anti-inflammatory	Chiang et al. (2004)
3,4-Di-O-caffeoylquinic acid	Phenylpropanoid	$C_{25}H_{24}O_{12}$	Antiviral	
			Antioxidant	
3,5-Di-O-caffeoylquinic acid	Phenylpropanoid	$C_{25}H_{24}O_{12}$	Antiviral	
			Antioxidant	

3-O-Rabinobioside	Saccharide	$C_{27}H_{38}O_{15}$	Antioxidant	Chiang et al. (2004)
3-β-D-Glucopyranosyl-1-hydroxy-6(E)-tetradecene-8,10,12-triyne	Polyyne	$C_{20}H_{26}O_{7}$	Antidiabetic	Dimo et al. (2003)
			Anti-inflammatory	Deba et al. (2008)
3, 6, 8-trichloro-5, 7, 3′, 4′-tetrahydroxyflavone	Flavonoid	–	α-Amylase inhibitor	Yang et al. (2012a)
4-O-(2″,3″-O-diacetyl-6″-O-p-coumaroyl-β-D-glucopyranosyl)-p-coumaric acid	Biscoumaric acid	–	α-Amylase inhibitor	Seelinger et al. (2008)
8,3′,4′-trihydroxyflavone-7-O-(6′‘-O-p-coumaroyl)-β-D-glucopyranoside	Flavonoid glycoside	$C_{21}H_{20}O_{11}$	Anti-inflammatory	Le et al. (2023)
8,3′,4′-trihydroxyflavone-7-O-β-D-glucopyranoside	Biscoumaric acid	$C_{21}H_{20}O_{11}$	α-Amylase inhibitor	Yang et al. (2012b)
4,5-Di-O-caffeoylquinic acid	Phenylpropanoid	$C_{25}H_{24}O_{12}$	Antiviral	Chiang et al. (2004)
			Antioxidant	Chiang et al. (2004)
Butein	Flavonoid	$C_{15}H_{12}O_{5}$	Anti-leishmanial	
			Cytotoxic	Seelinger et al. (2008)
Centaureidin	Flavonoid	$C_{18}H_{16}O_{8}$	Anti-listerial	Chang et al. (2000)
			Cytotoxic	Chang et al. (2000)
Centaurein	Flavonoid	$C_{24}H_{26}O_{13}$	Anti-listerial	Chang et al. (2000)
			Cytotoxic	Pattee (2002)
			Antiviral	Verma et al. (2001)
Chlorogenic acid	Phenolic	$C_{16}H_{26}O_{9}$	Antioxidant	
Ethyl caffeate	Phenylpropanoid	$C_{11}H_{12}O_{4}$	Anti-inflammatory	Chiang et al. (2004)
Heptanyl 2-O-β-xylofuranosyl-(1 → 6)-β-glucopyranoside	Miscellaneous	$C_{18}H_{44}O_{10}$	Antioxidant	
Heptadeca-2E,8E,10E,16-tetraene-4,6-diyne	Polyacetylenes	–	Anti-inflammatory	Redl and Erisman (1994)
Heptadeca-2E,8E,16-triene-4,6-diyne-10-ol	Polyacetylenes	–	Anti-inflammatory	
Heptadeca-2E,8Z,10E,16-tetraene-4,6-diyne	Polyacetylenes	–	Anti-inflammatory	

(continued)

Table 16.1 (continued)

Name	Classification	Molecular formula	Biological activities	References
Jacein	Flavonoid	$C_{24}H_{26}O_{13}$	Antioxidant	Chiang et al. (2004)
Linoleic acid	Fatty acid	$C_{18}H_{32}O_2$	Antiviral	Xia et al. (2009)
			Cytotoxic	
Luteolin	Flavonoid	$C_{15}H_{10}O_6$	Antiviral	Gachet et al. (2010)
			Cytotoxic	Kumari and Kakkar (2008)
			Anti-inflammatory	Tewtrakul et al. (2003)
			Anti-allergic	Nielsen (1986)
Phenylheptatriyne	Hydrocarbons	$C_{13}H_8$	Antifungal	Rybalchenko et al. (2010)
Quercetin 3,3′-dimethyl ether 7-O-α-L-rhamnopyranosyl-(1 → 6)-β-D-glucopyranoside	Flavonoid	$C_{21}H_{20}O_{11}$	Antimalarial	Andrade-Neto et al. (2008)
Quercetin 3,3′-dimethyl ether-7-O-β-D-glucopyranoside	Flavonoid	$C_{21}H_{20}O_{12}$	Antimalarial	Andrade-Neto et al. (2008)
Quercetin 3-O-rutinoside	Flavonoid	$C_{27}H_{38}O_{15}$	Antioxidant	Chiang et al. (2004)
Quercetin 3-O-β-D-galactopyranoside	Flavonoid	$C_{21}H_{20}O_{12}$	Anti-inflammatory	Nielsen (1986)
Safynol	Polyacetylenes	$C_{13}H_{12}O_2$	Anti-inflammatory	Redl and Erisman (1994)
Safynol-2-O-isobutyrate	Polyacetylenes	–	Anti-inflammatory	Redl and Erisman (1994)

products (AGEs) and was high in phenolics and 4-O-caffeoylquinic acid. In rat models of diabetes and oral glucose tolerance tests, the leaf extract (300 mg/kg) markedly decreased blood glucose levels and inhibited weight loss. It also prevented hyperglycemia in rats fed with a hypercaloric diet. According to these results, aqueous leaf extract from *B. gardneri* may be useful as a diabetes treatment (Felix et al. 2020; Table 16.1).

Bidens pilosa, in comparison with other plant extracts, demonstrated antidiabetic properties in reducing hyperglycemia. *B. pilosa* leaf extract effectively normalized diabetic-related electrolyte imbalances, similar to the conventional drug glibenclamide, by restoring blood glucose levels. In contrast, the other leaf extracts only moderately improved these metabolic issues. Thus, *B. pilosa* leaf extract shows significant antidiabetic activity (Langat et al. 2021).

16.2 Antidiarrheal Properties

Diarrhea is a prominent contributor to mortality among children below the age of five, and conventional medications have the potential to offer cost-effective treatments (Ogunro et al. 2023). Researchers assessed the efficacy of freeze-dried *B. biternata*, a plant commonly utilized in Ayurveda medicine, in treating diarrhea. The study utilized both in vivo and in vitro methods to evaluate the effects of *B. biternata* extracts on castor-oil-induced diarrhea, intestinal motility, enteropooling, and jejunal contractions in rats. The findings demonstrated that *B. biternata* effectively decreased fecal production, slowed down intestinal motility, and reduced intestinal fluid volume. The most notable effectiveness was reported at a dosage of 400 mg/kg. In addition, the extract reduced contractions in the jejunum, and these effects were altered by specific adrenergic and opioid blockers. The results indicate that *B. biternata* has notable antidiarrheal effects by regulating both the movement and secretion in the intestines, confirming its traditional medicinal use (Kinuthia 2017) (Table 16.1).

Another study investigated the antidiarrheal properties of chloroform, methanol, and aqueous extracts of *B. odorata* in mice with castor-oil-induced diarrhea. The chloroform extract (CBO) decreased the occurrence of diarrhea by 72.72% and suppressed the contraction of the ileum and the movement of substances through the intestines in rats. The active fraction of CBO (F4) effectively decreased diarrhea induced by castor oil and arachidonic acid but did not have the same effect on diarrhea caused by prostaglandin E2. FR5, the most potent subfraction, suppressed castor-oil-induced diarrhea by 92.1%. The GC-MS analysis detected the presence of oleic, palmitic, linoleic, and stearic acids in F4. The results corroborate the traditional usage of *B. odorata* as an antidiarrheal agent (Zavala-Mendoza et al. 2013).

16.3 Antihypertensive Activity

Long-term ingestion of fructose in rats commonly increases blood pressure, leading to the development of insulin resistance and hyperinsulinemia. A study investigated the effects of *B. pilosa* leaf methanolic extract on blood pressure and metabolic parameters in rats that were fed a diet high in fructose. Following a period of 3–6 weeks during which Wistar rats were exposed to a solution containing 10% fructose, there was a rise in plasma insulin levels, cholesterol levels, and systolic blood pressure (SBP). The extract of *B. pilosa* had preventive effects against hypertension and decreased high blood pressure and insulin levels. These data indicate that the antihypertensive benefits of *B. pilosa* may be attributed, at least in part, to enhanced insulin sensitivity (Dimo et al. 2002).

16.4 Antimalarial Activity

An investigation into the effectiveness of leaf extracts from *B. pilosa* in treating malaria utilized a mix of phytochemistry and bioassays. The study revealed that (*R*)-1,2-dihydroxytrideca-3,5,7,9,11-pentayne had antimalarial action against the *P. falciparum* NF54 strain of the parasite, with an IC_{50} value of 6.0 μg/mL (Kumari and Kakkar 2008). Furthermore, (*R*)-1,2-dihydroxytrideca-3,5,7,9,11-pentayne, extracted from the above-ground portions of *B. pilosa*, effectively suppressed the proliferation of the *P. falciparum* FCR-3 strain, with an IC_{50} value of 0.35 μg/mL. The in vivo effect of this chemical was evaluated in mice infected with the *P. berghei* NK-65 strain. The findings indicated that it reduced the average parasitemia in the RBC by 20.7%, decreasing it from 32.8% (in comparison to the control) to 12.1% after administering an intravenous injection of 0.8 mg/kg body weight/day for 4 days (Tobinaga et al. 2009). Additional research is required to investigate the antimalarial mechanism of both polyynes and clinical investigations.

The in vitro efficacy of crude root extracts made using 80% ethanol, against *Plasmodium falciparum* is attributed to the presence of polyacetylene and flavonoids. Following the separation of these substances using chromatography, both the extract and fractions underwent analysis and testing. In vivo, the ethanol extract, which includes both polyacetylene and flavonoids, decreased parasitemia by 36% on day five and 29% on day seven. The ether: methanol fraction exhibited a 38% reduction in parasitemia on day five, but its efficacy diminished by day seven. In summary, the ethanol extract's ability to increase survival rates can be attributed to the presence of both polyacetylene and flavonoids, which contribute to its effectiveness against malaria (De Oliveira and Vitória 2011).

The antimalarial research conducted on *B. pilosa* and other *Bidens* species demonstrated that both crude ethanol extracts and specific fractions from *B. pilosa* effectively suppressed the growth of *Plasmodium falciparum* by as much as 90% in laboratory conditions. The multiplication of parasites was suppressed by 65–91% when ethanol extracts from nine different *Bidens* species from outside Brazil were used. The active compounds discovered in the chloroform root fractions were phenyl acetylene and flavonoids, with a particular emphasis on 1-phenyl-1,3-diyn-5-en-7-ol-acetate. Species that had a greater abundance of aliphatic acetylenes demonstrated a higher level of effectiveness in combating malaria, whereas species with a lower quantity or absence of acetylenes were shown to be less potent (Brandão et al. 1997).

Species belonging to the genus *Bidens* that possess 6–14 aliphatic acetylenes exhibit significant levels of activity. Nevertheless, the extracts of *B. biternata*, which consist of only 3 acetylenes, demonstrated a mere 38% suppression of *Plasmodium falciparum* development in vitro. This suggests that the extracts are either inert or possess limited effectiveness in this particular setting (Zahara et al. 2015).

16.5 Immunomodulatory Activity

According to reports, *B. pilosa* is an immunomodulatory plant that effectively treats immunological illnesses, including allergies, arthritis, and T1D (Chang 2004; Chang et al. 2007) (Horiuchi and Seyama 2006).

The immunomodulatory effects of *B. pilosa* were investigated through the use of phytochemicals and T-cell activation tests. The study emphasized the importance of IFN-γ, a vital cytokine that is generated by T- and NK cells. It has shown that hot water extracts of *B. pilosa* enhanced the activity of the IFN-γ promoter, namely through the butanol fraction. The presence of centaurein and centaureidin in this fraction led to a considerable increase in the activity of the IFN-γ promoter. The mechanism of Centaurein involves the NFAT and NFκB pathways, which control the transcription of IFN-γ (Chang et al. 2007). Furthermore, extracts of *B. pilosa* hindered the process of transforming naïve $CD4^+$ helper T (Th0) cells into Th1 cells, instead favoring Th2 development. Cytopiloyne, a type of polyynes found in *B. pilosa*, inhibits the production of Th1 cytokines and promotes the production of Th2 cytokines, which is responsible for the plant's anti-inflammatory properties (Chang 2004).

In addition, Chang and colleagues demonstrated that the butanol extract of *B. pilosa* has an impact on autoimmune diabetes and airway inflammation in rats. This extract improved Th1-mediated diabetes by preventing β-cell death and the infiltration of leukocytes but worsened inflammation related to asthma. Both effects demonstrate *B. pilosa*'s capacity to alter the Th1/Th2 equilibrium (Chien et al. 2009b). Further investigation revealed that cytopiloyne reduces the production of IFN-γ and enhances the production of IL-4 in T-cells, elucidating its antidiabetic properties. Cytopiloyne regulates the development of T-cells and the release of cytokines, protecting against type 1 diabetes by shifting the balance from Th1 to Th2 polarization and decreasing immunological responses in the pancreas (Chiang et al. 2007). Cytopiloyne directly increases the expression of IL-4 in T-cells by upregulating GATA-3, while indirectly decreasing the expression of IFN-γ. Additionally, it stimulates the expression of Fas ligand in pancreatic β-cells and hinders the proliferation of T-cells, offering a comprehensive strategy for safeguarding against T1D and other autoimmune disorders mediated by Th1 cells (Abbas 1996).

The immunological regulation of *B. pilosa* is contingent upon its phytochemical makeup. The effects of different chemicals can vary, as seen by contradictory accounts on the exacerbation or improvement of allergies caused by their extracts (Horiuchi and Seyama 2006). The fraction of butanol comprises flavonoids that stimulate the expression of IFN-γ and polyynes that stimulate the expression of IL-4 and suppress the expression of IFN-γ in developing T-cells (Chang 2004). An essential aspect of comprehending the therapeutic characteristics of a substance is the identification of phytochemicals that are associated with its structure and bioactivity.

The methanolic extract of *B. pilosa* leaves and the polyacetylene PA-1 isolated from it exhibit significant immunomodulatory effects. The extract inhibits the proliferation of human and murine lymphocytes in vitro, with PA-1 being ten times more potent. In vivo, intraperitoneal administration of the extract in mice significantly reduced the size of the popliteal lymph node after zymosan-induced inflammation. These findings suggest that components of *B. pilosa* possess immunosuppressive properties, which may underlie its traditional use as an anti-inflammatory agent (Pereira et al. 1999).

16.6 Hypotensive and Vasodilatory Activities

The hypotensive properties of *B. pilosa* were examined by researchers utilizing different rat models and extracts (Dimo et al. 2002). The crude extract of *B. pilosa* exhibited a more pronounced reduction in systolic blood pressure in hypertensive rats (both salt-loaded and spontaneous hypertension) compared to normotensive rats. The extract was found to have an impact on the levels of urine sodium and potassium ions; however, the effect was not statistically significant. As a result, it was suggested that *B. pilosa* achieves its hypotensive effects by causing vasodilation (Dimo et al. 1999).

Subsequent research examined the effects of water and methylene chloride extracts from *B. pilosa* leaves on fructose-induced hypertensive rats. The findings revealed that both extracts could reduce blood pressure without affecting insulin levels. Intravenous administration of a neutral extract of *B. pilosa*, which is a combination of methanol and methylene chloride, resulted in a two-phase decrease in systolic blood pressure in both normal and hypertensive rats. The study also suggested that the hypotensive effects of extract of *B. pilosa* might be disrupted by atropine and propranolol, indicating a mechanism that involves the efficiency of the heart pump and the phases of vasodilation (Dimo et al. 2003).

The results of in vitro studies conducted on rat aorta, which were contracted using KCl and norepinephrine, demonstrated that NBP caused vasorelaxation that varied in intensity depending on the dosage. The extract decreased aortic contraction regardless of the presence of endothelial cells, and its ability to widen blood vessels was not influenced by ATP-dependent potassium channels but possibly by the inhibition of calcium. More precisely, NBP effectively suppressed contractions caused by KCl and $CaCl_2$, providing more evidence for its function as a calcium antagonist and vasodilator (Nguelefack et al. 2006). However, the exact molecule responsible for these activities in *B. pilosa* has not been determined yet. It is advisable to do a further study using bioactivity-guided approaches to separate and analyze the active chemicals that cause the plant's hypotensive and vasodilatory effects, as well as to understand how they work.

16.7 Uterotonic Activity

Bidens pilosa has also been investigated for its potential to induce labor in traditional medicine practices across the globe. The acute toxicity tests revealed that both the ethanol and aqueous extracts exhibited low toxicity. However, the ethanol extract had higher toxicity, with an LD_{50} value of 6.15 g/kg, compared to the aqueous extract which had an LD_{50} value of 12.30 g/kg. Immature mice subjected to in vivo experiments showed a rise in uterine wet weight and body weight gain that was dependent on the quantity of the substances. The ethanol extract demonstrated a greater impact on body weight, while the aqueous extract had a stronger effect on uterine weight. The aqueous extract demonstrated considerable uterine contractions in vitro, similar to the effects of estrogen and oxytocin. The results of this study provide evidence that supports the conventional usage of *Bidens pilosa* as a uterotonic agent for facilitating labor (Frida et al. 2008).

16.8 Wound Healing and Antiulcer Activity

Hassan et al. conducted studies that showed the ethanol extracts of *B. pilosa* have a comparable effect to neomycin sulfate in speeding up wound healing in Wistar rats. Histological examination demonstrated increased collagen synthesis, formation of new blood vessels, and improved tissue structure following the application of the substance to the affected area, indicating its potential as an alternative treatment for wounds (Rafa't Abdul Hassan 2011).

In addition, Tan et al. investigated the antiulcer effects of *B. pilosa* extracts in a rat model that was induced with a stomach-necrotizing solution containing HCl/ethanol. The methylene chloride extracts showed strong efficacy, reducing the formation of gastric lesions by 46.4% at a dosage of 500 mg/kg and suppressing them at a dosage of 750 mg/kg. The observed impact was partially influenced by the inhibition of prostaglandin formation, as demonstrated by the decreased level of protection when indomethacin was administered beforehand. Nevertheless, *B. pilosa* had restricted effectiveness in treating ulcers caused by 100% ethanol and had no impact on the release of stomach acid (Tan et al. 2000).

Phytochemical analysis has revealed the presence of flavonoids and acyclic chalcones in *B. pilosa*, which are recognized for their ability to prevent ulcers. This suggests that these substances may have a role in the therapeutic actions of *B. pilosa*. However, additional research is required to separate and describe the particular active substances that are responsible for their ability to heal wounds and treat ulcers. This emphasizes the necessity for more thorough investigations to confirm its traditional uses and possible medical purposes (Yamamoto et al. 1992).

The new compound beta-D-glucopyranosyloxy-3-hydroxy-6(E)-tetradecen-8, 10,12-triyne has been extracted and identified from the methanolic extract of the entire *B. pilosa* plants. Its identification was primarily done using IR and NMR techniques, along with the identification of a previously known polyyne. The novel chemical exhibited potent growth-inhibitory effects on both normal and altered human cell lines in vitro (Alvarez et al. 1996).

In rats, the flavonoid fraction of *B. aurea* greatly prevents stomach ulcers brought on by cold and confinement. This is a better form of protection than what ranitidine and omeprazole offer. The flavonoid extract increases gastric mucus content, proteins, and hexosamines, as well as the regions of neutral and sulfated glycoproteins (Table 16.1).

16.9 Conclusion

Traditionally, various *Bidens* species have been used in both Chinese medicine and Western herbalism. They address a range of health issues, including nausea, leprosy, fever, cough, and asthma. The bioactivities of *Bidens* species are associated with their diverse phytochemicals. These compounds include porphyrins, tannins, aliphatics, terpenoids, alkaloids, phenylpropanoids, cardiac glycosides, flavonoids, and aromatics. The bioactivities exhibited by these species are due to these phytochemicals. But in the future, more scientific studies are required. There is a need to further investigate the active compounds in *Bidens* species along with Clinical Trials to Validate efficacy and safety.

References

Abbas AK (1996) Die and let live: eliminating dangerous lymphocytes. Cell 84(5):655–657

Alvarez M, Cheibub JA, Limongi F, Przeworski A (1996) Classifying political regimes. Stud Comp Int Dev 31:3–36

Andrade-Neto VF, Brandão MGL, Nogueira F, Rosário VE, Krettli AU (2008) Ampelozyziphus amazonicus Ducke (Rhamnaceae), a medicinal plant used to prevent malaria in the Amazon Region, hampers the development of Plasmodium berghei sporozoites. Int J Parasitol 38(13):1505–1511

Baiwa FI, Yakubu A, Kanya GU (2018) Nutritional food composition analysis of some traditional foods in Jigawa, Nigeria. Int J Sci Res (IJSR) 7(11):676–680

Brandão A, Urmenyi T, Rondinelli E, Gonzalez A, de Miranda AB, Degrave W (1997) Identification of transcribed sequences (ESTs) in the Trypanosoma cruzi genome project. Mem Inst Oswaldo Cruz 92:863–866

Champagnat P (1951) Role of the terminal bud in the action exercised by the cotyledon of Bidens pilosus L. var. radiatus on its axillary bud. C R Seances Soc Biol Fil 145(17–18):1374–1376

Chang L (2004) Epidemiology and quality of life in functional gastrointestinal disorders. Aliment Pharmacol Ther 20:31–39

Chang C, Zhang Y, Li T (2000) Interannual and interdecadal variations of the East Asian summer monsoon and tropical Pacific SSTs. Part I: Roles of the subtropical ridge. J Clim 13(24):4310–4325

Chang J-S, Chiang L-C, Chen C-C, Liu L-T, Wang K-C, Lin C-C (2001) Antileukemic activity of Bidens pilosa L. var. minor (Blume) Sherff and Houttuynia cordata Thunb. Am J Chin Med 29(02):303–312

Chang T-C, Wentzel EA, Kent OA, Ramachandran K, Mullendore M, Lee KH et al (2007) Transactivation of miR-34a by p53 broadly influences gene expression and promotes apoptosis. Mol Cell 26(5):745–752

Chang C-H, Qiu J, O'Sullivan D, Buck MD, Noguchi T, Curtis JD et al (2015) Metabolic competition in the tumor microenvironment is a driver of cancer progression. Cell 162(6):1229–1241

Chiang Y-M, Chuang D-Y, Wang S-Y, Kuo Y-H, Tsai P-W, Shyur L-F (2004) Metabolite profiling and chemopreventive bioactivity of plant extracts from Bidens pilosa. J Ethnopharmacol 95(2–3):409–419

Chiang TC, Jeon BN, Li H (2007) Dynamic correlation analysis of financial contagion: evidence from Asian markets. J Int Money Finance 26(7):1206–1228

Chien S, Prochnow L, Cantarella H (2009a) Recent developments of fertilizer production and use to improve nutrient efficiency and minimize environmental impacts. Adv Agron 102:267–322

Chien S-C, Young PH, Hsu Y-J, Chen C-H, Tien Y-J, Shiu S-Y et al (2009b) Anti-diabetic properties of three common Bidens pilosa variants in Taiwan. Phytochemistry 70(10):1246–1254

De Oliveira JG, Vitória AP (2011) Papaya: nutritional and pharmacological characterization, and quality loss due to physiological disorders. An overview. Food Res Int 44(5):1306–1313

Deba F, Xuan TD, Yasuda M, Tawata S (2008) Chemical composition and antioxidant, antibacterial and antifungal activities of the essential oils from Bidens pilosa Linn. var. Radiata. Food Control 19(4):346–352

Dimo T, Nguelefack TB, Kamtchouing P, Dongo E, Rakotonirina A, Rakotonirina SV (1999) Hypotensive effects of a methanol extract of Bidens pilosa Linn on hypertensive rats. C R Acad Sci III 322(4):323–329

Dimo T, Rakotonirina SV, Tan PV, Azay J, Dongo E, Cros G (2002) Leaf methanol extract of Bidens pilosa prevents and attenuates the hypertension induced by high-fructose diet in Wistar rats. J Ethnopharmacol 83(3):183–191

Dimo T, Nguelefack T, Tan P, Yewah M, Dongo E, Rakotonirina S et al (2003) Possible mechanisms of action of the neutral extract from Bidens pilosa L leaves on the cardiovascular system of anaesthetized rats. Phytother Res 17(10):1135–1139

Felix JM, Bertoncello LA, do Amaral Alves MC, de Miranda Pedroso TF, Dos Santos RC, Formagio ASN et al (2020) Effects of Bidens gardneri Baker leaves aqueous extract in glucose homeostasis of rats. Biomed Pharmacother 127:110185

Frida L, Rakotonirina S, Rakotonirina A, Savineau J-P (2008) Complementary, medicines a in vivo and in vitro effects of Bidens pilosa L(Asteraceae) leaf aqueous and ethanol extracts on primed-oestrogenized rat uterine muscle. Afr J Tradit Complement Altern Med 5(1):79–91

Gachet MS, Kunert O, Kaiser M, Brun R, Munoz RA, Bauer R et al (2010) Jacaranone-derived glucosidic esters from Jacaranda glabra and their activity against Plasmodium falciparum. Nat Prod 73(4):553–556

Horiuchi M, Seyama Y (2006) Antiinflammatory and antiallergic activity of Bidens pilosa L var radiata SCHERFF. J Health Sci 52(6):711–717

Karis PO, Ryding O (1994) Tribe. Heliantheae. In: Bremer K (ed) Asteraceae. Cladistics & classification. Timber Press, Portland, pp 559–624

Kinuthia BK (2017) Export spillovers: comparative evidence from Kenya and Malaysia. J Afr Econ 26(1):24–51

Kumari A, Kakkar P (2008) Screening of antioxidant potential of selected barks of Indian medicinal plants by multiple in vitro assays. Biomed Environ Sci 21(1):24–29

Langat SK, Eyase F, Bulimo W, Lutomiah J, Oyola SO, Imbuga M et al (2021) Profiling of RNA viruses in biting midges (Ceratopogonidae) and related Diptera from Kenya using metagenomics and metabarcoding analysis. mSphere 6(5). https://doi.org/10.1128/msphere.00551-21

Le DT, Uram JN, Wang H, Bartlett BR, Kemberling H, Eyring AD et al (2015) PD-1 blockade in tumors with mismatch-repair deficiency. N Engl J Med 372(26):2509–2520

Le CVC, Tran PH, Ton THD, Le TA, Nguyen TL, Vo TKT et al (2023) Flavonoid glycoside constituents of the leaves of Solanum melongena collected in Thua Thien Hue and their anti-inflammatory activity. Ministry Sci Technol Vietnam 65(2):54–58

Li Y-L, Li J, Wang N-L, Yao X-S (2008) Flavonoids and a new polyacetylene from Bidens parviflora Willd. Molecules 13(8):1931–1941

Lin Y-x, Duan R-y, Tan Z-x, Ma Y-h, Wu H (2020) The complete chloroplast genome of the invasive and Cd-hyperaccumulator herb Bidens pilosa L. (Asteraceae). Mitochondrial DNA B Resour 5(1):436–438

Nguelefack T, Nana P, Atsamo A, Dimo T, Watcho P, Dongmo A et al (2006) Analgesic and anticonvulsant effects of extracts from the leaves of Kalanchoe crenata (Andrews) Haworth (Crassulaceae). J Ethnopharmacol 106(1):70–75

Nielsen FH (1986) Other elements: Sb, Ba, B, Br, Cs, Ge, Rb, Ag, Sr, Sn, Ti, Zr, Be, Bi, Ga, Au, In, Nb, Sc, Te, Tl, W. In: Trace elements in human and animal nutrition, vol 2. Academic, Orlando

Ogunro OB, Ofeniforo EB, Fakayode AE (2023) Quercetin-3-O-β-D-glucopyranoside-rich fraction demonstrated efficacy against infectious, secretory, and osmotic models of diarrhoeal rats. J Genet Eng Biotechnol 21(1):36

Orhan F, Çeker S, Anar M, Agar G, Arasoglu T, Gulluce M (2016) Protective effects of three luteolin derivatives on aflatoxin B 1-induced genotoxicity on human blood cells. Med Chem Res 25:2567–2577

Pattee HE (2002) Commemorative history of the agricultural and food chemistry division. ACS Publications, pp 3–6

Pereira RL, Ibrahim T, Lucchetti L, da Silva AJR, de Moraes VLG (1999) Immunosuppressive and anti-inflammatory effects of methanolic extract and the polyacetylene isolated from Bidens pilosa L. Immunopharmacology 43(1):31–37

Rafa't Abdul Hassan MJ (2011) Antimicrobial effect of bee honey on some pathogenic bacteria isolated from infected wounds in comparison to commonly used antibiotics. J Basrah Res 37(4):78–83

Redl R, Erisman BP (eds) (1994) Reducing distortion in peak-current-controlled boost power-factor correctors. Proceedings of 1994 IEEE Applied Power Electronics Conference and Exposition-ASPEC'94. IEEE

Rybalchenko N, Prykhodko V, Nagorna S, Volynets N, Ostapchuk A, Klochko V et al (2010) In vitro antifungal activity of phenylheptatriyne from Bidens cernua L against yeasts. Fitoterapia 81(5):336–338

Seelinger G, Merfort I, Schempp CM (2008) Anti-oxidant, anti-inflammatory and anti-allergic activities of luteolin. Planta Med 74(14):1667–1677

Tadesse M, Crawford DJ, Smith E (1995) Comparative capitular morphology and anatomy of Coreopsis L. and Bidens L.(Compositae), including a review of generic boundaries. Brittonia 47:61–91

Tan PV, Dimo T, Dongo E (2000) Effects of methanol, cyclohexane and methylene chloride extracts of Bidens pilosa on various gastric ulcer models in rats. J Ethnopharmacol 73(3):415–421

Tewtrakul S, Subhadhirasakul S, Puripattanavong J, Panphadung T (2003) HIV-1 protease inhibitory substances from the rhizomes of Boesenbergia pandurata Holtt. J Sci Technol 25(4):503–508

Tobinaga S, Sharma MK, Aalbersberg WG, Watanabe K, Iguchi K, Narui K et al (2009) Isolation and identification of a potent antimalarial and antibacterial polyacetylene from Bidens pilosa. Planta Med 75(06):624–628

Ubillas RP, Mendez CD, Jolad SD, Luo J, King SR, Carlson TJ et al (2000) Antihyperglycemic acetylenic glucosides from Bidens pilosa. Planta Med 66(01):82–83

Verma SC, Ladha JK, Tripathi AK (2001) Evaluation of plant growth promoting and colonization ability of endophytic diazotrophs from deep water rice. J Biotechnol 91(2–3):127–141

Wang S, Dai H, Ji D, Cui S, Jiang C, Skuza L et al (2023) Influencing factors of Bidens pilosa L hyperaccumulating cadmium explored by the real-Time uptake of Cd2+ influx around root apexes under different exogenous nutrient ion levels. Toxics 11(3):227

Wu J, Wan Z, Yi J, Wu Y, Peng W, Wu J (2013) Investigation of the extracts from Bidens pilosa Linn. var. radiata Sch. Bip for antioxidant activities and cytotoxicity against human tumor cells. J Nat Med 67:17–26

Xia Q, Guo Y, Zhang Z, Li D, Xuan Z, Li Z et al (2009) Complete resequencing of 40 genomes reveals domestication events and genes in silkworm (Bombyx). Science 326(5951):433–436

Yamamoto S, Yamamoto T, Shimada S, Kuramoto E, Yano O, Kataoka T et al (1992) DNA from bacteria, but not from vertebrates, induces interferons, activates natural killer cells and inhibits tumor growth. Microbiol Immunol 36(9):983–997

Yang Y, Hu C, Abu-Omar MM (2012a) Conversion of glucose into furans in the presence of AlCl3 in an ethanol–water solvent system. Bioresour Technol 116:190–194

Yang X-W, Huang M-Z, Jin Y-S, Sun L-N, Song Y, Chen H-S (2012b) Phenolics from Bidens bipinnata and their amylase inhibitory properties. Fitoterapia 83(7):1169–1175

Zahara K, Bibi Y, Tabassum S, Bashir T, Haider S, Araa A et al (2015) A review on pharmacological properties of Bidens biternata: a potential nutraceutical. Asian Pac J Trop Dis 5(8):595–599

Zavala-Mendoza D, Alarcon-Aguilar FJ, Pérez-Gutierrez S, Escobar-Villanueva MC, Zavala-Sánchez MA (2013) Composition and antidiarrheal activity of Bidens odorata Cav. Evid Based Complement Alternat Med 2013(1):170290

Potential of Genus *Bidens* in Pharmacology

17

Syed Muhammad Ahmad Shah, Sobia Nisa, and Maimoona Sabir

Abstract

Importance of medicinal plants with special emphasis on pharmacological and nutritional potential of worldwide distributed plants of genus *Bidens*. Among various species of genus *Bidens, Bidens pilosa* and *Bidens odorata* are well known for their folk use against multiple ailments. The therapeutic potential, efficacy, and underlying mechanisms are further strengthened by published scientific data in favor of the *Bidens* species. The pharmacological actions of plants included in genus *Bidens* like *Bidens pilosa* and *Bidens odorata* have antimicrobial, anti-inflammatory, antioxidant, anticancer, antimalarial, antidiabetic, immunomodulatory, hypotensive, and wound-healing effects are highlighted in this chapter. The active constituents, polyacetylenes, flavonoids, phenols, phytosterols, and fatty acids thought to be responsible for biological actions are meticulously discussed. The presence of unique bioactive compounds makes it a useful resource for the development of new therapeutics. Thus, *Bidens* is emerging as a rich source of new therapeutic compounds, which can be used in advanced pharmacotherapy and improve patients' responses. However, improved experimental methodologies, more human studies, clinical trials, research funding and opportunities, and high throughput screenings are still required to publish significant scientific data and to establish the therapeutic potential and efficacy of other species of *Bidens*.

Keywords

Bidens · Research · Mechanism · Therapeutic · Bioactive

S. M. A. Shah · S. Nisa · M. Sabir (✉)
Department of Microbiology, The University of Haripur, Haripur, Pakistan
e-mail: monasabir@uoh.edu.pk

Y. Bibi et al. (eds.), *The Genus Bidens*,
https://doi.org/10.1007/978-981-96-4257-1_17

17.1 Introduction

Genus *Bidens* includes medicinal flowering plants belonging to the family Asteraceae. These plants are also known by the name of beggarticks or bur-marigolds. There are several species used for their medicinal purposes: which includes *Bidens pilosa* also known as Blackjack or Spanish needles; *Bidens alba*, which is also known as *Bidens pilosa* var. Radiata, *Bidens tripartite* (three-lobbed beggarticks); and *Bidens frondosa* known as common beggarticks (Xuan and Khanh 2016). The pharmacological and biological actions of genus *Bidens* include the following: anti-inflammatory activity in which a very complex biological response occurs in the tissues in response to noxious stimuli or irritants; antimicrobial activity by the inhibition of the microbial growth or killing them completely; antioxidant activity in which they stabilize the harmful outcomes of free radicals in the body; anticancer activity in which they inhibit the growth, proliferation, and spread of cancer cells; antidiabetic activity in which they lower blood glucose levels in humans suffering from diabetes and improve the insulin sensitivity. Hepatoprotective effects are also reported in which extracts of *Bidens* protect liver cells from damage and diuretic activity by increasing the production and output of urine (Zahara et al. 2015).

Bidens pilosa L. belonging to genus *Bidens* is an edible herb. In many countries, it has been used traditionally for broad range of illnesses. Based on 218 sources, the chemical constituents, pharmacological effects, and toxicity of *Bidens pilosa* L. are reported over 40 years. The main chemical constituents include almost 301 compounds and encompasses a variety of flavonoids, aurones, polyacetylenes, flavone glycosides, fatty acids, polyacetylene glycosides, terpenes, and phenolic acids. Some other compounds like pheophytins, chalcones, phytosterols, and okanin glycosides have also been identified from this plant. Majority of these are considered bioactive having pharmacological potential (Xuan and Khanh 2016). Among these compounds, the most biologically active are polyacetylenes and derivatives of polyacetylenes, as they are isolated in larger quantities than other compounds. A compound called phenylheptatriyne is credited for its most pharmacological effects. Table 17.1 details all the major compounds obtained from *Bidens pilosa* L.

Bidens pilosa L. has extraordinary medicinal properties due to its varied composition and is found to be effective in almost 40 illnesses like pharyngitis, gastritis, asthma, wound management, and even in cancer. Besides flavonoids, porphyrins, terpenes, aromatic, and aliphatic compounds, nearly 201 metabolites have been found. Some flavonoids and polyynes like cytopiloyne, luteolin, and centaureidin from this plant have shown immunosuppressive, antioxidant, and antiallergic properties (Bartolome et al. 2013). Treatments available for cancer and autoimmune diseases, in recent years, cause undesirable side effects. New research focuses on development of strategies for therapy based on immunomodulation—most importantly by using natural products that have the potential to regulate immune responses. Various medicinal plants are under study to find new phytomedicines for symptom management and immune response regulation, with negligible side effects. *B. pilosa* in traditional medicine has been used for treatment of different inflammatory

Table 17.1 Compounds involved in the pharmacological and biological activities of genus *Bidens*

Polyynes			
1,2-Dihydroxy-5(*E*)-tridecene-7,9,11-triyne	$C_{13}H_{14}O_2$	Antiangiogenetic Antiproliferative	Bartolome et al. (2013), Yang (2014)
Cytopiloyne	$C_{19}H_{24}O_6$	Antiobese Antidiabetic	Kuo et al. (2021), Liang and Xu (2016)
1,3-Dihydroxy-6(*E*)-tetradecene-8,10,12-triyne	$C_{14}H_{16}O_2$	Antiangiogenetic	Bartolome et al. (2013), Yang (2014), Wu et al. (1997)
1,2-Dihydroxytrideca-5,7,9,11-tetrayne	$C_{13}H_{12}O_2$	Antiangiogenetic	Bartolome et al. (2013), Yang (2014), Wu et al. (1997)
R)-1,2-dihydroxytrideca-3,5,7,9,11-pentayne	$C_{13}H_8O_2$	Antimalarial Antibacterial	Yang (2014), Wu et al. (1997)
2-*β*-*D*-Glucopyranose-1-hydroxy-5*E*)-tridecene-7,9,11-triyne	$C_{19}H_{24}O_7$	Antidiabetic Anti-inflammatory Antimalarial Antibacterial	Bartolome et al. (2013), Yang (2014), Wu et al. (1997), Dimo et al. (2001), Nguelefack et al. (2005)
1-Phenylhepta-1,3,5-triyne	$C_{13}H_8$	Antimicrobial Antimalarial Cytotoxic Antifungal Nematicidal	Bartolome et al. (2013), Yang (2014), Rybalchenko et al. (2010)
3-*β*-*D*-Glucopyranosyl-1-hydroxy-6(*E*)- tetradecene-8,10,12-triyne	$C_{20}H_{26}O_7$	Antidiabetic Anti-inflammatory	Bartolome et al. (2013), Yang (2014), Dimo et al. (2001), Nguelefack et al. (2005)
Flavonoids			
Centaureidin	$C_{18}H_{16}O_8$	Anti-listerial Cytotoxic	Bartolome et al. (2013), Kuo et al. (2021), Chang et al. (2007)
Centaurein	$C_{24}H_{26}O_{13}$	Antiviral Anti-listerial Cytotoxic	Bartolome et al. (2013), Chang et al. (2007), Verma et al. (2001)
Luteolin	$C_{15}H_{10}O_6$	Antiviral Cytotoxic Anti-inflammatory Anti-allergic Anti-flavivirus Anti-HIV	Bartolome et al. (2013), Kuo et al. (2021), Gachet et al. (2010), Tewtrakul et al. (2003), Kumari et al. (2009)

(continued)

Table 17.1 (continued)

Butein	$C_{15}H_{12}O_5$	Cytotoxic Anti-leishmanial	Bartolome et al. (2013), Kuo et al. (2021), Seelinger et al. (2008)
Quercetin 3-*O*-*β*-*D*-galactopyranoside	$C_{21}H_{20}O_{12}$	Anti-inflammatory	Bartolome et al. (2013), Nielsen et al. (1998)
Quercetin 3,3′-dimethyl ether-7-O-*β*-*D*-glucopyranoside	$C_{21}H_{20}O_{12}$	Antimalarial	Bartolome et al. (2013), Andrade-Neto et al. (2004)
Quercetin 3,3′-dimethyl ether 7-O-*α*-*L*-rhamnopyranosyl-(1 → 6)-*β*-*D*-glucopyranoside	$C_{21}H_{20}O_{11}$	Antimalarial	Bartolome et al. (2013), Andrade-Neto et al. (2004)
Jacein	$C_{24}H_{26}O_{13}$	Antioxidant	Bartolome et al. (2013), Chiang et al. (2004)
Quercetin 3-O-rutinoside	$C_{27}H_{38}O_{15}$	Antioxidant	Bartolome et al. (2013), Chiang et al. (2004)
Fatty acids			
Linoleic acid	$C_{18}H_{32}O_2$	Antiviral Cytotoxic Anti-herpes Anti-Sendai Anti-influenza Anti-sindbis	Bartolome et al. (2013), Kuo et al. (2021), Ayyanar and Ignacimuthu (2005)
Palmitic acid	$C_{16}H_{32}O_2$	Antidiarrheal	Zavala-Mendoza et al. (2013)
Stearic acid	$C_{18}H_{36}O_2$	Antidiarrheal	Zavala-Mendoza et al. (2013)
Phenolics			
Chlorogenic acid	$C_{16}H_{26}O_9$	Antioxidant	Bartolome et al. (2013), Kuo et al. (2021), Chiang et al. (2004), Chiang et al. (2005)
Caffeic acid	$C_9H_8O_4$	Antioxidant Anti-inflammatory	Deba et al. (2008)
Phenylpropanoid			
3,5-Di-O-caffeoylquinic acid	$C_{25}H_{24}O_{12}$	Antiviral Antioxidant	Bartolome et al. (2013), Chiang et al. (2005), Chang et al. (2000)
4,5-Di-O-caffeoylquinic acid	$C_{25}H_{24}O_{12}$	Antiviral Antioxidant	Bartolome et al. (2013), Chiang et al. (2005), Chang et al. (2000)

(continued)

Table 17.1 (continued)

3,4-Di-O-caffeoylquinic acid (DiCQA)	$C_{25}H_{24}O_{12}$	Antiviral Antioxidant	Bartolome et al. (2013), Chiang et al. (2005), Chang et al. (2000)
Ethyl caffeate	$C_{11}H_{12}O_4$	Anti-inflammatory Immunosuppressive	Bartolome et al. (2013), Kuo et al. (2021), Chiang et al. (2005)

symptoms and conditions. There is clear proof available in literature supporting its traditional usage (Furman et al. 2019).

17.2 Pharmaceutical Formulations

17.2.1 Natural Antimicrobial

B. pilosa has been involved in antibacterial activity by making the cell wall of bacteria permeable, thereby disrupting the barriers and inhibiting the protein synthesis. It acts against both Gram-positive (e.g., *Staphylococcus aureus* and *Bacillus subtilis*) and Gram-negative bacteria (e.g., *Escherichia coli* and *Salmonella typhimurium*). Gram-positive bacteria are more susceptible to the polyphenolic compound's activity and essential oils due to their cell wall structure and Gram-negative bacteria show resistance due to the presence of an outer membrane. Still *Bidens pilosa* acts on it by permeating the membrane and disrupting the barriers (Rodriguez-Otero et al. 2023). The antifungal activity is shown by *Bidens tripartita* against the *Candida* species (e.g., *Candida albicans*) and *Aspergillus* species (e.g., *Aspergillus niger* and *Aspergillus fumigatus*). *B. tripartita* disrupts the cell membrane of *Candida* species and interferes with the synthesis of ergosterol, which is an important component of fungal cell membrane. The antifungal activity happens in *Aspergillus* species by the inhibition of spore germination and hyphal growth. Antiviral properties have also been observed in some studies that used the extracts of some *Bidens* species against viruses such as herpes simplex virus (HSV) and human immunodeficiency virus (HIV) by inhibiting the viral replication and reducing the cytopathic effects as well (Rodin et al. 2024).

Hexane extract of leaves of *Bidens bipinnata* has shown maximum activity, with inhibition being 46.89% in *E. coli*, 67.15% in *Shigella dysenteriae*, 66.09% in *Shigella flexneri*, 62.72% in *Shigella sonnei*, and 67.40% in *Staphylococcus Aureus*. This indicates the presence of active antibacterial substances. It has high levels of tannins and alkaloids which contribute to this substantial bactericidal activity. 16-Pregnenolone and 9-octadecenoic acid (Z)-, methyl ester are the two antibacterial compounds isolated from *Bidens bipinnata* and reported for the first time (Zahara et al., 2022). The essential oil extracted from the roots of *B. tripartita* was found to inhibit the growth of Gram-positive bacteria moderately, while it inhibited the growth of Gram-negative bacteria: *Moraxella catarrhalis* in MIC value of 3.12–2.1 mg/mL and *Neisseria gonorrhoeae* in MIC value of 3.1 mg/mL. Fungistatic effect against *Candida albicans* ATCC 10231 in MIC value of 3.1 mg/mL was shown by the same essential oil (Tomczykowa et al., 2011).

There are three steps involved in the mechanism of action used by the plants of genus *Bidens*. The first step is the disruption of cell wall and membrane. In this step, the flavonoids and essential oils get integrated into the cell membrane of a microorganism causing the membrane to be permeable which then causes leakage of cellular contents; and after that cell lysis and death occurs. The other constituent involved in this step is polyacetylenes insertion into the lipid bilayer of microbial membranes which causes the disruption of the cell membrane and leads to the cell death. The second step is inhibition of protein synthesis. In this step, flavonoids which are the essential compounds bind to the ribosomes of bacteria. This binding causes the blockage of the assembly of protein synthesis machinery which is crucial for the bacterial cell to grow further and replicate. The third step is interference of metabolic pathway. In this step phenolic acid and terpenoids are the essential compounds that inhibit the enzymes which are responsible for bacterial DNA replication and repair (e.g., gyrase) (Rodriguez-Otero et al. 2023).

17.2.2 Antioxidant Supplements

The responsible active constituents for the antioxidant activity in genus *Bidens* include flavonoids, phenolic acids, tannins, terpenoids, vitamins, and some other compounds. Flavonoids, including quercetin, luteolin, and apigenin, are the compounds that neutralize free radicals, thereby reducing oxidative stress and inhibiting the lipid peroxidation. Phenolic acids such as caffeic acid and chlorogenic acid possess antioxidant activities which include scavenging reactive oxygen species and protecting lipids and other proteins from oxidative harm. Tannins are also responsible for active constituents for antioxidant activity which are polyphenolic compounds that can precipitate proteins and have strong antioxidant activities. Terpenoids include sesquiterpenes and diterpenes. Their function is to protect cells from oxidative stress. Some *Bidens* species also contain vitamins which include vitamin C that help in neutralizing free radicals and provide protection against oxidative harm (Xuan and Khanh 2016). The ethanol extract of *B. tripartita* was found to be 352.51 close to that of Trolox which is 382.5. In chloroform sub-extract, the antioxidant capacity was found to be highest at 1532.45, while it was found inactive on DPPH radical scavenging activity test. Other extracts were found to be active ranging from 62.86 to 19.35% at 2–0.5 mg/mL concentrations. Among all, the ethyl acetate extract has 62.86 to 59.60% activity, being the highest, and showed a similar effect as that of ascorbic acid which has 70.88 to 65.62% activity in DPPH radical scavenging assay (Xuan and Khanh 2016).

Compared to aqueous extracts, the essential oils from leaves and flowers of *B. pilosa* have shown stronger activity. These essential oils were found to be capable of reducing DPPH to diphenylpicrylhydrazine. The aqueous extracts of the leaves were found to be more efficient in radical scavenging as compared to lower aqueous extracts. These essential oils contained a-pinene, ocimene, verbenol, and b-myrcene that are oxygenated terpenes and monoterpenes, a group of nonphenolic compounds (Zahara et al. 2015). In INS-1 cells, flavonoid-rich extract from *B. bippinata* (BBTF) was shown to reduce H_2O_2-induced ROS production. INS-1 cells pretreated with BBTF could reduce H_2O_2-induced cell apoptosis due to its antioxidant activity. In H_2O_2-treated INS-1 cells, pretreatment with BBTF has shown to increase the expression of Bcl-2 and reduce the expression of caspase-9, Bax, and cleaved caspase-3. The result has shown that, in pancreatic β-cells, the pretreatment with BBTF regulates the expressions of Bax and Bcl-2 resulting in reducing the downstream apoptotic effect. Thus, BBTF can significantly improve cell viability and reduces cell apoptosis and ROS production in H_2O_2-treated INS-1 cell (Chien et al. 2009).

17.2.3 Dermatological Applications

The leaf extract of *Bidens pilosa* can be squeezed in water and used for bathing. This helps in treating skin allergy, while application of the extract is used for treating wounds by tying on wounds/cuts (Furman et al. 2019). The methanol/water and petrol ether extracts of aerial parts of *B. pilosa* have shown to have antimicrobial activity. Various compounds extracted from *B. pilosa* have its use in traditional medicine for treatment of bacterial infection of GIT, wounds, and inflammations. Flavonoids extracted from leaf extracts *of B. pilosa* include chalcones and aurones. Flavonoids like friedelin, friedelan-3β-ol, and many others have anti-inflammatory capabilities. These flavonoids and many described acetylenes may rationalize the use of *B. pilosa* in traditional medicine for treating wounds. In countries like Colombia, this plant is ingested as teas, decoctions, and juice for treatment of respiratory infections. They can also be made into a poultice for direct application on skin burns and infected wounds (Chien et al. 2009).

17.2.4 Skin Health

In various parts of the world, *Bidens* species are used to improve skin health. In Philippines, leaves of the plants are used to treat boils, while, in Tonga, leaves are applied on inflamed eyes directly. Compared to stem and roots, flowers of these plants have shown unique DPPH activity. In DPPH, the butanol fraction has shown excellent antioxidant activity. The plant extract's antioxidant ability is reported due to its activation of NO and inhibition of ROS production in dermal endothelial cells (Rafa't Abdul Hassan 2011).

Chen et al. (2014) have reported the ethnobotanical usage of this plant to heal cut wounds due to a combination of anti-inflammatory, antioxidant, and antimicrobial activities. Moreover, the Angiotensin-converting enzyme and renin are inhibited by the ethanol extract from the parts of these plants, thereby improving wound bleeding. These investigations have provided a base for studying the potential of *B. pilosa* for synthesizing an effective anti-aging product. Supercritical CO_2 extract of the plant's aerial parts and especially phytol have shown same activity (Diamant et al. 2021).

The extract of *B. pilosa* plant has been already patented as a skin whitening agent under patent number: US2015/0359734A1. This is based on the fact that *B. pilosa* can degrade melanin and its transport due to its high retinoid-like and antioxidant activities. Ferulic acid, caffeic acid, and their derivatives from this plant are potential sources of anti-tyrosinase compounds that result in the whitening activity (Chen et al. 2016).

17.2.5 Mosquito Control

A compound called polyacetylene E-7-phenyl-2-heptene-4,6-diyn-1-ol was isolated from *Bidens pilosa* as an isopods repellent. The crude extract of plant has shown stronger repellent activity before purification. To show much stronger activity, this compound needed an eightfold dosage at 1.6 g Polyacetylene E-7-phenyl-2-heptene-4,6-diyn-1-ol emerged to be the main constituent of the repellent (Ishida et al. 2023). *B. pilosa* among the Asteraceae family is the most potent plant against malaria. It shows a strong inhibiting effect against parasitemia in vitro (Connelly et al. 2009). The strong antimalarial ability of *B. pilosa* is due to its production of polyacetylenes and flavonoids in abundance. To develop novel antimalarial and anti-leishmania drugs, these compounds are being investigated further. The potential therapeutic actions of *B. pilosa* can be used in the management of malaria and leishmania.

Monoterpene hydrocarbon and α-pinene are the main constituents present in the essential oils of *B. pilosa*. This essential oil of *B. pilosa* was found to be toxic for mature larvae of *Anopheles Gambiae,* and it would be effective due to high levels of α-pinene present in the plant (Roland et al. 2022). Larvicidal effect is shown by the methanolic extract from *B. pilosa* against the third instar larva of *Culex quinquefasciatus*. This extract showed 100% mortality rate at the concentration of 1000 ppm after 12 h of incubation (Singh et al. 2017). Sudarmani et al. (2021) has reported that the LC50 value of extract of *Bidens biternata* to be 45.135 ppm against the pupae and larvae of *Aedes aegypti*, while the synthesized silver nanoparticles (AgNPs) from the same plant extract has an LC50 value of 5.261 ppm. The same study has also confirmed the lethal effects of these nanoparticles and showed that the mortality of mosquito's larvae was dose-dependent.

17.3 Conclusion

In conclusion, the pharmacological potential of genus *Bidens*, in particular *B. pilosa* and *B. odorata*, highlights their important therapeutic benefits. These plants are known for their traditional use against a wide range of diseases, which is supported by scientific research confirming their efficacy. Important pharmacological properties of genus *Bidens* include antimicrobial, anti-inflammatory, antioxidant, anticancer, antimalarial, antidiabetic, immunomodulatory, and wound-healing effects. They have shown antimicrobial potential against many pathogenic organisms including bacteria, fungi, and viruses. These properties are attributed to active compounds extracted from these plants like polyacetylenes, flavonoids (quercetin, luteolin, apigenin), phenols (chlorogenic acid, caffeic acid), pheophytins, phytosterols, and various fatty acids. These compounds have anti-inflammatory effects involving suppression of pro-inflammatory cytokines and enzymes, while exhibiting antioxidant properties against oxidative stress. Moreover, compounds extracted from genus *Bidens* can induce apoptosis and inhibit tumor growth highlighting their potential in cancer treatment. They have shown their ability by enhancing insulin sensitivity and regulation of glucose metabolism highlighting their antidiabetic properties.

The pharmacological potential of *B. pilosa* and *B. odorata* makes these plants a potential source for the development of new therapeutic medicines. However, there are some challenges in the establishment of libraries of such compounds and their screening in larger quantities. There is a need for improved technologies to make full use of such compounds. Further research is needed to confirm the potential of genus *Bidens*; in particular, human studies, high throughput screenings, and clinical trials are much needed. This genus could significantly contribute to advanced therapeutics with such improvements. Overall, the broad range of pharmacological properties of genus *Bidens* emphasizes their potential as an important source of natural medicines. It offers therapeutic benefits across a diverse spectrum of health conditions without many side effects which is accepted by both traditional use and scientific research.

References

Andrade-Neto VF, Brandão MG, Oliveira FQ, Casali VW, Njaine B, Zalis MG et al (2004) Antimalarial activity of *Bidens pilosa* L.(Asteraceae) ethanol extracts from wild plants collected in various localities or plants cultivated in humus soil. Phytother Res 18(8):634–639

Ayyanar M, Ignacimuthu S (2005) Traditional knowledge of Kani tribals in Kouthalai of Tirunelveli hills, Tamil Nadu, India. J Ethnopharmacol 102(2):246–255

Bartolome AP, Villaseñor IM, Yang W-C (2013) *Bidens pilosa* L.(Asteraceae): botanical properties, traditional uses, phytochemistry, and pharmacology. Evid Based Complement Alternat Med 2013(1):340215

Chang MH, Wang GJ, Kuo YH, Lee CK (2000) The low polar constituents from *Bidens pilosa* L. var. minor (Blume) Sherff. J Chin Chem Soc 47(5):1131–1136

Chang S-L, Chiang Y-M, Chang CL-T, Yeh H-H, Shyur L-F, Kuo Y-H et al (2007) Flavonoids, centaurein and centaureidin, from *Bidens pilosa*, stimulate IFN-γ expression. J Ethnopharmacol 112(2):232–236

Chen G, Wang Y, Li M, Xu T, Wang X, Hong B et al (2014) Curcumol induces HSC-T6 cell death through suppression of Bcl-2: involvement of PI3K and NF-κB pathways. Eur J Pharm Sci 65:21–28

Chen Y, Luo J, Zhang Q, Kong L (2016) Identification of active substances for dually modulating the renin–angiotensin system in *Bidens pilosa* by liquid chromatography–mass spectrometry–based chemometrics. J Funct Foods 21:201–211

Chiang Y-M, Chuang D-Y, Wang S-Y, Kuo Y-H, Tsai P-W, Shyur L-F (2004) Metabolite profiling and chemopreventive bioactivity of plant extracts from *Bidens pilosa*. J Ethnopharmacol 95(2–3):409–419

Chiang YM, Lo CP, Chen YP, Wang SY, Yang NS, Kuo YH et al (2005) Ethyl caffeate suppresses NF-κB activation and its downstream inflammatory mediators, iNOS, COX-2, and PGE2in vitro or in mouse skin. Br J Pharmacol 146(3):352–363

Chien S-C, Young PH, Hsu Y-J, Chen C-H, Tien Y-J, Shiu S-Y et al (2009) Anti-diabetic properties of three common *Bidens pilosa* variants in Taiwan. Phytochemistry 70(10):1246–1254

Connelly W, Shenton F, Lethbridge N, Leurs R, Waldvogel H, Faull R et al (2009) The histamine H4 receptor is functionally expressed on neurons in the mammalian CNS. Br J Pharmacol 157(1):55–63

Deba F, Xuan TD, Yasuda M, Tawata S (2008) Chemical composition and antioxidant, antibacterial and antifungal activities of the essential oils from Bidens pilosa Linn. var. Radiata. Food Control 19(4):346–352

Diamant S, Pierce SJ, Rohner CA, Graham RT, Guillemain d'Echon A, Guillemain d'Echon T et al (2021) Population structure, residency, and abundance of whale sharks in the coastal waters off nosy be, north-western Madagascar. Aquat Conserv Mar Freshw Ecosyst 31(12):3492–3506

Dimo T, Azay J, Tan PV, Pellecuer J, Cros G, Bopelet M et al (2001) Effects of the aqueous and methylene chloride extracts of *Bidens pilosa* leaf on fructose-hypertensive rats. J Ethnopharmacol 76(3):215–221

Furman D, Campisi J, Verdin E, Carrera-Bastos P, Targ S, Franceschi C et al (2019) Chronic inflammation in the etiology of disease across the life span. Nat Med 25(12):1822–1832

Gachet MS, Lecaro JS, Kaiser M, Brun R, Navarrete H, Muñoz RA et al (2010) Assessment of anti-protozoal activity of plants traditionally used in Ecuador in the treatment of leishmaniasis. J Ethnopharmacol 128(1):184–197

Ishida M, Yamaji Y, Isoda W, Abe H, Sato M, Kariya K et al (2023) (E)-7-phenyl-2-heptene-4, 6-diyn-1-ol from *Bidens pilosa* as a repellent against isopods. Biosci Biotechnol Biochem 87(8):833–838

Kumari P, Misra K, Sisodia BS, Faridi U, Srivastava S, Luqman S et al (2009) A promising anticancer and antimalarial component from the leaves of *Bidens pilosa*. Planta Med 75(01):59–61

Kuo TF, Yang G, Chen TY, Wu YC, Tran Nguyen Minh H, Chen LS et al (2021) *Bidens pilosa*: nutritional value and benefits for metabolic syndrome. Food Front 2(1):32–45

Liang X, Xu Q (2016) Separation and identification of phenolic compounds in *Bidens pilosa* L. by ultra high performance liquid chromatography with quadrupole time-of-flight mass spectrometry. J Sep Sci 39(10):1853–1862

Nguelefack T, Dimo T, Mbuyo EN, Tan P, Rakotonirina S, Kamanyi A (2005) Relaxant effects of the neutral extract of the leaves of *Bidens pilosa* Linn on isolated rat vascular smooth muscle. Phytother Res 19(3):207–210

Nielsen SF, Christensen SB, Cruciani G, Kharazmi A, Liljefors T (1998) Antileishmanial chalcones: statistical design, synthesis, and three-dimensional quantitative structure– activity relationship analysis. J Med Chem 41(24):4819–4832

Rafa't Abdul Hassan M (2011) Antimicrobial effect of bee honey on some pathogenic bacteria isolated from infected wounds in comparison to commonly used antibiotics. Ann Burns Fire Disasters 37(4):78–83

Rodin MN, Baeva VM, Bokov DO, Voronov VM, Samylina IAJF (2024) Comparative study of morphological and anatomical features of *Bidens* L. species. Farmaciya (Pharmacy) 73(2):25–33

Rodriguez-Otero P, Ailawadhi S, Arnulf B, Patel K, Cavo M, Nooka AK et al (2023) Ide-cel or standard regimens in relapsed and refractory multiple myeloma. N Engl J Med 388(11):1002–1014

Roland TJG, Mbida JA, Ngaha R, Christelle AD, Gaelle MT, Pasma NM et al (2022) Larvicidal activities of some plants essential oils of the Cameroonian flora against Anopheles gambiae SL, major vector of malaria in the city of Douala. J Entimol Zool Stud 10(4):7–12

Rybalchenko N, Prykhodko V, Nagorna S, Volynets N, Ostapchuk A, Klochko V et al (2010) In vitro antifungal activity of phenylheptatriyne from *Bidens cernua* L. against yeasts. Fitoterapia 81(5):336–338

Seelinger G, Merfort I, Wölfle U, Schempp CM (2008) Anti-carcinogenic effects of the flavonoid luteolin. Molecules 13(10):2628–2651

Singh G, Passsari AK, Singh P, Leo VV, Subbarayan S, Kumar B et al (2017) Pharmacological potential of *Bidens pilosa* L and determination of bioactive compounds using UHPLC-QqQ LIT-MS/MS and GC/MS. BMC Complement Alternat Med 17:1–16

Sudarmani D, Sundareswari C, Samuel P (2021) Bioefficacy of Sesbania grandiflora leaves silver nanoparticles against Aedes aegypti larvae. J Adv Sci Res 12(04 Suppl 1):176–184

Tewtrakul S, Miyashiro H, Nakamura N, Hattori M, Kawahata T, Otake T et al (2003) HIV-1 integrase inhibitory substances from Coleus parvifolius. Phytother Res 17(3):232–239

Verma A, Su A, Golin AM, O'Marrah B, Amorosa JK (2001) The lateral view: a screening method for knee trauma. Acad Radiol 8(5):392–397

Wu H, Chen H, Hua X, Shi Z, Zhang L, Chen J (1997) Clinical therapeutic effect of drug-separated moxibustion on chronic diarrhea and its immunologic mechanisms. J Tradit Chin Med= Chung i tsa Chih Ying wen pan 17(4):253–258

Xuan TD, Khanh TD (2016) Chemistry and pharmacology of *Bidens pilosa*: an overview. J Pharm Investig 46:91–132

Yang W-C (2014) Botanical, pharmacological, phytochemical, and toxicological aspects of the antidiabetic plant *Bidens pilosa* L. Evid Based Complement Alternat Med 2014(1):698617

Zahara K, Bibi Y, Tabassum S, Bashir T, Haider S, Araa A et al (2015) A review on pharmacological properties of *Bidens biternata:* a potential nutraceutical. Asian Pac J Trop Dis 5(8):595–599

Zavala-Mendoza D, Alarcon-Aguilar FJ, Pérez-Gutierrez S, Escobar-Villanueva MC, Zavala-Sánchez MA (2013) Composition and antidiarrheal activity of *Bidens odorata* Cav. Evid Based Complement Alternat Med 2013(1):170290

Part V

Phytochemical and Pharmacological Profiles of Key Bidens Species

Metabolomic, Phytochemical, and Pharmacological Properties of *Bidens pilosa* L.

18

Salaudeen Abdulwasiu Olawale, Alfiah Hayati, and Vuanghao Lim

Abstract

People's lifestyles are changing from active to sedentary as a result of urbanization; they are replacing organic home-cooked meals with processed fast foods and snacks that are high in sugar, salt, and fat. This increases the risk of metabolic syndromes, including diabetes, obesity, and other noncommunicable illnesses linked to food. Therefore, it is vital to harness the potential of plants like *Bidens pilosa* L. (BP; Asteraceae) to combat malnutrition, food scarcity, and health issues. About 230–240 varieties of the *Bidens* genus exist. *Bidens pilosa* L. is a common perennial plant found throughout tropical and temperate regions. It is an edible medicinal plant with a wide range of bioactivities that may help prevent and treat several illnesses. Historically, BP has been utilized in medicinal therapy, phytoremediation of heavy metal-contaminated soils, and in agriculture as a biopesticide and herbicide, as well as a food flavoring ingredient. Bioactives derived from BP are resurfacing as a promising alternative medicine for drug formulations and herbal therapy to treat over 40 ailments in humans and animals without any evident side effects. Even though phytochemical and biological

S. A. Olawale
Advanced Medical and Dental Institute, Universiti Sains Malaysia, Bertam, Kepala Batas, Penang, Malaysia

Applied Mathematics Programme (Chemistry) of National Mathematical Centre Abuja, Abuja, Nigeria

A. Hayati
Department of Biology, Faculty of Science and Technology, Universitas Airlangga, Surabaya, Indonesia

V. Lim (✉)
Advanced Medical and Dental Institute, Universiti Sains Malaysia, Bertam, Kepala Batas, Penang, Malaysia
e-mail: vlim@usm.my

Y. Bibi et al. (eds.), *The Genus Bidens*,
https://doi.org/10.1007/978-981-96-4257-1_18

investigations of BP have advanced significantly in recent years, thorough and critical assessments of this plant are either outdated or have a narrow focus. While there has been some progress, more comprehensive research is needed to fully comprehend and confirm the conventional applications of BP and find therapeutic benefits for the specific chemicals extracted from the substance. In summary, the broad range of its applications for various purposes is undetermined, which is why this book chapter aims to evaluate BP's pharmacological, metabolomic, and phytochemical features as well as its uses, health advantages, and/or hazards.

Keywords

Bidens pilosa · Metabolomics · Phytochemical and pharmacological · Anticancer

18.1 Introduction

Bidens pilosa L. (BP), also known as blackjack (Mtenga and Ripanda 2022), is a tropical/subtropical weed (Chiang et al. 2004) that is used as an edible vegetable (Bilanda et al. 2017). As a member of the Asteraceae family (see Fig. 18.1), it has yellow blooms and reaches a height of 30–100 cm. It is used in phytoremediation of heavy metal-polluted soils (Yan et al. 2020), food, and traditional medicine (Yang

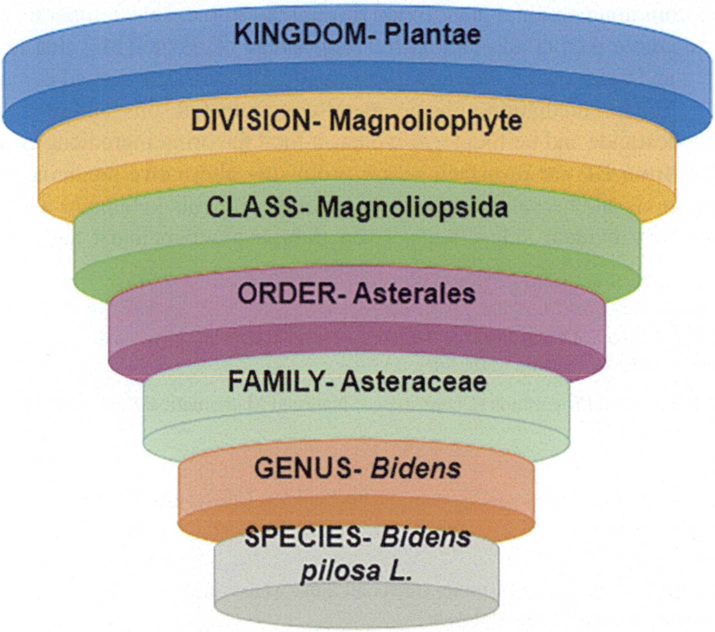

Fig. 18.1 Taxonomy of *Bidens pilosa* L. (Adapted from (Bartolome et al. 2013) licensed under a Creative Commons Attribution 4.0 International License)

2014), as a plant biopesticide, in weed control (Xuan et al. 2005), etc. The plant has a long history in ethnobotany. It is reportedly a classic element in tea preparations and sauce dishes. It has been used for treating conditions such as hyperglycemia, malaria, tumors, microbial infections, rheumatoid arthritis, and inflammation (Lans et al. 2020).

According to reports in ethnoveterinary medicine, BP can effectively cure a variety of cattle and poultry illnesses. For example, BP is employed to enhance horse nutrition in Trinidad and British Columbia, Canada (Lans et al. 2020). Chang et al. (2016) reported that BP improved chicken development by minimizing cecal damage, lowering protozoan infection, and modulating gut microbes (Chang et al. 2016). When administered for 6 weeks, BP was shown to reduce the degree of liver fibrosis and damage in rats (Yuan et al. 2008), indicating that it may be applied to cure liver illnesses. *Bidens pilosa* L. extracts in methylene chloride and aqueous medium corrected fructose-mediated hypertension and hypertriglyceridemia in rats, suggesting that the plant extracts exerted consequences of hypotension independent of insulin level regulation (Dimo et al. 2001).

Bidens pilosa L.'s varied phytochemicals, including flavonoids, aromatics, aliphatics, terpenoids, alkaloids, phenylpropanoids, porphyrins, and cardiac glycosides, are linked to its bioactivities (Bartolome et al. 2013). The genus *Bidens* has a large variety of bioactive chemicals; however, the precise phenolic compounds are yet to be identified. Generally speaking, food additives, pharmaceuticals, flavors, colognes, cosmetics, and a host of other goods use phytochemicals or plant secondary metabolites. It is challenging to access most medicinal plants—including BP—because they are not currently farmed for commercial purposes, which makes them a major natural source of secondary metabolites (Smith et al. 2021). Plants continue to be the most dependable source of secondary metabolites, even though bacteria, fungi, and a few marine species, including corals, sponges, and tunicates, also generate them (Thirumurugan et al. 2018).

The purpose of this chapter is to provide a summary of BP's primary chemical components, an in-depth analysis of its major pharmacological and rare toxicological effects, as well as its uses in folk medicine and metabolomics studies. Relevant reports from previous years have also been taken into consideration.

18.2 Metabolomics Studies

In addition to some omics technologies like transcriptomics, proteomics, and genomes, metabolomics investigations in therapeutic herbs seek to offer a thorough and precise analysis of every low-molecular-weight compound (Siddaiah et al. 2020). All variables, including genetic, physiological, and environmental, in addition to other factors, influence variations in plant metabolism (Hall et al. 2002). As a result, the metabolomics method has gained paramount significance in plant research, allowing for the in-depth examination of their constituent parts, evaluation of their organoleptic and dietary qualities, and investigation of their toxicological and functional features. Even the most minor alterations in a complex biological system may be detected and assessed using metabolomics analysis.

The ability of metabolomics research to fully identify and quantify every metabolite depends on the techniques and tools it employs (Hong et al. 2016). Metabolomics is a complex undertaking because of the intricacy of the many molecular prevalences and metabolic characteristics. The phrases "metabolomics" or "metabolite profiling" can also be used to refer to three other kinds of methods: untargeted, targeted, and semi-targeted (Piasecka et al. 2019).

There are two primary methods for gathering metabolomics data: targeted and untargeted analyses. In targeted and untargeted metabolomics, the steps of sample preparation and collection, as well as the mobile and stationary phases, are comparable. Targeted metabolomics studies, however, generally concentrate on a small number of metabolites from one or more pathways. In contrast, the goal of untargeted metabolomics is to examine an organism's whole metabolome (Waris et al. 2022).

A metabolite profiling study by Nxumalo et al. (2023) to ascertain the compositions of BP by liquid chromatography-mass spectrometry (LC-MS) indicated 15 identified metabolites and 6 unknown polyphenols (Nxumalo et al. 2023). In another study, the metabolite makeup and main components of the BP butanol portion were examined by Chiang et al. using LC-MS (Chiang et al. 2004). Using reverse-phase high-performance liquid chromatography (HPLC), the total metabolite profile of the butanol layer was ascertained. Trifluoroacetic acid (TFA) was introduced to the HPLC analysis' mobile phase to inhibit any possible ionic dissociation of the plant's phytocompounds on the column, as the majority among the metabolites were found to have phenolic hydroxyl groups, typically found in acidic solutions. Using a variety of spectroscopic investigations, the metabolite assay and main components of the butanol fraction were examined and described. A UHPLC–QTOF mass spectrometer was used to perform untargeted metabolomic profiling of various BP plant materials, encompassing the qualitative and quantitative composition of phenolic chemicals. The metabolite profiles of the four plant portions exhibited distinct characteristics when all annotated compounds were clustered using principal component analysis (PCA) and normalized peak intensity (Angelini et al. 2021). The stem had a higher concentration of monoglyceride fraction than the other plant sections. Future research could apply internal standards for all ontology groups to manage the variations in the ionization and extraction methods, thereby improving our comprehension of metabolic routes and considering the intricacy and diverse overall properties of the compounds analyzed.

A few reviews of BP have been published, although not all types of metabolites have been included in the phytochemical data. This species has produced about 198 chemicals that have been characterized so far (Lima Silva et al. 2011). Some of the secondary metabolites are presented in Table 18.1.

Table 18.1 Selected metabolites derived from *Bidens pilosa* L

S/N	Class of compound	Name of compound	Structure/Molecular formula	Plant portion
	SUH	Heneicosane	$CH_3(CH_2)_{19}CH_3$	AP
	SUH	Pentacosane	$CH_3(CH_2)_{23}CH_3$	AP
	SUCA	Eicosanoic acid (Arachidic acid)	$CH_3(CH_2)_{18}CO_2H$	AP
	SUA	2-tetracosan-1-ol	$CH_3(CH_2)_{22}CH_2OH$	AP
	SUA	1-hentriacontanol	$CH_3(CH_2)_{29}CH_2OH$	NF
	UACAE	2-butenedioic acid	$C_4H_4O_4$	AP
	SP	1,2-benzenediol (Pyrocatechin)	$C_6H_6O_2$	EP
	SAA	4-hydroxy-3-methoxybenzaldehyde (vanillin)	$C_8H_8O_3$	AP
	SBA	4-hydroxy-benzoic acid	$C_7H_6O_3$	EP
	SPP	2-methoxy-4(2-propen-1-yl)-phenol (eugenol)	$C_{10}H_{12}O_2$	LF/RT
	CCE	Ethyl ester, 3-(3,4-dihydroxyphenyl)-2-propenoic acid	$C_{11}H_{12}O_4$	EP
	Aurones	2-[(3,4-dihydroxyphenyl)-methylene]-6-hydroxy-3(2H) benzofuranone (Sulfuretin)	$C_{15}H_{10}O_5$	AP
	Chalcones	1-(2,4-dihydroxyphenyl)-3-(3,4-dihydroxy-phenyl)-2-propen-1-one (Butein)	$C_{15}H_{12}O_5$	AP
	Flavanones	2-(3,4-dihydroxyphenyl)-2,3-dihydro-7,8-dihydroxy-4H-1-benzopyran-4-one (Okanin, iso)	$C_{15}H_{12}O_6$	
	Diterpenes	(2E,7R,11R)-3,7,11,15-tetramethyl-2-hexadecen-1-ol (phytol)	$C_{40}H_{20}O$	EP
	NSNP	Thymidine	$C_{10}H_{14}N_2O_5$	NF

AP Aerial part, *LF* Leaf, *EP* Entire plant, *RT* Root, *NF* Not found. *SUA* Saturated unbranched alcohols, *SUH* Saturated unbranched hydrocarbons, *SUCA* Saturated unbranched carboxylic acids, *SPP* Simple phenylpropanoids, *NSNP* Nitrogen- and Sulfur-containing Natural Products, *SP* Simple phenols, *SAA* Simple aryl aldehydes, *SBA* Simple benzoic acids, *CCE* Coumaric and caffeoyl esters, and *UACAE* Unbranched aliphatic carboxylic acid esters

18.3 Phytochemical Studies

Bidens pilosa L. is an exceptional phytochemical source, especially for flavonoids and polyynes. It is common knowledge that plant flavonoids have antioxidant, anti-inflammatory, anticancer, and other bioactivities. Since BP was discovered in 1753, there has been a rising interest in its fundamental research and practical applications. This is primarily because it is widely used in foods, beverages, and medications. The usage and utilization of BP for medicinal purposes have been reported in about 116 articles (Bartolome et al. 2013).

Table 18.2 Isolated polyynes from *Bidens pilosa* L

S/N	Plant part	Common name	IUPAC name	Ref.
1	Whole/ aerial	Pilosol A	7-Phenyl-4,6-heptadiyn-2-ol	Chang et al. (2000), Wang et al. (2010)
2	Whole	1-Phenylhepta-1,3-diyn-5en	(5*E*)-5-Heptene-1,3-diyn-1-yl-benzene	Chang et al. (2000)
3	Aerial	(*R*)-1,2-Dihydroxytrideca-3,5,7,9,11-pentayne	(*R*)-3,5,7,9,11-Tridecapentayne-1,2-diol	Tobinaga et al. (2009)
4	Leaves	Pentadiene	1-Tridecene-3,5,7,9,11-Pentayne	Harborne (1974), Sarg et al. (1991)
5	Roots	1,11-Tridecadiene-3,5,7,9-tetrayn-13-acetate	2,12-Tridecadiene 4,6,8,10-tetrayn-1-ol,1-acetate	Harborne (1974)
6	Whole	1,2-Dihydroxytrideca-5,7,9,11-tetrayne	5,7,9,11-Tridecatetrayne-1,2-Diol	Wu et al. (2007), Wang et al. (2007)

Bidens pilosa L. contains approximately 201 substances, including 8 porphyrins, 25 terpenoids, 13 aromatics, 19 phenylpropanoids, 70 aliphatics (36 polyynes), 60 flavonoids, along with 6 more phytochemicals, making it an extraordinary source of phytochemicals (Lima Silva et al. 2011). The details of some polyynes found in BP are provided in Table 18.2.

Polyacetylenes and flavonoids, characteristic metabolite groups in the *Bidens* genus, are the most common types of chemicals among those described (Christensen and Lam 1991). They are also the most reported class of chemicals in terms of biological activity (Chang et al. 2007a). The isolation of hydrocarbons (Chang et al. 2000), phenylpropanoids (Horiuchi and Seyama 2008), terpenoids (Grombone-Guaratini et al. 2005), and steroids (Chang et al. 2000) have also been documented in several prior investigations.

Among the 60 flavonoids found in BP, the biological activities of only seven have been examined. Further research is required as the bioactivities of the other 53 flavonoids are not well characterized. Some identified flavonoids and their biological actions are presented in Table 18.3.

Sesquiterpenes, acetylacetone (Deba et al. 2008), phenylheptadiynol (Wu et al. 2007), phenylpropanoid glucosides, pheophytins (Lee et al. 2008), polyacetylenes (Wu et al. 2007), polyacetylenic glycosides (Chiang et al. 2007), flavonoids, derivatives of p-coumaric acid, flavonoid glycosides, aurons, auron glycosides and derivatives of caffeoylquinic acid (Tobinaga et al. 2009) are among the chemical constituents of BP.

An increasing body of research indicates that BP's phytochemical complexity may be responsible for its range of bioactivities (Yang 2014).

Table 18.3 The biological activities of some of *Bidens pilosa* L.'s chemical components

S/N	Name	Molecular formula	Biological activities	Ref.
1	Centaureidin	$C_{18}H_{16}O_8$	Anti-listerial	Chang et al. (2007a)
2	Jacein	$C_{24}H_{26}O_{13}$	Antioxidant	Chiang et al. (2004)
3	Quercetin 3-O-rutinoside	$C_{27}H_{38}O_{15}$	Antioxidant	Chiang et al. (2004)
4	Quercetin 3,3′-dimethyl ether-7-O-*β*-D-glucopyranoside	$C_{21}H_{20}O_{12}$	Antimalarial	Oliveira et al. (2004)
5	Butein	$C_{15}H_{12}O_5$	Anti-leishmanial	Li (2016)
6	Quercetin 3-O-*β*-D-galactopyranoside	$C_{21}H_{20}O_{12}$	Anti-inflammatory	Nielsen et al. (1998)
7	Quercetin 3,3′-dimethyl ether-7-O-*α*-L-rhamnopyranosyl-(1→6)--D-glucopyranoside	$C_{21}H_{20}O_{11}$	Antimalarial	Andrade-Neto et al. (2004)

18.4 Pharmacological Properties

Researchers worldwide are paying attention to the natural compounds derived from historically used medicinal plants to treat human illnesses. BP has been used in folk medicine for various ailments for many years. The Indigenous People in many areas around the world have a long history of using this plant as an herbal medicine. This current section aims to present BP's pharmacological potential (Singh et al. 2017).

Traditional uses for BP include treating a wide range of illnesses. Over 40 different types of ailments have allegedly been treated with various formulations from the entire plant or its other parts (Bartolome et al. 2013). The entire plant, including the roots and aerial portion, has been linked to the following biological activities: antimicrobial (Geissberger and Séquin 1991), antifungal (Ashafa and Afolayan 2009), IFN-α pro5moter (Chang et al. 2007a), anti-mycobacterium (van Puyvelde et al. 1994), anticancer, anti-angiogenic, antimalarial, antidiabetic, and antibacterial (Pharma et al. 2010). Besides, it also contains antioxidative, immunomodulatory, anti-inflammatory and anti-allergic, anti-secretory and anti-ulcer in the stomach (Pharma et al. 2010), antihypertensive (Dimo et al. 2001) and antitumor (Chang et al. 2007b; Kumari et al. 2009; Sundararajan et al. 2006) properties.

18.4.1 Anticancer Activity

Researchers have validated and documented that phytochemicals from BP possess antitumor properties against various cancerous cells. Folk literature suggests that

BP may be effective against tumors; several rigorous research studies conducted in vitro have confirmed that BP extracts and other isolated constituents have cancer-preventive properties. Bioassay-guided isolation and fractionation techniques have been employed in numerous research studies to isolate novel chemicals from BP (Bartolome et al. 2013).

Extracts from BP whole plant were obtained using n-hexane, chloroform, and methanol extract. Utilizing in vivo models, several extracts, and fractions were screened for antipyretic and anticancer properties utilizing the in vitro comet test (Pharma et al. 2010). Moreover, BP extracts in n-hexane, chloroform, and methanol, as well as their ethyl acetate and aqueous fractions, were tested for toxicity against Hela and KB cells using in vitro comet assays. The ethyl acetate portion of the methanol extract had the most vigorous activity against Hela and KB cells, with CTC50 of 965.2 and 586.2μg/mL, respectively. These results imply that these fractions or BP extracts may be worthy of further research, even with their mild toxicity (Sundararajan et al. 2006).

Chang et al. conducted cytotoxicity studies employing a colorimetric test based on XTT to assess the antileukemic properties of *Houttuynia cordata* Thunb (*H. cordata*) and BP L. var. minor (Blume) Sherff. In the study, BP L. var. minor exhibited greater efficacy in suppressing most leukemic cells than *H. cordata*. Hence, BP L. var. minor (Blume) Sherff was proposed as a potentially useful medicinal herb for leukemia treatment (Chang et al. 2001).Compound 1,2-Dihydroxytrideca-5,7,9,11-tetrayne at 2.5μg/mL and 1,3-Dihydroxy-6(E)-tetradecene-8,10,12-triyne found in BP are known to inhibit tube formation in HUVECs, though the former has a more pronounced effect than the later. Additionally, compound 1,2-Dihydroxy-5(E)-tridecene-7,9,11-triyne entirely prevents the migration and development of endothelial cells in collagen gels at a dosage of 2.5μg/mL (Wu et al. 2004).

The overall results of an investigation by Singh et al. (2017) supported the ecological relevance of human health as well as the bioactive capacity of BP's methanolic leaf extract. The findings provide evidence to encourage further research on the plant under study, which may result in its application in the treatment of cancer, oxidative stress, and antimicrobial therapy (Chang et al. 2004).

18.4.2 Anti-inflammatory Activity

Bioactive substances found in BP, such as phenolic acids, terpenoids, and flavonoids, have been demonstrated to impede the synthesis or function of pro-inflammatory mediators, including leukotrienes, prostaglandins, and cytokines. *Bidens pilosa* L. may aid in reducing inflammation by lowering the concentrations of certain inflammatory chemicals. According to some research, extracts from BP may prevent the initiation of inflammatory signaling mechanisms such as nuclear factor kappa-light-chain-enhancer of activated B cells and mitogen-activated protein kinase (MAPK). These routes are essential for coordinating the body's inflammatory reactions, and BP may have anti-inflammatory properties because it inhibits these pathways.

An essential enzyme for bodily functions, cyclooxygenase-2 (COX-2), transforms arachidonic acid into prostaglandin (PGE2). It serves as an inflammatory marker and becomes activated by various external triggers, suggesting its potential role in inflammatory illnesses (Yoshida et al. 2006). Based on its ability to suppress inflammation, Xin et al. used bioassay-guided fractionation to isolate isookanin, a phytochemical of the flavonoid class, from BP. The anti-irritant properties of the isolate were evaluated. It was demonstrated that the isolate inhibits development of PTGS2 and NOS2, reducing the synthesis of pro-inflammatory intermediaries (nitric oxide and PGE2). Additionally, the isolate reduced the phosphorylation of p38 MAPK and JNK in the MAPK signaling pathway, along with the formation of activator protein 1. Furthermore, in LPS-induced THP-1 cells, there was suppression of Pro-inflammatory mediators. These findings imply that it has anti-inflammatory and medicinal properties (Xin et al. 2021).

Yoshida et al. investigated how aqueous extracts of BP's aerial parts affected dermal cells, HDF development of COX-2 and PGE2, including how they activated MAPKs in reaction to the inflammatory cytokine IL-1β. This study demonstrated that IL-1β promoted the production of COX-2 and activated MAPKs like ERK1/2, p38, and JNK to varying degrees. p38 was the primary regulator of COX-2 expression in HDFs after IL-1β stimulation. SB203580, a p38 inhibitor, consistently inhibited its expression. The ability of BP extracts to reduce inflammation was investigated using this cell platform. In addition to inhibiting COX-2 expression and PGE2 generation, the extract also moderately inhibited ERK1/2 and decreased the initiation of p38 and JNK, depending upon the dose (Yoshida et al. 2006).

Based on formalin, acetic acid, and tail-flick pain tests, scientists have observed that BP's leaf extracts demonstrate analgesic properties in rats. *Bidens pilosa* L. exhibits potent analgesic properties and could potentially alleviate the discomfort from oral lesions in people with HIV/AIDS, thus supporting trado-medical usage

Table 18.4 List of several anti-inflammatory compounds present in *Bidens pilosa*

S/N	Compound	Chemical formula	Class	Ref.
1	Quercetin 3-O-β-D-galactopyranoside	$C_{21}H_{20}O_{12}$	Flavonoid	Nielsen et al. (1998)
2	Ethyl caffeate	$C_{11}H_{12}O_4$	Phenylpropanoid	Chiang et al. (2005)
3	Luteolin	$C_{15}H_{10}O_6$	Flavonoid	Tewtrakul et al. (2003)
4	2-O-β-Glucosyltrideca-11(E)-en-3,5,7,9-tetrayn-1,2-diol	$C_{19}H_{20}O_7$	Polyyne	Tan et al. (2000)
5	2-β-D-Glucopyranosyloxy-1-hydroxy-5(E)-tridecene-7,9,11-triyne	$C_{19}H_{24}O_7$	Polyyne	Nguelefack et al. (2005)

(Ezeonwumelu et al. 2018). All things considered, evidence exists that BP has anti-inflammatory qualities and compounds, as shown in Table 18.4. However, further investigation—including human clinical trials—is required to comprehensively understand its modes of action and therapeutic potential in inflammatory disorders. Furthermore, to prevent any interactions or harmful effects, it is imperative that consumers speak with a healthcare provider before using BP or other herbal medicines, particularly if they have underlying medical issues or are on medication.

18.4.3 Antidiabetic Activity

The scientific community worldwide has paid considerable attention to ethnopharmaceutical herbs such as BP for managing diabetes mellitus because of their potential antidiabetic qualities. A third of the 1200 plants that have been proclaimed as antidiabetic treatments have undergone scientific evaluations to manage type 2 diabetes, T2D (Singh et al. 2011). Among these, BP is frequently utilized for both human and animal consumption and medication (Pozharitskaya et al. 2010). Regarding its antidiabetic properties, research has indicated that BP and its antidiabetic polyynes may successfully mitigate and cure diabetes (both types 1 and 2) with different etiologies (Chang et al. 2013). The etiology of type 1 diabetes (T1D), being an autoimmune-mediated death of pancreatic β-cells, results in hyperglycemia, insulin insufficiency, and related problems. Type 1 diabetes currently has no known cure. Adopting specific characteristics of Th cell polarization regulates the course of T1D.

In the research conducted by Ajagun-Ogunleye et al., 30 rats were placed into six groups, with each group further subdivided into extract and non-extract treatment groups. Three groups were created from the extract-treated group, and each group was administered 200 mg/kg, 400 mg/kg, or 800 mg/kg dose of BP extract, respectively. Two liver function enzymes were biochemically analyzed after week 4. The blood glucose levels were tracked for a month using a standard glucometer. According to the research findings, BP caused hypoglycemia for 2 weeks before losing it in the following weeks. The glucose levels of T1D rats treated with 200 mg/kg recovered better (355.25–164.5 mg/dl), with the rats receiving 400 mg/kg treatment showing the second-best rates of recovery (Ajagun-Ogunleye et al. 2015).

According to another research, butanol fraction of BP prevented T-cell multiplication, diminished Th1 cells and cytokines, and elevated Th2 cells and cytokines in nonobese diabetic (NOD) mice, therefore preventing T1D (Ajagun-Ogunleye et al. 2015). The butanol fraction's IC_{50} value was 200 μg/mL. It was discovered that at 180 μg/mL, the butanol fraction could lead to 50% mortality of Th1 cells, explaining some of the inhibition. Furthermore, via downregulating Th1 cells or upregulating Th2 cells, which have effects opposite to those of Th1 cells, this research revealed that the butanol fraction of BP may avert diabetes in NOD mice in vivo (Chang et al. 2004). In mice that were neither normoglycemic nor mildly diabetic (200–340 mg/dL), the extract of dried BP boiled with 15% water/ethanol for 5 min significantly reduced blood sugar levels; however, it did not affect mice with chronic diabetics. This suggests the relevance of insulin as a moderator of the plant extracts' hypoglycemic effects. In mildly diabetic mice, sound hypoglycemic effects were observed after employing aqueous-based extracts (Alarcon-Aguilar et al. 2002).

Growing data from epidemiological research infer that hereditary and external influences are the main causative factors of T2D. Insulin action, insulin synthesis, or both are impaired due to these causes, which also reduce *β*-cell function and insulin resistance. Thus, this impairment is concomitant with the development of hyperglycemia, a primary pathogenic characteristic of T2D (Laakso 2001).

In another research study (Hsu et al. 2009), 6–8 week-old diabetic db/db mice with blood sugar values of 350–400 mg/dL after meals were evaluated using BP aqueous extract (BPWE). Like the insulin-releasing effects of oral antidiabetic glimepiride, a single BPWE dosage reduced blood sugar values from 374 mg/dL to 144 mg/dL. The glucose-lowering effect of BPWE was linked to an increased level of serum insulin, suggesting that BPWE increases the production of insulin to decrease blood sugar levels. However, compared to glimepiride, BPWE had a unique dynamics of insulin production (Hsu et al. 2009).

Among the flavonoids that have been isolated, butein reportedly has promising properties for treating diabetes mellitus (Lim et al. 2010). Growing *Bidens* species in various climatic conditions might alter their phytochemical composition. For instance, several species of BP plants contain 2-b-D-Glucopyranosyloxy-1-hydroxytrideca-5,7,9,11-tetrayne, or cytopiloyne, a potent polyacetylene that may be used to manage T1D (Hsu et al. 2009; Ubillas et al. 2000). Interestingly, compounds α-Bergamotene, Germacrene A, and α-Gurjunene were absent in the volatile oils of BP cultivated in Japan, while they were present in the essential oils extracted from the plants grown in Argentina in amounts of 11.2%, 39.5%, and 3.3%, respectively (Dang Khanh et al. 2009).

Conclusively, extracts from BP have been demonstrated to offer potential benefits in blood sugar regulation. Various studies have shown that its extracts lower blood glucose levels, indicating the plant's potential as an antidiabetic treatment.

18.4.4 Antioxidant Activity

Since BP contains various bioactive components, including flavonoids, phenolic compounds, and other antioxidants, it has been investigated for its possible antioxidant qualities. This section provides an overview of BP's potential antioxidant activity. Free radicals can harm cellular components through a sequence of chemical interactions (Muchuweti et al. 2007), resulting in the onset and advancement of neurological illnesses, cancer, cardiovascular disease, and aging (Chiang et al. 2004). Macrophages can create superoxide anions, free radicals, and nitric oxide (NO) to eliminate bacteria. However, various diseases are linked to an overabundance of free radicals in pathological settings.

In tests for lipid peroxidation, β-carotene, and 1,1-diphenyl-2-picrylhydrazyl (DPPH), the methanolic extracts of BP aerial portions demonstrate exceptional antioxidant capabilities. *Bidens pilosa* L. has been established to contain 15 typical

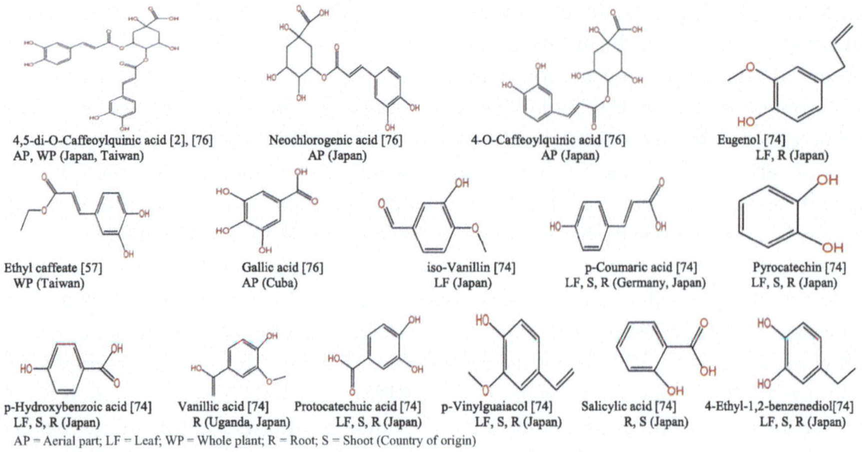

Fig. 18.2 Isolated flavonoids and their derivatives from *Bidens Pilosa* L. (Adapted from Xuan and Khanh 2016)

Table 18.5 Antioxidant activity of *Bidens Pilosa* L.'s water extracts and essential oils

Extract	IC_{50} (μg/mL)
Essential oils from leaves	47
Essential oils from flowers	50
Extract from leaves	61
Extract from flowers	172

Adapted from Deba et al. (2008)

phenolic acids, including the compounds in Fig. 18.2. 5-O-Methylhoslundin is the most abundant compound in the leaves, stems, and roots, with values of 117.4, 298.7, and 350.3μg/g, and p-Coumaric acid is the next abundant compound in the leaves, stems, and roots with values of 18.5, 32.9, and 29.6μg/g, respectively (Deba et al. 2008). Eugenol (6.1, 6.2, and 37.1μg/g, respectively) is the following most abundant compound. These compounds are likely responsible for BP's antioxidant properties (Muchuweti et al. 2007; Deba et al. 2007).

Muchuweti and colleagues conducted a supplementary investigation to ascertain BP's methanolic extract's antioxidant properties and phenolic profile. They calculated BP's methanolic extract to have a phenolic concentration of 1102.8 ± 2.2 mg/g (Muchuweti et al. 2007). This extract contained hydroxybenzaldehyde, ferulic acid, caffeic acid, coumaric acid, and vanillin. Additionally, the extract exhibited effectiveness for DPPH radical scavenging activity (Yuan et al. 2008). The anti-inflammatory properties of BP may also be responsible for this action (Yoshida et al. 2006). The findings of the DPPH free radical scavenging assay are summarized in Table 18.5.

18.4.5 Antimalarial Activity

One of the most serious illnesses, malaria, affects around 219 million people in underdeveloped nations globally. One of the main challenges in controlling malaria is the emergence of resistance against chemical antimalarial medications (Nadia et al. 2020). *Bidens pilosa* L. has been studied for its antimalarial properties, particularly in ethnomedicine, where it is used to manage symptoms of malaria. According to Connelly (2009), BP exhibited substantial suppression opposing parasitemia colonies grown in vitro, making it one of the most promising and effective botanicals against malaria (Connelly 2009).

An investigation of BP's leaf extract's potential to prevent malaria through phytochemistry and biological tests revealed it effectiveness against the *Plasmodium falciparum* NF54 strain of malaria parasites (Kumari et al. 2009). Furthermore, (*R*)-1,2-dihydroxytrideca-3,5,7,9,11-pentayne extracted from aerial portions of BP, prevented the development of the *P. falciparum* FCR-3 strain. The in vivo effects of this chemical were investigated in mice harboring the NK-65 strain of *Plasmodium berghei* infection. Administered as an intravascular shot for a period of 4 days, (*R*)-1,2-dihydroxytrideca-3,5,7,9,11-pentayne was established to reduce the mean red blood cells' parasitemia from 32.8% of the control group to 12.1% (Tobinaga et al. 2009). In mice diagnosed, with *P. berghei*, the ethanol-based extracts of the root produce a greater reduction than extracts of the entire plant and its shoot (Andrade-Neto et al. 2004). In vitro, the development of *P. falciparum* is 86% suppressed by the root's chloroform fractions (Andrade-Neto et al. 2004).

Oliveira et al. (2004) established that raw root extracts produced by percolation with 80% ethanol are efficacious in vitro against *P. falciparum;* this property is a consequence of the flavonoids and polyacetylene present in BP. The extract was subjected to ether and ether methanol (1:1) column chromatography, yielding two layers high in flavonoids and polyacetylene. Antimalarial tests were conducted in vivo, and HPLC/DAD analysis was used to evaluate the extract and its fractions. The ethanol extract demonstrated many peaks for polyacetylene and flavonoids using HPLC. The peaks for polyacetylene in the ether to methanol ratio and flavonoids in the ether fraction were missing. In vivo experiments demonstrated that ethanol extracts reduced parasitemia by 29% on the seventh day and by 36% on the fifth. Compared to those who received fractions, the animals administered with 80% ethanol extract had a greater survival rate. The results demonstrated that polyacetylene and flavonoids are necessary for the in vivo action of the ethanolic extract.

In their work, Nadia et al. (2020) created extracts by macerating the powder of BP leaf in ethyl acetate. The extract and the in vitro antiplasmodial fraction obtained by HPLC were concentrated and vaporized until dry. The efficacy of BP plant components as an antimalarial was evaluated by the use of the Peter and Rane test in vivo in contrast to mice harboring *P. berghei* infections. While the antimalarial efficacy of a selected raw extract was evaluated at two different concentrations, the selected fraction 12 from the ethyl acetate extract was also investigated at higher concentrations. Blood was collected from test subjects (mice) to evaluate their hematological features. Both fraction 12 and the pure ethyl acetate extract exhibited complete in vivo parasite suppression efficacy at 125 mg/kg and 500 mg/kg, respectively. The intraerythrocytic phase of *P. berghei*, or parasitemia, decreased significantly in mice given 250 and 500 mg/kg treatment doses and subsided completely in the animals administered 500 mg/kg by the eighth day. There was much more antiplasmodial activity in the BP fraction 12 ethyl acetate extract. The mice's hematological parameters changed as a result of the therapy by the fifth day of the trial. Over the course of the 30-day trial, all three groups—the groups treated with fraction 12 (125 mg/kg), crude extract (500 mg/kg), and chloroquine (5 mg/kg)—survived, but the negative control group experienced 100% fatalities (Nadia et al. 2020).

Brandão et al. (1997) administered a dose of 50µg/mL, where the crude ethanol extracts of the entire plant, leaves, and roots, as well as the BP chloroform and butanol fractions, inhibited the in vitro development of *P. falciparum* by up to 90%. The fractions partially reduced the *P. berghei* parasitemia in mice in vivo. Given that BP appeared to be a viable antimalarial drug, the study also described the compounds causing this action. A significant component of the root-based chloroform fractions, 1-phenyl-1,3-diyn-5-en-7-ol-acetate, caused 86% of the suppression of parasite development in vitro (Brandão et al. 1997). Clinical research as well as more investigations into the antimalarial mechanism of BP's active components are required.

18.4.6 Antibacterial Activity

Numerous investigations have revealed that BP possesses potent antibacterial properties.

The emergence of several antibiotic-resistant bacteria is posing a worldwide health risk and making it more challenging to treat illnesses. Penicillin, for example, is frequently used to treat the intracellular Listeria, a bacterium transmitted through food; nevertheless, penicillin-resistant bacteria have recently been discovered (Chang et al. 2007b).

Singh et al. (2017) findings highlighted BP's crude methanolic leaf extract's antibacterial activity. The extract displayed noteworthy antibacterial efficacy within the range of 9.1–18.2 mm. Compared with conventional ampicillin (30µg/mL), BP had the strongest antibacterial efficacy in treating *E. coli*, with an inhibition zone of 18.2 mm (10 mg/mL). At 10 mg/mL, the extract significantly suppressed the development of *S. aureus, M. luteus*, and *P. aeruginosa* (inhibition zones of 15.66, 14.66, and 14 mm, respectively), while it was less effective against *C. albicans* (an inhibition zone of 9.1 mm) at the same dose. At greater doses, the extract did, however, exhibit modest action. The extract exhibited noteworthy efficacy against certain bacterial infections, demonstrating minimum inhibitory concentration (MIC) levels between 80 and 870µg/mL. The extract exhibited the highest level of action against *E. coli* (80µg/mL), *S. aureus* (110µg/mL), and *P. aeruginosa* (220µg/mL). The plant extract may evolve into an antibacterial agent since the extract's MIC dramatically inhibited the growth of harmful bacterial strains (Singh et al. 2017) evaluated the in vitro antibacterial properties of nine extracts derived from root, stem, flower, and leaves of BP using ethanol:water (7:3, v/v). The extracts were tested against 60 strains of oxacillin-resistant *S. aureus* (ORSA), as well as *S. aureus* ATCC6538. The average inhibitory zone diameters of BP leaves were found to be substantially higher than those of chlorhexidine (0.12%) against ORSA, and the extracts showed greater activity against *S. aureus* ATCC ($p < 0.05$) (da Silva et al. 2014).

According to the study by Deba et al. (2008) against six bacterial strains, the oils derived from BP demonstrated noteworthy antibacterial properties. Compared to Gram-positive bacteria, the flower's essential oils exhibited a substantially stronger inhibitory action against Gram-negative bacteria. The study highlighted that BP essential oils and aqueous extracts have antibacterial properties, suggesting that they might be a natural remedy for bacterial infections (Deba et al. 2008).

Mohamed and Mathew (2021) examined the antibacterial activity of BP leaf extracts in methanol against *S. aureus and E. coli*. Based on the detected zone of inhibition, the antibacterial activity of the leaf extracts was evaluated at 20%, 50%, and 70% concentrations using the Agar well diffusion technique. Antibacterial activity was detected at all concentrations; however, the maximum zones of inhibition values (18.5 mm and 32 mm, respectively) for *E. coli and S. aureus* were detected at 70% concentration. The findings demonstrate the antibacterial property of BP leaf extracts against the tested bacteria and indicate that they may be developed further, potentially leading to the discovery of active ingredients that may be investigated for the treatment of illnesses linked to *S. aureus and E. coli* (Mohamed and Mathew 2021).

18.4.7 Wound-Healing Activity

Many investigations have been conducted to evaluate BP's capacity for wound healing. In most studies, using plant materials or extracts has demonstrated greater or equivalent effectiveness than the positive controls. However, since there is no information available on the methodology utilized to establish the dosages used in these tests, there is still a significant gap regarding the standardization of doses to be employed.

Fokou et al. (2023) applied mechanical pressure extraction on freshly harvested leaves to extract BP juice. A phytochemical screening process was used on the resulting extract. A noninfected wound model created by skin excision of a 2 × 2 cm region was used to measure its healing activity. Juice from BP was extracted with a 60% yield. Flavonoids, polyphenols, tannins, and saponins were detected in the juice using phytochemical screening. When compared to BIAFINE® (trolamine 0.0670 g/100 g), a protective and soothing agent that can be purchased at pharmacies, it was demonstrated that all dosages of BP juice examined accelerated the rate of wound healing. Like cyclophosphamide, a standard immunosuppressant recommended for inflammatory wounds, BP juice at all tested dosages lowered the phagocytic index and increased leukocyte and platelet counts following cyclophosphamide-induced neutropenia. The findings show that fresh BP leaf juice has immunomodulatory and wound-healing properties in vivo (Fokou et al. 2023).

A study by Namunana et al. (2018) sought to create and develop a medicinal oil from BP; ethanolic extracts from the plant's leaves were made and tested for alkaloids and flavonoids to determine the quality and effectiveness of the prepared medicinal oil in healing wounds. The herbal ointment was formulated using the fusion method. There were no discernible alterations in the physical, chemical, or microbial features. Compared to the other therapies, the herbal oil demonstrated the quickest rate of wound closure and the shortest timeframes for epithelialization and healing, within a day with no alterations to the skin were visible. Therefore, an herbal oil made from BP is a potent wound-healing cure that may be used as an alternative to prescription medication for wound care (Namunana et al. 2018).

In a study by Hassan et al. (2011), Wistar albino rats were used as excision wound models to assess the wound-healing abilities of ethanolic extracts of *Ocimum suave (O. suave)* and BP. Nine ($n = 9$) rats in each group were treated to evaluate the extracts' capacity to cure wounds. Four sets of animals received topical treatments twice a day using distilled water, neomycin sulfate, BP, or *O. suave*. Throughout the experiment, the wound shrinkage, epithelialization, and full recovery rates were measured. Compared to the negative control, wounds treated using neomycin sulfate, BP, and *O. suave* extracts showed quicker rates of wound shrinkage ($p < 0.05$) on the third, sixth, and ninth days. Better collagenation, angiogenesis, and wound tissue organization were also observed upon histological evaluation of the lesion on the seventh day. For both BP and *O. suave*, epithelialization and overall healing time were similar to neomycin sulfate ($p > 0.05$). Hence, BP and *O. suave* extracts have the potential to replace neomycin in wound healing (Hassan 2011).

Conclusively, extracts from BP can speed up wound healing with no noticeable scar. It is suggested that extracts be formulated into stable dose forms for clinical trials and marketed as cost-effective medications for wound healing.

18.4.8 Toxicological Studies

It is necessary to research the toxicological features of BP to ascertain its safety for use as medicine for humans. Research on BP's systemic toxicity, such as its acute, subacute, chronic, and subchronic toxicities in animals and humans, is currently scarce and inadequate despite the plant's use as a component in food intended for human consumption. Toxicities that are acute or subchronic have only been studied in rats and mice thus far (Bartolome et al. 2013). Nevertheless, no extended animal toxicity investigation of BP has been conducted yet.

In the study by Liang et al. (2020), mice were used to examine the oral toxicity of BP after 24 weeks at dosages of 0%, 0.5%, 2.5%, 5%, and 10% of diet. The death rate, body size, organ weight, diet and water intake, hematological markers, serum biochemistry, urinalysis, genotoxicity, and organ histopathology of animals of both sexes were measured. Except for body size and diet intake, which were considerably lessened in mice given 10% BP than in mice that were fed controls, no discernible changes were observed in the aforementioned characteristics between the two groups. Additionally, hens who were given BP for 28 days showed comparable outcomes. The above findings indicate that when administered at a dose of 5% or less of the diet, BP does not have any adverse effects on mice or hens (Liang et al. 2020).

Tcheutchoua et al. (2022) performed a toxicological study to conduct acute and subchronic toxicity investigations. For the acute investigation, rats of both sexes received one injection each of 50:50 combination of BP and *C. citratus*. According to the study, the combination of BP and *C. citratus* do not negatively impact the kidneys, liver, lungs, or spleen when taken as an aqueous extract because of their low toxicity (Tcheutchoua et al. 2022).

Furthermore, it has been documented that BP extracts in both aqueous and ethanolic forms are acutely hazardous to mice (Frida et al. 2007). Mice aged 5–6 weeks, with body weights between 28 and 35 g, were injected with the two extracts at varying dosages intraperitoneally. The 50% lethal doses (LD_{50}) for ethanol and aqueous extracts in mice were 6.15 g/kg BW and 12.30 g/kg BW, respectively (Frida et al. 2007).

A thorough toxicological investigation has not yet been carried out for humans. Moreover, it is uncertain how BP interacts with other medications. Clinical investigations and further safety verification should be executed before considering BP for therapeutic usage (Bartolome et al. 2013).

18.5 Conclusion

The composition, bioactivity, and possible modes of action of BP and its phytochemicals have received attention from scholars. While there has been some progress, more thorough research is still needed to fully understand and validate the traditional usage of each of the specific chemicals identified in BP, as well as to create therapeutic applications. Pharmacological and phytochemical investigations have been conducted worldwide throughout the previous decades to validate BP's usage in traditional folk medicine and there have been some successful human clinical trials to validate BP's effectiveness. Even though these potential pharmacological qualities are encouraging, it is vital to highlight that additional clinical studies are required to completely understand BP's impact. Finally, it is critical to consult medical specialists before utilizing any herbal therapies because they can have negative effects or interfere with medication, particularly in cases of specific medical issues.

References

Ajagun-Ogunleye MO et al (2015) Hypoglycemic and high dosage effects of *Bidens pilosa* in Type-1 diabetes mellitus. J Diabetes Mellitus 05(03):146–154. https://doi.org/10.4236/jdm.2015.53018

Alarcon-Aguilar FJ, Roman-Ramos R, Flores-Saenz JL, Aguirre-Garcia F (2002) Investigation on the hypoglycaemic effects of extracts of four Mexican medicinal plants in normal and Alloxan-diabetic mice. Phytother Res 16(4):383–386. https://doi.org/10.1002/ptr.914

Andrade-Neto VF et al (2004) Antimalarial activity of *Bidens pilosa* L. (Asteraceae) ethanol extracts from wild plants collected in various localities or plants cultivated in humus soil. Phytother Res 18(8):634–639. https://doi.org/10.1002/ptr.1510

Angelini P et al (2021) Metabolomic profiling, antioxidant and antimicrobial activity of Bidens pilosa. PRO 9(6):903. https://doi.org/10.3390/pr9060903

Ashafa AOT, Afolayan AJ (2009) Screening the root extracts from Bidens pilosa L. var. radiata (Asteraceae) for antimicrobial potentials. J Med Plants Res 3(8):568–572. [Online]. Available: http://www.academicjournals.org/JMPR

Bartolome AP, Villaseñor IM, Yang WC (2013) *Bidens pilosa* L. (Asteraceae): botanical properties, traditional uses, phytochemistry, and pharmacology. Evid Based Complement Alternat Med 2013. https://doi.org/10.1155/2013/340215

Bilanda DC et al (2017) Bidens pilosa Ethylene acetate extract can protect against L-NAME-induced hypertension on rats. BMC Complement Altern Med 17(1):479. https://doi.org/10.1186/s12906-017-1972-0

Brandão MGL, Krettli AU, Soares LSR, Nery CGC, Marinuzzi HC (1997) Antimalarial activity of extracts and fractions from Bidens pilosa and other Bidens species (Asteraceae) correlated with the presence of acetylene and flavonoid compounds. J Ethnopharmacol 57(2):131–138. https://doi.org/10.1016/S0378-8741(97)00060-3

Chang M, Wang G, Kuo Y, Lee C (2000) The low polar constituents from *Bidens Pilosa* L. *var. Minor* (Blume) Sherff. J Chin Chem Soc 47(5):1131–1136. https://doi.org/10.1002/jccs.200000152

Chang J-S, Chiang L-C, Chen C-C, Liu L-T, Wang K-C, Lin C-C (2001) Antileukemic activity of *Bidens pilosa* L. *var. minor* (Blume) Sherff and *Houttuynia cordata* Thunb. Am J Chin Med (Gard City N Y) 29(02):303–312. https://doi.org/10.1142/S0192415X01000320

Chang S-L et al (2004) Polyacetylenic compounds and butanol fraction from *Bidens pilosa* can modulate the differentiation of helper T cells and prevent autoimmune diabetes in non-obese diabetic mice. Planta Med 70(11):1045–1051. https://doi.org/10.1055/s-2004-832645

Chang CL-T et al (2007a) Cytopiloyne, a polyacetylenic glucoside, prevents type 1 diabetes in nonobese diabetic mice. J Immunol 178(11):6984–6993. https://doi.org/10.4049/jimmunol.178.11.6984

Chang S-L et al (2007b) Flavonoids, centaurein and centaureidin, from Bidens pilosa, stimulate IFN-γ expression. J Ethnopharmacol 112(2):232–236. https://doi.org/10.1016/j.jep.2007.03.001

Chang CLT, Chen Y-C, Chen H-M, Yang N-S, Yang W-C (2013) Natural cures for type 1 diabetes: a review of phytochemicals, biological actions, and clinical potential. Curr Med Chem 20(7):899–907

Chang CLT, Chung C-Y, Kuo C-H, Kuo T-F, Yang C-W, Yang W-C (2016) Beneficial effect of Bidens pilosa on body weight gain, food conversion ratio, gut bacteria and coccidiosis in chickens. PLoS One 11(1):e0146141. https://doi.org/10.1371/journal.pone.0146141

Chiang Y-M, Chuang D-Y, Wang S-Y, Kuo Y-H, Tsai P-W, Shyur L-F (2004) Metabolite profiling and chemopreventive bioactivity of plant extracts from Bidens pilosa. J Ethnopharmacol 95(2–3):409–419. https://doi.org/10.1016/j.jep.2004.08.010

Chiang Y et al (2005) Ethyl caffeate suppresses NF- κ B activation and its downstream inflammatory mediators, iNOS, COX-2, and PGE $_2$ *in vitro* or in mouse skin. Br J Pharmacol 146(3):352–363. https://doi.org/10.1038/sj.bjp.0706343

Chiang Y-M, Chang CL-T, Chang S-L, Yang W-C, Shyur L-F (2007) Cytopiloyne, a novel polyacetylenic glucoside from Bidens pilosa, functions as a T helper cell modulator. J Ethnopharmacol 110(3):532–538. https://doi.org/10.1016/j.jep.2006.10.007

Christensen LP, Lam J (1991) Acetylenes and related compounds in heliantheae. Phytochemistry 30(1):11–49. https://doi.org/10.1016/0031-9422(91)84096-B

Connelly P (2009) Horrible weed or miracle herb? A review of Bidens Pilosa. J Austr Tradit Med Soc 15(2):77–79. [Online]. Available: http://www.sutherland.nsw.gov.au/ssc/rwpattach.nsf/viewasattach

da Silva JJ et al (2014) In vitro screening antibacterial activity of Bidens pilosa linné and annona crassiflora mart. Against oxacillin resistant Staphylococcus aureus (ORSA) from the aerial environment at the dental clinic. Rev Inst Med Trop Sao Paulo 56(4):333–340. https://doi.org/10.1590/S0036-46652014000400011

Dang Khanh T et al (2009) Allelopathic plants: 20. Hairy Beggarticks (Bidens pilosa L.). Allelopathy J 24(2):243–254. [Online]. Available: http://www.allelopathy-journal.com/allelopathy.aspx?target=ijor:aj&volume=24&issue=2&article=002

Deba F, Xuan TD, Yasuda M, Tawata S (2007) Herbicidal and fungicidal activities and identification of potential phytotoxins from *Bidens pilosa* L. var. *radiata* Scherff. Weed Biol Manag 7(2):77–83. https://doi.org/10.1111/j.1445-6664.2007.00239.x

Deba F, Xuan TD, Yasuda M, Tawata S (2008) Chemical composition and antioxidant, antibacterial and antifungal activities of the essential oils from Bidens pilosa Linn. var. Radiata. Food Control 19(4):346–352. https://doi.org/10.1016/j.foodcont.2007.04.011

Dimo T et al (2001) Effects of the aqueous and methylene chloride extracts of Bidens pilosa leaf on fructose-hypertensive rats. J Ethnopharmacol 76(3):215–221. https://doi.org/10.1016/S0378-8741(01)00229-X

Ezeonwumelu JOC et al (2018) Analgesic appraisal of Bidens pilosa (Asteraceae) leaf extracts used in management of oral lesion pain in HIV/AIDS patients in Rodents. Pharmacol Pharm 09(06):175–192. https://doi.org/10.4236/pp.2018.96014

Fokou JBH et al (2023) Wound healing and immunomodulatory effects of juice from Bidens pilosa fresh leaves. Int J Herb Med 11(6):28–34. https://doi.org/10.22271/flora.2023.v11.i6a.911

Frida L, Rakotonirina S, Rakotonirina A, Savineau J-P et al (2007) Afr J Tradit Complement Altern Med 5(1):79–91

Geissberger P, Séquin U (1991) Constituents of Bidens pilosa L.: do the components found so far explain the use of this plant in traditional medicine? Acta Trop 48(4):251–261. https://doi.org/10.1016/0001-706X(91)90013-A

Grombone-Guaratini MT, Silva-Brandão KL, Solferini VN, Semir J, Trigo JR (2005) Sesquiterpene and polyacetylene profile of the Bidens pilosa complex (Asteraceae: Heliantheae) from Southeast of Brazil. Biochem Syst Ecol 33(5):479–486. https://doi.org/10.1016/j.bse.2004.11.005

Hall R, Beale M, Fiehn O, Hardy N, Sumner L, Bino R (2002) Plant metabolomics. Plant Cell 14(7):1437–1440. https://doi.org/10.1105/tpc.140720

Harborne JB (1974) Naturally occurring acetylenes. Phytochemistry 13(1):315. https://doi.org/10.1016/S0031-9422(00)91337-2

Hassan A (2011) Wound healing potential of the ethanolic extracts of Bidens pilosa and Ocimum suave. Afr J Pharm Pharmacol 5(2):132–136. https://doi.org/10.5897/AJPP10.266

Hong J, Yang L, Zhang D, Shi J (2016) Plant metabolomics: an indispensable system biology tool for plant science. Int J Mol Sci 17(6):767. https://doi.org/10.3390/ijms17060767

Horiuchi M, Seyama Y (2008) Improvement of the antiinflammatory and antiallergic activity of Bidens pilosa L. var. radiata SCHERFF treated with enzyme (Cellulosine). J Health Sci 54(3):294–301. https://doi.org/10.1248/jhs.54.294

Hsu Y-J, Lee T-H, Chang CL-T, Huang Y-T, Yang W-C (2009) Anti-hyperglycemic effects and mechanism of Bidens pilosa water extract. J Ethnopharmacol 122(2):379–383. https://doi.org/10.1016/j.jep.2008.12.027

Kumari P et al (2009) A promising anticancer and antimalarial component from the leaves of *Bidens pilosa*. Planta Med 75(01):59–61. https://doi.org/10.1055/s-0028-1088362

Laakso M (2001) Insulin resistance and its impact on the approach to therapy of type 2 diabetes. Int J Clin Pract Suppl 121:8–12

Lans C, Sant C, Georges K (2020) Ethnoremedies used for horses in british columbia and trinidad and tobago. In: Ethnoveterinary medicine. Springer, Cham, pp 57–71. https://doi.org/10.1007/978-3-030-32270-0_4

Lee T, Lu C, Kuo Y, Lo J, Lee C (2008) Unexpected novel pheophytin peroxides from the leaves of *Bidens pilosa*. Helv Chim Acta 91(1):79–84. https://doi.org/10.1002/hlca.200890016

Li TSC (2016) Chinese and related North American herbs. CRC Press. https://doi.org/10.1201/9781420031881

Liang Y-C et al (2020) Toxicity study of Bidens pilosa in animals. J Tradit Complement Med 10(2):150–157. https://doi.org/10.1016/j.jtcme.2019.04.002

Lim SS, Jung SH, Ji J, Shin KH, Keum SR (2010) Synthesis of flavonoids and their effects on aldose reductase and sorbitol accumulation in streptozotocin-induced diabetic rat tissues. J Pharm Pharmacol 53(5):653–668. https://doi.org/10.1211/0022357011775983

Lima Silva F, Fischer DCH, Fechine Tavares J, Sobral Silva M, Filgueiras de Athayde-Filho P, Barbosa-Filho JM, Compilation of secondary metabolites from Bidens pilosa L. (2011) Molecules 16(2):1070–1102. https://doi.org/10.3390/molecules16021070

Mohamed SA, Mathew C (2021) Antimicrobial activity of Bidens pilosa leaves extracts against staphylococcus aureus and Escherichia coli. Tanz Vet J 38:20–24. https://doi.org/10.4314/tvj.v38i1.4s

Mtenga DV, Ripanda AS (2022) A review on the potential of underutilized Blackjack (Bidens Pilosa) naturally occurring in sub-Saharan Africa. Heliyon 8(6):e09586. https://doi.org/10.1016/j.heliyon.2022.e09586

Muchuweti M, Mupure C, Ndhlala A, Murenje T, Benhura MAN (2007) Screening of antioxidant and radical scavenging activity of Vigna ungiculata, Bidens pilosa and Cleome gynandra. Am J Food Technol 2(3):161–168. https://doi.org/10.3923/ajft.2007.161.168

Nadia NAC et al (2020) Antimalarial activity of ethyl acetate extract and fraction of *Bidens pilosa* against *Plasmodium berghei* (ANKA). J Parasitol Res 2020:1–8. https://doi.org/10.1155/2020/8832724

Namunana S et al (2018) Formulation, development and validation of a wound healing herbal ointment from extracts of Bidens pilosa and Aloe barbadensis. J Pharm Pharmacol Res 02(02). https://doi.org/10.26502/jppr.0008

Nguelefack TB, Dimo T, Mbuyo EPN, Tan PV, Rakotonirina SV, Kamanyi A (2005) Relaxant effects of the neutral extract of the leaves of Bidens pilosa Linn on isolated rat vascular smooth muscle. Phytother Res 19(3):207–210. https://doi.org/10.1002/ptr.1646

Nielsen SF, Christensen SB, Cruciani G, Kharazmi A, Liljefors T (1998) Antileishmanial chalcones: statistical design, synthesis, and three-dimensional quantitative structure–activity relationship analysis. J Med Chem 41(24):4819–4832. https://doi.org/10.1021/jm980410m

Nxumalo KA, Aremu AO, Fawole OA (2023) Metabolite profiling, antioxidant and antibacterial properties of four medicinal plants from Eswatini and their relevance in food preservation. S Afr J Bot 162:719–729. https://doi.org/10.1016/j.sajb.2023.10.008

Oliveira FQ, Andrade-Neto V, Krettli AU, Brandão MGL (2004) New evidences of antimalarial activity of Bidens pilosa roots extract correlated with polyacetylene and flavonoids. J Ethnopharmacol 93(1):39–42. https://doi.org/10.1016/j.jep.2004.03.026

Pharma D et al (2010) Scholars Research Library An updated review on Bidens Pilosa L. [Online]. Available: www.derpharmachemica.com

Piasecka A, Kachlicki P, Stobiecki M (2019) Analytical Methods for Detection of Plant Metabolomes Changes in Response to Biotic and Abiotic Stresses. Int J Mol Sci 20(2):379. https://doi.org/10.3390/ijms20020379

Pozharitskaya ON et al (2010) Anti-inflammatory activity of a HPLC-fingerprinted aqueous infusion of aerial part of Bidens tripartita L. Phytomedicine 17(6):463–468. https://doi.org/10.1016/j.phymed.2009.08.001

Sarg TM, Ateya AM, Farrag NM, Abbas FA (1991) Constituents and biological activity of Bidens pilosa L. grown in Egypt. Acta Pharm Hung 61(6):317–323

Siddaiah C et al (2020) Metabolite Profiling of alangium salviifolium bark using advanced LC/MS and GC/Q-TOF technology. Cells 10(1):1. https://doi.org/10.3390/cells10010001

Singh J, Cumming E, Manoharan G, Kalasz H, Adeghate E (2011) Medicinal chemistry of the antidiabetic effects of momordica charantia: active constituents and modes of actions. Open Med Chem J 5(Suppl 2):70–77. https://doi.org/10.2174/1874104501105010070

Singh G et al (2017) Pharmacological potential of Bidens pilosa L. and determination of bioactive compounds using UHPLC-QqQLIT-MS/MS and GC/MS. BMC Complement Altern Med 17, no. 1:492. https://doi.org/10.1186/s12906-017-2000-0

Smith HH, Idris OA, Maboeta MS (2021) Global trends of green pesticide research from 1994 to 2019: a bibliometric analysis. J Toxicol 2021:1–11. https://doi.org/10.1155/2021/6637516

Sundararajan P, Dey A, Smith A, Doss AG, Rajappan M, Natarajan S (2006) Studies of anticancer and antipyretic activity of Bidens pilosa whole plant. Afr Health Sci 6(1):27–30. https://doi.org/10.5555/afhs.2006.6.1.27

Tan PV, Dimo T, Dongo E (2000) Effects of methanol, cyclohexane and methylene chloride extracts of Bidens pilosa on various gastric ulcer models in rats. J Ethnopharmacol 73(3):415–421. https://doi.org/10.1016/S0378-8741(00)00290-7

Tcheutchoua YC et al (2022) Acute and subchronic toxicity studies on the aqueous extract of the plant mixture (Bidens pilosa and Cymbopogon citratus Aerial Parts) in rat model. J Toxicol 2022:1–13. https://doi.org/10.1155/2022/1998433

Tewtrakul S et al (2003) HIV-1 integrase inhibitory substances from *Coleus parvifolius*. Phytother Res 17(3):232–239. https://doi.org/10.1002/ptr.1111

Thirumurugan D, Cholarajan A, Raja SSS, Vijayakumar R (2018) An introductory chapter: secondary metabolites, in Secondary metabolites—sources and applications. INTECH. https://doi.org/10.5772/intechopen.79766

Tobinaga S et al (2009) Isolation and identification of a potent antimalarial and antibacterial polyacetylene from *Bidens pilosa*. Planta Med 75(06):624–628. https://doi.org/10.1055/s-0029-1185377

Ubillas RP et al (2000) Antihyperglycemic acetylenic glucosides from Bidens pilosa. Planta Med 66(1):82–83. https://doi.org/10.1055/s-0029-1243117

van Puyvelde L, Ntawukiliyayo JD, Portaels F, Hakizamungu E (1994) *In vitro* inhibition of mycobacteria by Rwandese medicinal plants. Phytother Res 8(2):65–69. https://doi.org/10.1002/ptr.2650080202

Wang H, Lu S, Hua LI, Yao Z (2007) EDTA-enhanced phytoremediation of lead contaminated soil by Bidens maximowicziana. J Environ Sci 19(12):1496–1499. https://doi.org/10.1016/S1001-0742(07)60243-5

Wang R, Wu Q-X, Shi Y-P (2010) polyacetylenes and flavonoids from the aerial parts of *Bidens pilosa*. Planta Med 76(09):893–896. https://doi.org/10.1055/s-0029-1240814

Waris M, Koçak E, Gonulalan EM, Demirezer LO, Kır S, Nemutlu E (2022) Metabolomics analysis insight into medicinal plant science. TrAC Trends Anal Chem 157:116795. https://doi.org/10.1016/j.trac.2022.116795

Wu L-W et al (2004) Polyacetylenes function as anti-angiogenic agents. Pharm Res 21(11):2112–2119. https://doi.org/10.1023/B:PHAM.0000048204.08865.41

Wu L-W et al (2007) A novel polyacetylene significantly inhibits angiogenesis and promotes apoptosis in human endothelial cells through activation of the CDK inhibitors and Caspase-7. Planta Med 73(7):655–661. https://doi.org/10.1055/s-2007-981527

Xin Y-J et al (2021) Anti-inflammatory activity and mechanism of Isookanin, isolated by bioassay-guided fractionation from Bidens pilosa L. Molecules 26, no. 2:255. https://doi.org/10.3390/molecules26020255

Xuan TD, Khanh TD (2016) Chemistry and pharmacology of Bidens pilosa: an overview. J Pharm Investig 46(2):91–132. https://doi.org/10.1007/s40005-016-0231-6

Xuan TD, Shinkichi T, Khanh TD, Chung IM (2005) Biological control of weeds and plant pathogens in paddy rice by exploiting plant allelopathy: an overview. Crop Prot 24(3):197–206. https://doi.org/10.1016/j.cropro.2004.08.004

Yan A, Wang Y, Tan SN, Mohd Yusof ML, Ghosh S, Chen Z (2020) Phytoremediation: a promising approach for revegetation of heavy metal-polluted land, Front. Plant Sci 11. https://doi.org/10.3389/fpls.2020.00359

Yang W-C (2014) Botanical, pharmacological, phytochemical, and toxicological aspects of the antidiabetic plant *Bidens pilosa* L. Evid Based Complement Alternat Med 2014:1–14. https://doi.org/10.1155/2014/698617

Yoshida N, Kanekura T, Higashi Y, Kanzaki T (2006) *Bidens pilosa* suppresses interleukin-1β-induced cyclooxygenase-2 expression through the inhibition of mitogen activated protein kinases phosphorylation in normal human dermal fibroblasts. J Dermatol 33(10):676–683. https://doi.org/10.1111/j.1346-8138.2006.00158.x

Yuan L-P et al (2008) Protective effects of total flavonoids of Bidens pilosa L. (TFB) on animal liver injury and liver fibrosis. J Ethnopharmacol 116(3):539–546. https://doi.org/10.1016/j.jep.2008.01.010

Bidens biternata: Chemistry and Pharmacology

19

Xue Yang, Zi-Chang Jia, Chunmei Gan, Xiaoqiong Zhao, and Mo-Xian Chen

Abstract

Bidens biternata is a genus of flowering plants in the Asteraceae family with a long history in numerous countries and regions. In this chapter, we first explore the historical significance of *Bidens biternata*, highlighting its traditional medicinal uses in different cultures. We then discuss the main chemical constituents of the plant, identify the main bioactive compounds that contribute to its therapeutic potential, and study the biological activities of these compounds, focusing on their anti-inflammatory, antibacterial, antiviral, antitumor, and antioxidant properties. In addition to exploring the individual properties of *Bidens biternata*, we compared it with those of other species of the same genus. This comparative analysis revealed similar or unique pharmacological properties between *Bidens biternata* and other plants of the same genus. Our study contributes to the understanding of the chemical composition and pharmacological properties of *Bidens biternata*, provides valuable insights into contemporary healthcare practices, and provides a theoretical basis for future research and potential applications of *Bidens biternata* in the development of novel therapeutic drugs. By combining traditional knowledge with modern scientific research, we emphasize the relevance of *Bidens biternata* in modern medicine. With its rich history and diverse pharmacological properties, *Bidens biternata* holds promise as an important source of new therapies and medicines for contemporary healthcare.

Authors Xue Yang and Zi-Chang Jia have equally contributed to this chapter.

X. Yang · Z.-C. Jia · C. Gan · X. Zhao · M.-X. Chen (✉)
State Key Laboratory of Green Pesticide, Key Laboratory of Green Pesticide and Agricultural Bioengineering, Ministry of Education, Center for R&D of Fine Chemicals of Guizhou University, Guiyang, China

Y. Bibi et al. (eds.), *The Genus Bidens*,
https://doi.org/10.1007/978-981-96-4257-1_19

Keywords

Biternata · Compounds · Pharmacology · Chemistry · Flavonoids

19.1 Introduction

19.1.1 Definition of the *Bidens biternata*

Bidens biternata is an annual herbaceous plant in the genus *Bidens* of the family Asteraceae, commonly known as Biternate Beggarticks or Spanish needles. *Bidens biternata* is widely distributed in Asia, Africa, and the southwestern Pacific islands, and in China, it is mainly found in the eastern, central, southern, and southwestern regions. *Bidens biternata* usually grows in moist or wet environments such as swamps, marshes, and riverbanks (Li et al. 2012a). It has erect stems, serrated leaves, and flowers that are a combination of white and yellow in color, surrounded by three to five ligulate flowers appearing white on the periphery, and tubular flowers that are yellow on the inside. The fruits are black striped with barbed hairs on them, and these barbs can be attached to fur or clothes when people or animals pass by and carried to distant places which helps in reproduction (Chen et al. 2021).

The overall morphological characteristics of *Bidens biternata* are shown in Fig. 19.1.

Fig. 19.1 Image of *Bidens biternata*. (**a**) Flower of *Bidens biternata*; (**b**) leaves of *Bidens biternata*; (**c**) fruit pods of *Bidens biternata*. (Figure comes from the following website: https://powo.science.kew.org/taxon/urn:lsid:ipni.org:names:184771-1)

19.1.2 Importance of Studying *Bidens biternata* Chemistry and Pharmacology

Chemical and pharmacological studies of *Bidens biternata* are of great significance in revealing its medicinal value and developing new natural drugs. *Bidens biternata* is widely used in traditional medicine, and the plant is rich in a variety of bioactive components such as flavonoids, alkaloids, volatile oils, phenols, and terpenoids, which make it play an important role in anti-inflammatory, antioxidant, antimicrobial, and immunomodulatory properties (Shen et al. 2015a; Zahara et al. 2021). Despite such a wide range of applications, the specific molecular mechanisms involved in their action are not well understood. With the continuous development of modern pharmacological research techniques, this drawback has been effectively remedied. First, the analysis of their chemical composition, structure, and properties based on modern pharmacological research has made it easier to further analyze the mechanism and drug development. In addition, in-depth pharmacological studies have revealed the mechanism of interaction between the bioactive components of *Bidens biternata* and molecular targets in organisms, which is essential for promoting the modernization and standardization of traditional medicines, improving their efficacy and safety, and promoting the integration of traditional and modern medicine (Zhu et al. 2022; Jia et al. 2024). Moreover, research on the pharmacological effects of *Bidens biternata* is important for public health and global health. Many regions rely on traditional plant medicines to treat common ailments such as infection, inflammation, and pain. Scientific studies can validate the safety and efficacy of these traditional uses and inform public health policy development.

In conclusion, chemical and pharmacological studies of *Bidens biternata* are important to explore its medicinal value and promote drug development. With the continuous improvement of modern pharmacological research techniques, these studies will continue to play an important role in the future.

19.1.3 The Historical Background of *Bidens biternata* in China

Bidens biternata, commonly known as Jin Zhan Yin Pan in Chinese, as a kind of traditional Chinese medicine, *Bidens biternata* has a wide range of practical applications. According to "A Supplement to the Compendium of Materia Medica" and "Records of Medicinal Herbs Collected in Lingnan," the whole herb of *Bidens biternata* can be used medicinally and is effective in clearing heat and removing toxins, dispersing blood stasis and activating blood circulation. Specifically, its leaves, stems, and roots can be used to treat colds, fever, gastroenteritis, rheumatic pain, and other diseases (Li and Xia 1996), and its medicinal value and efficacy have been fully recognized and applied in traditional Chinese medicine. In recent years, the chemical composition and pharmacological effects of *Bidens biternata* have been systematically explored, and its medicinal mechanism has been preliminarily revealed to provide a scientific basis for the further development and utilization of bitter ginseng.

In addition, besides the exploration of its pharmacological effects, some studies have also focused on its ecological properties, and the distribution of *Bidens biternata* in different environments affects the community structure and its interactions with other plant species. In addition, *Bidens biternata* has some positive impacts on the environment. Its dense root system makes it highly adaptable, which can effectively restore the soil environment and stabilize damaged ecosystems. On the one hand, it can effectively stabilize the soil and reduce soil erosion; on the other hand, it can also enrich heavy metals in the soil, improve the soil environment, and participate in the ecological restoration of the soil (Xiong et al. 2023; Xu et al. 2023). In addition, *Bidens biternata* can absorb some organic pollutants during its growth, improve water quality, and protect aquatic organisms (Du 2021). In addition, it also provides habitats and food sources for various insects and small animals and promotes the maintenance and stabilization of biodiversity (Wu et al. 2021). In conclusion, *Bidens biternata* also plays an important role in maintaining ecological balance and protecting the environment.

19.2 Bioactive Compounds and Metabolic Regulation Pathways in *Bidens*

19.2.1 Key Bioactive Compounds in *Bidens biternata*

As a medicinal plant, *Bidens biternata* contains a wide range of bioactive constituents including flavonoids (quercetin, chrysin, and astragaloside), volatile oils (geraniol, stigmasterol, and geranial), phenols (2,4-di-tert-butylphenol and chlorogenic acid), terpenes (monoterpenes and sesquiterpenes), amino acids, and sugars. Detailed information and the corresponding chemical structure formulas are shown in Fig. 19.2 and Table 19.1.

19.2.2 Isolation and Identification Methods

The isolation and characterization of the active bioactive components of *Bidens biternata* can be divided into four steps in detail which are extraction, separation, purification, and structural characterization, respectively.

Among them, the extraction is often carried out by solvent extraction method, the process usually choose methanol, ethanol, or water as the extraction solvent for reflux extraction, can also be used to ultrasonic-assisted extraction, microwave-assisted extraction in order to shorten the extraction time to improve efficiency (Peng et al. 2012; Wei and Xiong 2012). Separation of active ingredients is often carried out by column chromatography; the principle is based on the adsorption-desorption capacity difference or partition coefficient difference between the stationary phase and the mobile phase of each component of the mixture, and the separation is realized by continuous elution (Li et al. 2016). The purification of the product is often carried out by preparative thin-layer chromatography (TLC),

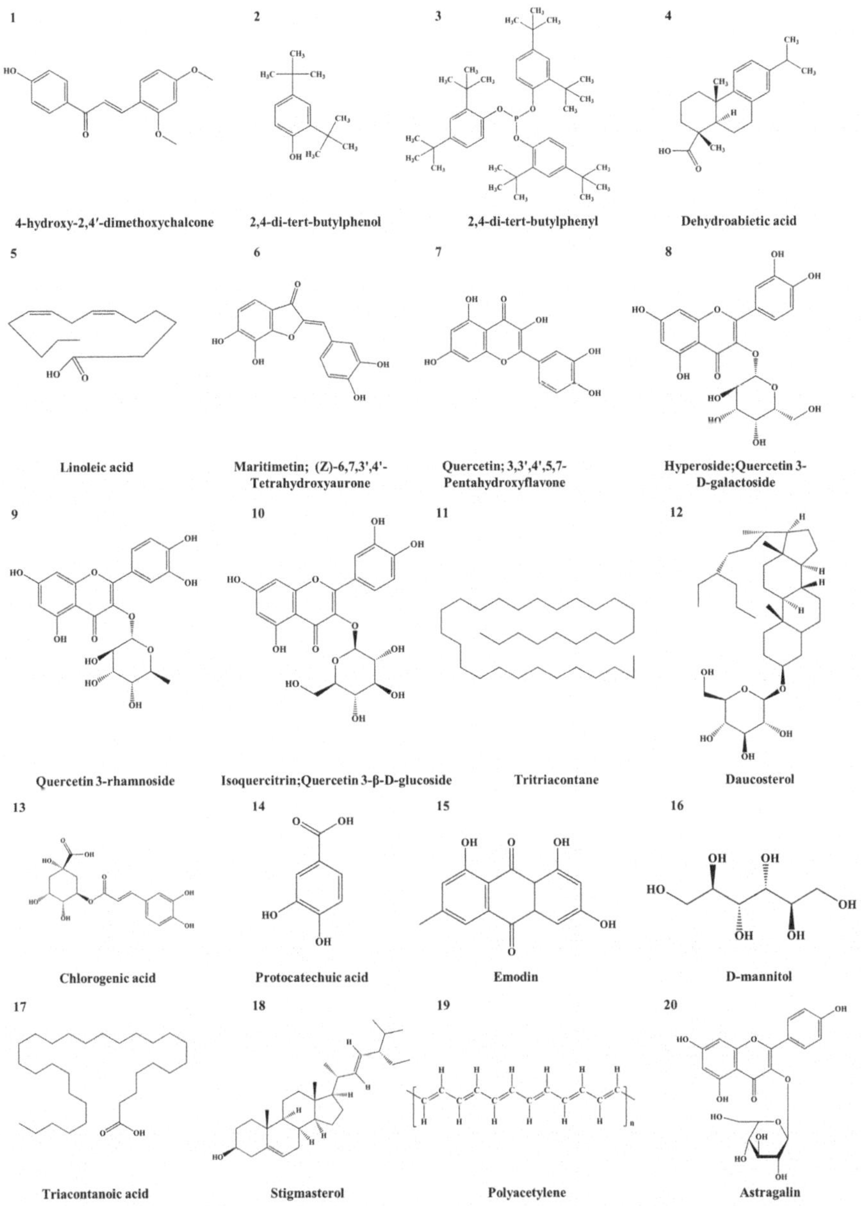

Fig. 19.2 Bioactive components in *Bidens biternata* and their chemical structural formulae

high-performance liquid chromatography (HPLC) and semi-preparative high-performance liquid chromatography (HPLC), etc. The principle is similar to that of column chromatography in that it is separated and purified through the interaction between the stationary phase and the mobile phase (Zhou 2010). Structural identification is commonly carried out by analytical techniques such as mass spectrometry

Table 19.1 Differences in key compounds and applications between *Bidens biternata* and other medicinal plants

No.	Compound type	Chemical constituents	Molecular formula	Extracts	Parts	References
1	Flavonoids	4-Hydroxy-2,4′-dimethoxychalcone	$C_{17}H_{16}O_4$	Ethanol	Leaves	Zahara et al. (2022)
2	Phenols	2,4-di-*tert-butylphenol*	$C_{14}H_{22}O$	Ethanol	Leaves	Zahara et al. (2022)
3	Phenols	2,4-di-*tert-butylphenyl*	$C_{42}H_{63}O_3P$	Ethanol	Leaves	Zahara et al. (2022)
4	Organic acids	Dehydroabietic acid	$C_{20}H_{28}O_2$	Hexane	Leaves	Zahara et al. (2021)
5	Fatty acid	Linoleic acid	$C_{18}H_{32}O_2$	Hexane	Leaves	Zahara et al. (2021)
6	Flavonoids	Maritimetin; (Z)-6,7,3′,4′-tetrahydroxyaurone	$C_{15}H_{10}O_6$	Ethanol	Leaves	Li et al. (2016), Jiang et al. (2008), and Du et al. (2013)
7	Flavonoids	Quercetin; 3,3′,4′,5,7-pentahydroxyflavone	$C_{15}H_{10}O_7$	Ethanol	Leaves	Li et al. (2016), Du et al. (2013), and Wagner et al. (1976)
8	Flavonoids	Hyperoside; quercetin 3-D-galactoside	$C_{21}H_{20}O_{12}$	Ethanol	Leaves	Li et al. (2016) and Ma et al. (2012)
9	Flavonoids	Quercetin 3-rhamnoside	$C_{21}H_{20}O_{11}$	Ethanol	Leaves	Li et al. (2016) and Jiang et al. (2008)
10	Flavonoids	Isoquercitrin; quercetin 3-β-D-glucoside	$C_{21}H_{20}O_{12}$	Ethanol	Leaves	Li et al. (2016), Jiang et al. (2008), and Ma et al. (2012)
11	–	Tritriacontane	$C_{33}H_{68}$	Ethanol	Leaves	Li et al. (2011)
12	Flavonoids	Daucosterol	$C_{35}H_{60}O_6$	Ethanol	Leaves	Li et al. (2011)
13	Phenols	Chlorogenic acid; 3-O-caffeoylquinic acid	$C_{16}H_{18}O_9$	Ethanol	Leaves	Zhou (2010)
14	Organic acids	Protocatechuic acid; 3,4-dihydroxybenzoic acid	$C_7H_6O_4$	Ethanol	Leaves	Li et al. (2011)
15	Anthraquinone	Emodin	$C_{15}H_{10}O_5$	Ethanol	Leaves	Li et al. (2011)

16	–	D-mannitol	$C_6H_{14}O_6$	Ethanol	Leaves	Li et al. (2011)
17	Organic acids	Triacontanoic acid	$C_{30}H_{60}O_2$	Ethanol	Leaves	Li et al. (2012b)
18	Volatile oils	Stigmasterol	$C_{29}H_{48}O$	Ethanol	Leaves	Li et al. (2012b)
19	Polyacetylene	Polyacetylene	$(C_2H_2)n$	Ethanol	Leaves	Li et al. (2012b)
20	Flavonoids	Astragalin	$C_{21}H_{20}O_{11}$	Ethanol	Leaves	Ma et al. (2012)
21	Organic acids	4-O-(2″-O-acetyl-6″-O-p-coumaroyl-β-D-glucopyranosyl)-β-coumaric acid	–	Ethanol	Leaves	Li et al. (2012b)

(MS), nuclear magnetic resonance (NMR), and infrared spectroscopy (IR), and the chemical structure of a compound is determined by comparing the spectral data obtained with the spectral information of known compounds (Li et al. 2012b).

19.2.3 Metabolic Pathway of Key Chemical Compounds in *Bidens biternata*

The chemical constituents of *Bidens biternata* include flavonoids, alkaloids, volatile oils, phenols, terpenoids, and others. The anabolism of these compounds in plants involves several pathways. The anabolic pathways of several major compounds in *Bidens biternata* are summarized below.

19.2.3.1 Metabolic Pathways of Flavonoids

Flavonoids are secondary metabolites of polyphenols in plants, which have a wide range of biological activities including antioxidant, anti-inflammatory, and anticancer activities. Their biosynthesis begins with phenylalanine, which is gradually converted into various derivatives under the action of various enzymes such as phenylalanine deaminase, cinnamic acid-4-hydroxylase, chalcone synthetase, and flavonoid 3′,5′-hydroxylase.

19.2.3.2 Metabolic Pathway of Alkaloids

Alkaloids are a group of nitrogen-containing natural secondary metabolites with analgesic, antibacterial, anticancer, and other biological activities. They are usually synthesized from amino acids (e.g., lysine, tryptophan, and phenylalanine) by various enzymatic reactions such as decarboxylation, methylation, and oxidation.

19.2.3.3 Metabolic Pathways of Volatile Oils

Volatile oils are composed of terpenoids, aromatic, aliphatic, and sulfur-containing nitrogen compounds, which have antibacterial, anti-inflammatory, antioxidant, and other biological activities. The metabolic pathway of volatile oil is centered on terpene synthesis, which begins with the mevalonate pathway and the methylerythritol phosphate pathway. These pathways produce isopentenyl pyrophosphate (IPP) and dimethylallyl pyrophosphate (DMAPP) as precursors for terpene synthesis to form different types of terpenes, such as monoterpenes, sesquiterpenes, and diterpenes, catalyzed by different terpene synthases.

19.2.3.4 Metabolic Pathways of Phenolic Compounds

Phenolic compounds are formed by the combination of benzene ring with hydroxyl group and have biological activities such as antioxidant, anti-inflammatory, and anticancer. The metabolism of phenolic compounds is centered on the phenylalanine pathway, with phenylalanine and tyrosine as precursors, and is cross-regulated through the phenylalanine metabolic pathway and the manganic acid pathway.

19.2.3.5 Metabolic Pathways of Other Organic Compounds

The metabolism of other organic compounds, such as amino acids and organic acids, involves a variety of biosynthetic pathways, e.g., glycolysis/pentose-phosphate pathway and tricarboxylic acid cycle.

19.3 Traditional Uses and Modern Pharmacological Research in *Bidens biternata*

19.3.1 Historical Use of Traditional Medicines in *Bidens biternata*

Bidens biternata has a wide range of medicinal and therapeutic potentials and is widely used in the practical application of traditional medicine for the treatment of hepatitis, colds, coughs, diarrhea, and dysentery (Shahwar et al. 2011; Wang 2015). For example, studies have shown that *Bidens biternata* extract promotes intestinal peristalsis and secretory activity (Kinuthia et al. 2016). Moreover, research indicates that the ethanol extract of *Bidens biternata* exhibits cytotoxicity and efficacy against various cancers, including cervical cancer (HELA), hepatocellular carcinoma (HEPG), and adenocarcinoma human alveolar base epithelial cells (A549), indicating significant anticancer potential (Zahara et al. 2022; Almosnid et al. 2018; Marwan 2018). Additionally, *Bidens biternata* shows robust antioxidant activity, effectively inhibiting various pathogenic bacteria, such as Fusarium spp. and Alternaria spp., and demonstrating broad-spectrum antimicrobial activity (Zahara et al. 2019; Dai et al. 2025). Furthermore, *Bidens pilosa* and related plants exhibit promising antimalarial properties (Brandao et al. 1997), with studies indicating that *Bidens biternata* is resistant to pathogenic bacteria and viruses (House 2013). Thus, understanding plant chemical compositions and pharmacological effects will reveal their potential health benefits, facilitating drug discovery and development.

19.3.2 Modern Pharmacological Studies Validating Traditional Uses in *Bidens biternata*

Bidens biternata has been widely used in traditional medicine for the treatment of inflammation and pain relief, but the exact mechanisms involved are unknown. This has been remedied in modern pharmacological studies, where the mechanism of the anti-inflammatory effects of the flavonoids and polyphenols in *Bidens biternata* has been elucidated based on in vitro studies and animal model analysis, and these compounds reduce inflammation and pain mainly by inhibiting the release of inflammatory mediators, such as TNF-α, IL-1β, and IL-6, as well as by inhibiting the NF-κB signaling pathway (Lin et al. 2013). In addition, the active ingredient in *Bidens biternata* may be used to treat certain tumors or cancer-related symptoms. For example, *Bidens biternata* extracts can inhibit the proliferation of various cancer cells, induce apoptosis, and inhibit tumor angiogenesis by modulating the Bax/Bcl-2 ratio, activating caspase-3 and caspase-9, and inhibiting the expression of vascular endothelial growth factor (VEGF) (Zahara et al. 2022; Zhu et al. 2022). Moreover, *Bidens biternata* extracts significantly reduced blood glucose levels in experimental diabetic animal models for the treatment of diabetes mellitus (Deng et al. 2017). Among them, flavonoids and polyphenols are considered to be the main active components of its hypoglycemic effect, which mainly act in various ways such as improving insulin sensitivity, promoting insulin secretion, and improving glucose metabolism (Wang 2015). In addition, the flavonoids and polyphenols in *Bidens biternata* can enhance the activity of immune cells such as macrophages and natural killer cells, regulate the balance of cytokines, and enhance the immune response and anti-inflammatory capacity of the body in order to improve immunity and prevent disease (Wang et al. 2014). Finally, several studies have shown that *Bidens biternata* has a strong inhibitory effect on a variety of bacteria, fungi, and viruses. First, *Bidens biternata* extracts have been shown to have strong inhibitory effects against common pathogenic bacteria such as *Candida albicans* and *Escherichia coli*. Second, the antiviral potential of *Bidens biternata* has also been demonstrated, and its antiviral mechanism involves the inhibition of viral RNA synthesis and protein assembly (Zahara et al. 2018, 2021). The pharmacological mechanism of the bioactive compounds in *Bidens biternata* is shown in Fig. 19.3.

19.4 Comparative Studies with Other Medicinal Plants in *Bidens biternata*

19.4.1 Comparison of the Chemical Composition of *Bidens biternata* with That of the Same Genus

Besides *Bidens biternata*, many other types of *Bidens* plants have medicinal properties. Some of the more common ones, such as *Bidens biternata*, *Bidens pilosa*, *Bidens bipinnata*, and *Bidens tripartita*, are widely used in traditional medicine in different regions. Compared to them, *Bidens biternata* has many similarities in terms of active ingredients and medicinal values, but also has its own characteristics. The similarities lie in the fact that all four species *Bidens biternata*, *Bidens pilosa*, *Bidens bipinnata*, and *Bidens tripartita* are rich in flavonoids such as quercetin and lignans as well as polyphenolic compounds such as chlorogenic and caffeic acids, which have significant anti-inflammatory, antioxidant, and antitumor activities. The differences are that both *Bidens biternata* and *Bidens pilosa* contain polyacetylene compounds with unique antitumor and antimicrobial activities (Liu et al. 2024). *Bidens tripartita* contains coumarin constituents with diuretic and antipyretic effects. Sesquiterpenes in *Bidens pilosa* and volatile oil constituents in *Bidens bipinnata* have unique antitumor and antimicrobial activities (Deng et al. 2017).

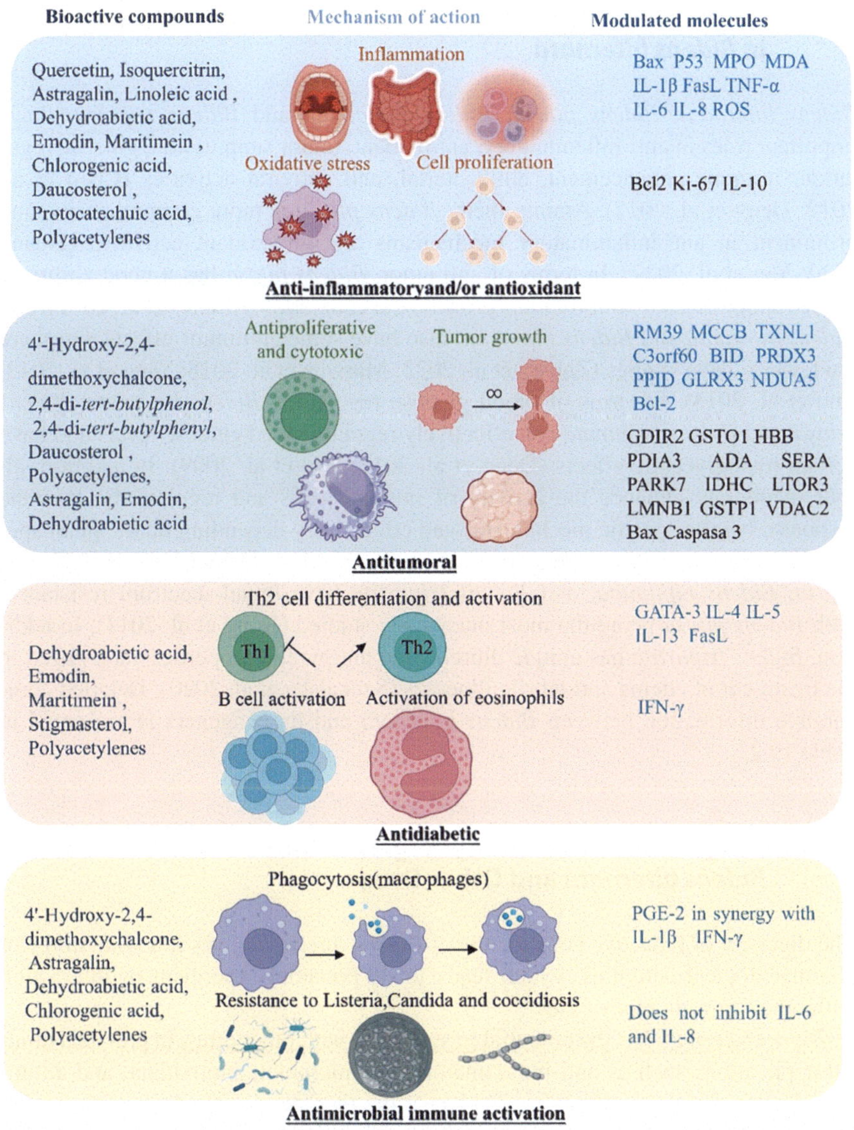

Fig. 19.3 The pharmacological mechanisms of bioactive compounds in *Bidens biternata*. Anti-inflammatory and/or antioxidant: *Bidens biternata* bioactive compounds, such as flavonoids and phenols, are capable of scavenging free radicals, thereby reducing oxidative stress, which has been linked to the prevention of several diseases. They also inhibit the release of pro-inflammatory cytokines, thereby reducing the inflammatory response; antitumoral: Compounds such as 4-hydroxy-2,4′-dimethoxy-chalcone in *Bidens biternata* show significant anticancer activity. These compounds inhibit tumor growth and spread by inducing cancer cell apoptosis (programmed cell death); antidiabetic: Some components of *Bidens biternata* can regulate blood sugar levels and improve insulin resistance, thus exerting antidiabetic effects. These compounds work by affecting glucose metabolism and insulin signaling pathways; antimicrobial immune activation: Flavonoids, phenols, and other bioactive compounds in *Bidens biternata* have significant antibacterial and antifungal activity and are effective in fighting infection. In addition, these compounds can activate the immune system and enhance the body's immune defense

19.4.2 Contrasting Pharmacological Activities and Applications in *Bidens biternata*

Bidens biternata, *Bidens pilosa*, *Bidens bipinnata*, and *Bidens tripartita* have important roles in anti-inflammatory, antioxidant, tumor suppression, glucose regulation, immune enhancement, antibacterial, and antiviral activities (Chen et al. 2012; Deng et al. 2017). Among them, *Bidens pilosa* is more comprehensive and prominent in anti-inflammatory mechanisms and antioxidant activities (Huang 2012; Yan et al. 2022). In terms of antitumor, *Bidens pilosa* has a good ability to induce apoptosis and inhibit tumor growth (Cai et al. 2023; Arantes et al. 2021), and *Bidens biternata* and *Bidens bipinnata* also have some antitumor effects, but there are fewer related studies (Zahara et al. 2022; Marwan et al. 2018; Yang et al. 2013; Zhu et al. 2013). In terms of blood glucose regulation, *Bidens biternata*, *Bidens pilosa*, and *Bidens bipinnata* can effectively regulate blood glucose levels and have certain hypoglycemic effects (Deng et al. 2017; Hsu et al. 2009). In addition, all four plants can enhance the activity of immune cells and regulate the immune response, but the specific mechanisms and effects vary depending on the plant species. In terms of antimicrobial and antiviral activities, *Bidens biternata*, *Bidens pilosa*, *Bidens bipinnata*, and *Bidens tripartita* have broad-spectrum resistance, with *Bidens pilosa* being the most intensively studied (Kong et al. 2011). In addition, *Bidens tripartita* has unique diuretic and antipyretic properties and is used in the treatment of edema and febrile illnesses (Szekalska et al. 2020). Detailed comparative information between *Bidens biternata* and its congeners is presented in Table 19.2.

19.4.3 Insights into Potential Synergies or Interactions Between *Bidens biternata* and Other Drugs

The diversity of bioactive constituents of *Bidens biternata* gives it a wide range of pharmacological activities, which leads to its potential synergism or interaction with other drugs in many cases.

Bidens biternata has the potential to synergize with other drugs in practical clinical applications such as anti-inflammatory, antimicrobial, antioxidant, and antitumor. Such synergistic effects may enhance the therapeutic effect on diseases, but may also lead to the development of bacterial resistance, cytotoxicity, and other adverse reactions (Yang et al. 2024). Therefore, while *Bidens biternata* may be effective in relieving flare-ups when used in combination with a variety of other drugs, and there may be subsequent clinical applications for these combination therapies, additional clinical trials and in-depth studies are needed to validate their safety and efficacy.

Table 19.2 Differences in key compounds and applications between *Bidens biternata* and other medicinal plants

Item	*Bidens biternata*	*Bidens pilosa*	*Bidens bipinnata*	*Bidens tripartita*	References
Key compounds	4-Hydroxy-2,4′-dimethoxychalcone, 2,4-Di-tert-butylphenol (first found in *B. biternata*)	Flavonoids, phenolic compounds, terpenoids, alkanes	Sesquiterpenes, acetylenes, lactones, flavonoids	Flavonoids, phenolic compounds, terpenoids, naphthoquinones	Zahara et al. (2022), Xia et al. (2014), Rodríguez-Mesa et al. (2023), Bartolome et al. (2013),Yang et al. (2021), Wolniak et al. (2007), Shen et al. (2015b), and Wang et al. (2014)
Anticancer activity	Significant activity against cervical cancer (HeLa), liver cancer (HepG), and human lung adenocarcinoma (A549)	Inhibits cancer cell proliferation and induces apoptosis through various mechanisms	Inhibits tumor cell proliferation	Inhibits cancer cell proliferation and induces apoptosis	Zahara et al. (2022), Yang et al. (2013), Zhu et al. (2013), Rodríguez-Mesa et al. (2023), and Jia et al. (2013)
Anti-inflammatory	Effective anti-inflammatory compounds, reduces inflammatory response	Reduces inflammation by inhibiting the release of inflammatory mediators and cytokines	Contains various anti-inflammatory compounds, reduces inflammation	Inhibits inflammatory responses, reduces tissue damage	Yan et al. (2022), Wolniak et al. (2007), Gu et al. (1998), Shi et al. (1992), Pozharitskaya et al. (2010), and Kinuthia et al. (2016)
Antibacterial	Inhibits various pathogenic bacteria	Broad-spectrum inhibition of various bacteria and fungi	Inhibits the growth of bacteria and fungi	Inhibits the growth of various pathogens	Zahara et al. (2021), Bartolome et al. (2013), Jia et al. (2013), and Tomczykowa et al. (2011)

(continued)

Table 19.2 (continued)

Item	*Bidens biternata*	*Bidens pilosa*	*Bidens bipinnata*	*Bidens tripartita*	References
Antioxidant	Contains strong antioxidant compounds, reduces oxidative stress	Rich in antioxidants, protects cells from oxidative damage	Contains various antioxidants, reduces free radical damage to cells	Rich in antioxidants, reduces cellular oxidative damage	Bartolome et al. (2013), Jia et al. (2013), and Orhan et al. (2016)
Hepatoprotective	Protects liver cells from toxin damage	Hepatoprotective, reduces inflammation, promotes liver cell regeneration	Protects the liver by reducing oxidative stress and inhibiting inflammation	Protects liver cells, promotes liver regeneration	Bartolome et al. (2013), Jia et al. (2013), and Guo et al. (2021)
Applications	Traditional medicine, development of anticancer drugs	Traditional medicine, development of anti-inflammatory, anticancer, and antibacterial drugs	Traditional medicine, development of anti-inflammatory and antioxidant drugs	Traditional medicine, development of anti-inflammatory, anticancer, and antibacterial drugs	Bartolome et al. (2013) and Hu et al. (2021)

19.5 Conclusion and Outlook

Bidens biternata has been widely used in traditional medicine, and its anti-inflammatory, antibacterial, antioxidant, and antitumor effects have been confirmed by many studies. However, studies related to its modern pharmacology are still in their infancy (Deng et al. 2017). In future studies, the specific regulatory mechanisms of the active components of *Bidens biternata* should be revealed at the molecular level. In addition, drug compatibility studies as well as systematic clinical trials are important to evaluate the effects of combining *Bidens biternata* with other drugs and the safety of actual clinical applications.

In conclusion, the integration from traditional medical applications to modern clinical trials is a complex process that requires the assistance of a variety of approaches, including basic science research, clinical trials, and drug compatibility studies. Based on the continuous advancement of modern technological tools, it is believed that *Bidens biternata* will soon be applied in modern medicine, offering new hope to patients and healthcare providers.

References

Almosnid NM, Zhou XL, Jiang LH, Ridings A, Knott D, Wang S, Wei F, Yuan JQ, Altman E, Gao Y, Miao JH (2018) Evaluation of extracts prepared from 16 plants used in Yao ethnomedicine as potential anticancer agents. J Ethnopharmacol 211:224–234

Arantes DAC, Acgd S, Lima EM, Alonso ECP, Marreto RN, Mendonça EF, Valadares MC, Batista AC (2021) Biological effects of formulation containing curcuminoids and *Bidens pilosa* L. in oral carcinoma cell line. Braz Oral Res 35:e063

Bartolome AP, Villaseñor IM, Yang WC (2013) *Bidens pilosa* L. (Asteraceae): botanical properties, traditional uses, phytochemistry, and pharmacology. Evid Based Complement Alternat Med 2013:340215

Brandao MG, Krettli AU, Soares LS, Nery CG, Marinuzzi HC (1997) Antimalarial activity of extracts and fractions from *Bidens pilosa* and other Bidens species (Asteraceae) correlated with the presence of acetylene and flavonoid compounds. J Ethnopharmacol 57(2):131–138

Cai J, Shi SY, Cheng F, Wei M, Zou K, Yu XQ, Chen JF (2023) Polyacetylene isomers isolated from *Bidens pilosa* L. suppress the metastasis of gastric cancer cells by inhibiting Wnt/β--catenin and hippo/YAP signalling pathways. Molecules 28(4):1837

Chen JZ, Ge SL, Xiao YF (2012) Study on the antibacterial activity of extracts from *Bidens pilosa*, a plant of the Asteraceae family. Northern Hortic 21:83–85

Chen YT, Ma L, Lu TY, Wang Y, Gao GL, Hao JH (2021) Identification of weed species in the Bidens genus in China. J Changshu Inst Technol 35(02):87–91

Dai C, Liu Y, Lv F, Cheng P, Qu S (2025) An alternative approach to combat multidrug-resistant bacteria: new insights into traditional Chinese medicine monomers combined with antibiotics. Adv Biotechnol 3(1):6

Deng XJ, Pan WS, Zhang T, Zeng XY, Li H, Wang T (2017) Research Progress on the pharmacological effects of *Bidens pilosa*. China Pharm 28(013):1860–1864

Du YM (2021) Antimony pollution and remediation in soil and water environment of Guiyu electronic waste dismantling area. Guangzhou University, Guangzhou

Du B, Su BZ, Jiang HQ (2013) Antioxidant study on flavonoid components of *Bidens biternata*. Pharm Res 32(07):384–386

Gu HZ, Song JB, Zhu YF (1998) Modern research on *Bidens pilosa*. Shi Zhen Guo Yao Yan Jiu 02:88–90

Guo ML, Ma XY, Gong YQ, Feng ML, Zhao YW, Duan LX (2021) Protective effect of ethyl acetate extract from *Bidens bipinnata* on hepatocyte damage induced by endoplasmic reticulum stress. Zhongguo Zhong Yao Za Zhi 46(15):3893–3899

House M (2013) Evaluation of extracts used in traditional Chinese medicine for antiviral potential against herpes simplex virus type 1. Doctoral dissertation. Middle Tennessee State University

Hsu YJ, Lee TH, Chang CL, Huang YT, Yang WC (2009) Anti-hyperglycemic effects and mechanism of *Bidens pilosa* water extract. J Ethnopharmacol 122(2):379–383

Hu HG, Li ZG, Zhong XQ (2021) Authentication of Bidens genus plants and analysis of development prospects. J Jiangxi Univ Tradit Chinese Med 33(05):1–8

Huang GH (2012) Research progress on the pharmacological effects of *Bidens pilosa*. China Pharm 23(27):2578–2581

Jia J, Chen Z, Chen ZJ, Liu ZC, Li SH, Wang G (2013) Research progress on the chemical components, pharmacological effects, and application in animal husbandry of *Bidens pilosa*. China Anim Husband Vet Med 40(08):199–203

Jia ZC, Yang X, Wu YK, Li M, Das D, Chen MX, Wu J (2024) The art of finding the right drug target: emerging methods and strategies. Pharmacol Rev 76(5):896–914

Jiang HQ, Wang JP, Liu YH, Gong LL, Liu HY (2008) Separation and identification of chemical components of *Bidens pilosa*. Food Drug 09:15–17

Kinuthia DG, Muriithi AW, Mwangi PW (2016) Freeze dried extracts of *Bidens biternata* (Lour.) Merr. and Sheriff. show significant antidiarrheal activity in in vivo models of diarrhea. J Ethnopharmacol 193:416–422

Kong WZ, Xiong YH, Hu XL (2011) Research on the chemical components and pharmacological effects of *Bidens pilosa*. Med Inform: Mid-Month Edition 24(3):3

Li S.Z., and Xia K. Z., Compendium of materia medica by Li Shizhen. 1996

Li B, Liu X, Xiong J, Lai XW, Wang SJ, Chen J, Zhang ZL (2011) Separation of chemical components of *Bidens biternata*. Jiangxi Tradit Chin Med 42(10):51–53

Li Y, Ma XP, Li FY, Hou W, Yang CL, Cheng ZH (2012a) Characteristics of plant communities and their species diversity in the riparian zone of Haicheng River. Marine Lake and Marsh Bulletin 3:10

Li Y, Jiang HQ, Zhang QH (2012b) Separation and identification of chemical components of *Bidens biternata*. Food Drug 14(07):270–273

Li B, Zhan ZN, Xiong J, Liu HD, Wang ZW, Chen XT, Lai XW, Chen J, Liu B (2016) Separation of chemical components of *Bidens biternata* II. Jiangxi Tradit Chinese Med 47(4):64–66

Lin MY, Chen FH, Ge JF, Tang TJ, Ni WL (2013) Study on the protective effect of *Bidens pilosa* total flavonoids against acute inflammation and its possible mechanism. Chin J Clin Pharmacol Ther 18(06):614–620

Liu C, Han Y, Liu J, Zhang T, Zou ZM (2024) A new chalcone glycoside compound from the whole plant of *Bidens pilosa*. Acta Pharm Sin 59(06):1757–1764

Ma TT, Xie J, Zhang QL, Xu H, Li J, Chen FH (2012) Analysis of fingerprint and bioactive components of *Bidens biternata* by HPLC. Zhong Yao Cai 35(6):892–896

Marwan AN, Zhou X, Jiang L, Ridings A, Knott D, Wang S, Wei F, Yuan J, Altman E, Gao Y, Miao J (2018) Evaluation of extracts prepared from 16 plants used in Yao ethnomedicine as potential anticancer agents. J Ethnopharmacol 211:224–234

Orhan N, İçöz ÜG, Altun L, Aslan M (2016) Anti-hyperglycaemic and antioxidant effects of *Bidens tripartita* and quantitative analysis on its active principles. Iran J Basic Med Sci 19(10):1114–1124

Peng XQ, Zhang QL, Zhan X, Zhang Q, Gao WC, Chen FH (2012) Establishment of HPLC finger print map for *Bidens biternata* extract. In: Proceedings of the Anhui Pharmacological Society 2012 annual academic conference

Pozharitskaya ON, Shikov AN, Makarova MN, Kosman VM, Faustova NM, Tesakova SV, Makarov VG, Galambosi B (2010) Anti-inflammatory activity of an HPLC-fingerprinted aqueous infusion of aerial part of *Bidens tripartita* L. Phytomedicine 17(6):463–468

Rodríguez-Mesa XM, Contreras Bolaños LA, Mejía A, Pombo LM, Modesti Costa G, Santander González SP (2023) Immunomodulatory properties of natural extracts and compounds derived from *Bidens pilosa* L.: literature review. Pharmaceutics 15(5):1491

Shahwar D, Ullah S, Raza MA, Sana U, Yasmeen A, Ghafoor S, Ahmad N (2011) Acetylcholine esterase and antioxidant potential of some members of Asteraceae and Euphorbiaceae. J Med Plants Res 5(32):7011–7016

Shen YW, Lin LQ, Lin XH, Yao H (2015a) Research progress on the chemical composition and pharmacological activities of *Bidens bipinnata*. J Fujian Med Univ 49(01):58–61

Shen AZ, Li X, Hu W, Chen FH (2015b) Total flavonoids of *Bidens bipinnata* L. ameliorate experimental adjuvant-induced arthritis through induction of synovial apoptosis. BMC Complement Altern Med 15(1):437

Shi JY, Feng SQ, Song GY, Liu KC (1992) PAGE analysis of seeds from *Bidens pilosa* in Shandong. Chin Pharm J 12:717–719+759

Szekalska M, Sosnowska K, Tomczykowa M, Winnicka K, Kasacka I, Tomczyk M (2020) In vivo anti-inflammatory and anti-allergic activities of cynaroside evaluated by using hydrogel formulations. Biomed Pharmacother 121:109681

Tomczykowa M, Leszczyńska K, Tomczyk M, Tryniszewska E, Kalemba D (2011) Composition of the essential oil of *Bidens tripartita* L. roots and its antibacterial and antifungal activities. J Med Food 14(4):428–433

Wagner H, Chari VM, Sonnenbichler J (1976) 13C-NMR-spektren natürlich vorkommender flavonoide. Tetrahedron Lett 17(21):1799–1802

Wang YP (2015) Basic clinical medicine institute of CACMS and standardisation TCM China. Chin Pharm 04(73):132–135

Wang XY, Chen GR, Deng ZY, Zhao J, Ge JF, Li N, Chen FH (2014) Chemical component research on *Bidens pilosa*. China J Chin Mater Med 39(10):7

Wei GB, Xiong W (2012) Study on the inhibitory effect of *Bidens biternata* extract on α-glucosidase. Shandong Chem Indus 41(09):1–3

Wolniak M, Tomczykowa M, Tomczyk M, Gudej J, Wawer I (2007) Antioxidant activity of extracts and flavonoids from *Bidens tripartita*. Acta Pol Pharm 64(5):441–447

Yang X, Li M, Jia ZC, Liu Y, Wu SF, Chen MX, Hao GF, Yang Q (2024) Unraveling the secrets: Evolution of resistance mediated by membrane proteins. Drug Resist Updat 77:101140

Traditional Uses and Modern Pharmacological Investigations of *Bidens frondosa* and *Bidens tripartita*

20

Shahparan Islam Shawon, Rashmia Nargis Reyda, and Nazmul Qais

Abstract

Plants have established the cornerstone of complex conventional treatment for millennia. Medicinal plants are used as treatment for a wide range of disease categories. The genus *Bidens*, a member of the family Asteraceae, is extensively documented as an ethnomedicinally significant genus of plants due to its reported diversified biological activities. There are around 230 species which are distributed worldwide. A species of this genus named *Bidens frondosa* native to North America is used as aqueous, alcoholic, methanol extracts and subfractions of these extracts, as well as individual components, to treat a variety of illnesses. Several studies have shown that this species has significant anti-inflammatory, antibacterial, antioxidant, antimalarial, anticancer, antiviral, and antidiarrheal properties. According to reports, the presence of acetylene, polyenic D-glucosides, chalcones, aurones, flavones, and terpene structures is responsible for these biological activities. Again, significant amounts of active constituents, including aurones, chalcones, flavonoids, volatile oil, acetylene and polyacetylene, xanthophylls, sterols, phenolcarboxylic acids, coumarins, polysaccharides, aromatic hydroxy aldehydes, caffeine, and tannin, are found in *Bidens tripartita*, a different species. Known by most as three-part beggar-ticks, this plant has been used extensively in traditional medicine for its astringent, diuretic, antioxidant,

S. I. Shawon · N. Qais (✉)
Department of Clinical Pharmacy and Pharmacology, Faculty of Pharmacy, University of Dhaka, Dhaka, Bangladesh
e-mail: nqais@du.ac.bd

R. N. Reyda
Department of Clinical Pharmacy and Pharmacology, Faculty of Pharmacy, University of Dhaka, Dhaka, Bangladesh

East West University, Dhaka, Bangladesh

Y. Bibi et al. (eds.), *The Genus Bidens*,
https://doi.org/10.1007/978-981-96-4257-1_20

immunostimulating, anti-inflammatory, antiallergic, hepatoprotective, antitumor, antiulcerogenic, antimycotic, antithrombotic, narcotic, sedative, and possibly nociceptive properties. *B. tripartita* can be a potential target of multiple drug development, whereas *B. frondosa* can be useful as an alternative to *B. tripartita* or *Bidens pilosa* for active constituent isolation as it has structural similarity with them.

Keywords

Frondosa · *Bidens* · Tripartita · Polyacetylene · Xanthophylls

20.1 Introduction

Traditional medical systems all throughout the world have been using medicinal herbs for thousands of years. Medicinal plants are often utilized as raw materials for extracting active ingredients, which are then used in the development of various drugs (Hassan 2012). They play a significant role in the healthcare practices of many cultures and have contributed significantly to modern pharmacology and medicine. According to estimates from the United Nations World Health Organization, up to 80% of people worldwide use herbal medicine for primary healthcare (Organization WH 2004).

Bidens is a genus of flowering plants that includes over 230 species, commonly known as beggar-ticks, stickseeds, or Spanish needles. These plants have been used in traditional medicine in various cultures. Here are some medicinal uses of *Bidens* species: anti-inflammatory, antimicrobial, digestive health, antioxidant, wound healing, respiratory health, antimalarial, antidiabetic, anticancer, and liver protection (Xuan and Khanh 2016).

We have chosen two plants (species) *Bidens frondosa* and *Bidens tripartita* from the *Bidens* genus for this chapter. The purpose of the work was to systematize scientific data on the traditional uses and modern pharmacological investigations of *B. frondosa* and *B. tripartita.*

20.2 *Bidens frondosa* (Leafy Beggar-Ticks/Devil's Pitchfork)

Bidens frondosa (Family Asteraceae) is an annual herbaceous plant native to North America, particularly prevalent in the eastern and central United States. It is recognized for its barbed seeds that cling to clothing and fur, facilitating its spread (gobotany.nativeplanttrust.org, n.d.). Although originally from North America, it has extended its range to Europe and Asia, where it is often regarded as an invasive species. This plant typically thrives in wetlands, along riverbanks, in partial shade or full sun, and in wet ditches, disturbed soil, and other moist, disturbed areas (Jebaraj et al. 2024).

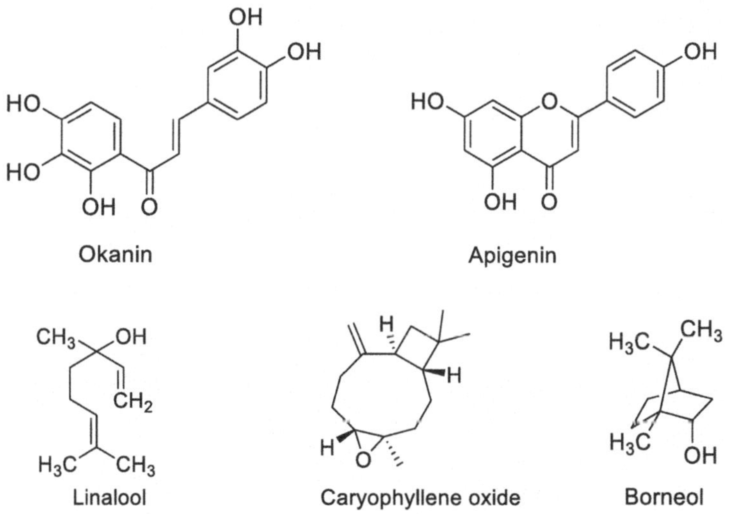

Fig. 20.1 Some active constituents isolated from *B. frondosa*

20.2.1 Phytoconstituents

20.2.1.1 Antioxidant Components

According to Ahn and Kim (2016), notable antioxidant compounds include:

Quercetin 3 O β D glucopyranoside.
Luteolin-7-O-β-D-glucopyranoside.
7,8,3′,4′-tetrahydroxyflavanone-3.
Okanin (Fig. 20.1).
Okanin-4’-O-β-D-glucopyranoside.

20.2.1.2 Anti-inflammatory Components

Le et al. (2015) identified several anti-inflammatory compounds, including:

(3E,5E,11E)-tridecatriene-7,9-diyne-1,2,13-triol-2-O-β-D-glucopyranoside.
8,3′,4′-trihydroxyflavone-7-O-(6″-O-p-coumaroyl)-β-D-glucopyranoside.
3-(4-hydroxy-3-methoxyphenyl)-3-methoxypropane-1,2-diol.
Wilfordiol B.
Apigenin.
4-hydroxy-2-furaldehyde.

20.2.1.3 Antibacterial Components

Rahman et al. (2011) reported the presence of various antibacterial compounds in essential oils, including:

Monoterpenes

- β-Caryophyllene (14.3%).

Rosefuran (13.5%).
Verbenone (11.5%).
Cis-tagetone (5.3%).
Linalool (3.2%).
Trans-tagetone (2.5%).
Dihydrotagetone (2.9%).
Limonene (1.4%).
Eucalyptol (1.3%).
Piperitenone (1.1%).

Sesquiterpenes

- Spathulenol (6.02%).

Caryophyllene oxide (5.49%).
α-Humulene (4.02%).
Epiglobulol.

Alicyclic diterpene alcohol

- Phytol (1.32%).

20.2.1.4 Insecticidal Components

Li et al. (2017) discovered 32 compounds in the essential oil of *Bidens frondosa* leaves, with key constituents responsible for insecticidal properties being:

Caryophyllene oxide (20.50%).
Borneol (17.66%).
4-Terpineol (14.56%).
α-Terpineol (6.28%).
β-Cedrene (6.94%).
Spathulenol (4.07%).
Caryophyllene (3.97%).

20.2.1.5 Others

Additionally, Karikome et al. (1992) isolated five acylated glucosides of chalcone from the fresh leaves of *Bidens frondosa:*

Okanin 4-O-(6″-O-acetyl-2″-O-caffeoyl-β-D-glucopyranoside).
Okanin 4-O-(2–0-caffeoyl-6″-O-p-coumaroyl-β-D-glucopyranoside).
4-O-Methylokanin 4-O-(6–0-acetyl-2-0-caffeoyl-β-D-glucopyranoside).
4-O-Methylokanin 4'-O-(6″-O-p-coumaroyl-β-D-glucopyranoside).
4-O-Methylokanin 4-O-(6″-O-acetyl-β-D-glucopyranoside).

20.2.2 Traditional Uses

20.2.2.1 Urinary Tract Issues

Bidens frondosa has been used traditionally to treat urinary tract infections, inflammation, and bleeding.

20.2.2.2 Pain and Inflammation

The plant is also used in traditional medicine for pain relief and to reduce inflammation.

20.2.2.3 Skin Conditions

Some cultures use *Bidens frondosa* for treating skin irritations and wounds.

20.2.3 Pharmacological Activities

The pharmacological activity of *Bidens frondosa* has been studied only in preclinical studies and to a lesser extent than in other *Bidens* species (*B. tripartita, B. pilosa*).

20.2.3.1 Anti-inflammatory Activity

Research has identified various anti-inflammatory compounds in *Bidens frondosa* that may be beneficial for conditions like arthritis, inflammatory bowel disease, and pain. *B. frondosa* is found to significantly decrease the levels of inflammatory cytokines TNF-α, IL-6, IL-1, and IL-10 compared to the control cell group, confirming the successful inhibition of inflammation in RAW 264.7 monocytes induced by LPS.

Compounds exhibited significant inhibitory effects are (Le et al. 2015):

(3E,5E,11E)-tridecatriene-7,9-diyne-1,2,13-triol-2-O-β-d-glucopyranoside.
2′-butoxyethylconiferin.
8,3′,4′- trihydroxyflavone-7-O-(6″-O-p-coumaroyl)-β-d-glucopyranoside.
3-(4-Hydroxy-3-methoxyphenyl)propane-1,2-diol.
4-hydroxy-2 -furaldehyde.

20.2.3.2 Antimicrobial Activity

The essential oil of *Bidens frondosa* is proved to have antibacterial effects against all Gram-positive and Gram-negative bacteria. Specifically, it strongly inhibited the growth of *Staphylococcus aureus, Listeria monocytogenes,Pseudomonas aeruginosa, Bacillus subtilis*. (Rahman et al. 2011).

20.2.3.3 Antioxidant Activity

Bidens frondosa extracts may have antioxidant properties, potentially helping with conditions associated with oxidative stress. Compounds extracted from *B. frondosa* such as quercetin-3-O-β-D-glucopyranoside, luteolin-7-O-β-D-glucopyranoside, 7,8,3′,4′-tetrahydroxyflavanone, okanin-4-O-β-D-glucopyranoside, and okanin were evaluated for their antioxidant activity. Among these,

7,8,3′,4′-tetrahydroxyflavanone and okanin demonstrated significant radical scavenging effects on DPPH (2,2-diphenyl-1-picrylhydrazyl) and showed strong superoxide quenching activities originating from riboflavin and xanthine (Ahn and Kim 2016).

20.2.3.4 Antimalarial Activity

One of the main ingredients 1-phenyl-1,3-diyn-5-en-7-ol-acetate was found in the chloroform fractions from the roots which inhibited parasite growth in vitro by 86% (Brandão et al. 1997).

20.2.3.5 Antidiabetic Activity

Studies on normoglycemic, glucose-loaded, and diabetic animals revealed significant antidiabetic properties. In streptozotocin-induced diabetic animals, an ethanolic extract of *B. frondosa* showed an acute antidiabetic effect, reducing blood glucose levels by 42% at a dose of 500 mg/kg. Additionally, the extracts demonstrated significant inhibitory activity against α-glucosidase and α-amylase enzymes, with inhibition rates ranging from 64.29% to 75.22% at a concentration of 2 mg/mL (Icoz et al. 2017).

20.2.3.6 Tyrosinase Inhibitory Activity

Silver nanoparticles synthesized using aqueous extract of *B. frondosa* have been discovered to have substantially increased tyrosinase inhibitory action (Abbas et al. 2017).

20.2.3.7 Insecticidal Properties

The essential oil found in *Bidens frondosa* and its isolates demonstrate contact toxicity against booklice. Caryophyllene oxide and borneol, isolated from *B. frondosa*, showed significant contact toxicity when tested against booklice (*Liposcelis bostrychophila*) compared to the essential oil, with no overlap in the 95% fiducial limit. Booklice were more vulnerable to the essential oil than to β-cedrene, although 4-terpineol showed the same degree of contact toxicity as the oil. It was also discovered that 4-terpineol and borneol were fumigant to booklice (*L. bostrychophila*) (Li et al. 2017).

20.3 *Bidens tripartita*

Bidens tripartita, locally called as three-lobed beggar-ticks, three-part beggar-ticks, leafy-bracted beggar-ticks, or trifid bur-marigold, is a prevalent and widely distributed flowering plant in the sunflower family, Asteraceae. This annual herb typically reaches a height of 30–60 cm (12–24 inches). It has compound leaves with three lobes, which gives it the name “tripartita.” The plant produces yellow flower characteristic of the Asteraceae family, featuring disk florets at the center and ray florets on the edges. *B. tripartita* is native to Europe, Asia, and parts of North Africa. It commonly thrives in wetlands, marshes, riverbanks, and other moist environments. It can adapt to various soil moisture levels, ranging from submerged conditions to periodically wet or damp soils.

20.3.1 Phytoconstituents

Bidens tripartita is rich in various biologically active compounds, including coumarins, flavonoids, tannins, phenolcarboxylic acids, essential oils, polysaccharides, polyacetylenes, and aromatic hydroxy aldehydes (Fig. 20.2). These substances contribute to its pharmacological properties and therapeutic potential.

Gas chromatography mass spectrometry (GC-MS) analysis of methanolic leaf extract of *B. tripartita* conducted by Sisein et al. (2022) revealed the presence of phenols, flavonoids, and alkaloids among a total of 44 identified compounds. Notable constituents included caryophyllene, spathulenol, patchoulane, and various methyl esters of fatty acids. This comprehensive analysis sheds light on the diverse chemical composition of *B. tripartita,* highlighting its potential bioactivity. The compounds identified are:

Caryophyllene.
4-(2,4,4-trimethyl-cyclohexa-1,5-dienyl)-but-3-ene-2-one 1-methyl-4-(5-methyl-1-methylene-4-hexenyl-cyclohexane.
Spathulenol.
Patchoulane.
1-chloro-octadecane.
7-hydroxyfarnesin.
Methyl palmitoleate.
Methyl-cis-6-octadecenoate.
N-hexadecanoic acid methyl ester.

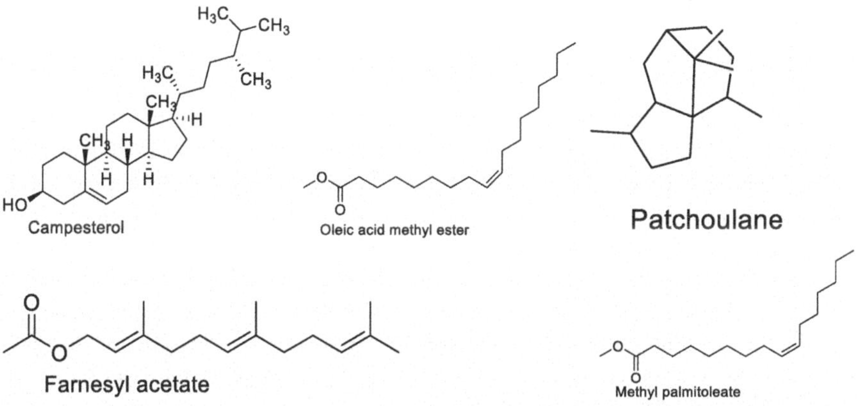

Fig. 20.2 Some active constituents isolated from *B. tripartita*

11,58 Dibutyl phthalate.
2-methyl-4-(2,6,6-trimethyl-1-cyclohexen-1-yl)-2-butenal.
1-(7-hydroxy-5-methoxy-2,2-dimethyl-2H-1-benzopyran-6-yl)-ethanone.
1-(cyclohexylmethyl)-2-methyl-cyclohexane.
Octadecenoic acid methyl ester.
N-(2-methoxyethoxycarbonyl)-pentadecyl ester.
Methyl-9-cis, 11-trans-octadecadienoate.
Cis-11-eicosenoic acid methyl ester.
2-methyl tetracosane.
N-butyl-benzenesulfonamide.
Heptadecyl-3-chloropropanoate.
N-nonadecane.
6,10,14-trimethyl-2-pentadecanone.
Z-9-dodecenyl acetate.
Heptadecane.
Trans, trans farnesyl acetate.
9,12-octadecadienoic acid methyl ester.
Oleic acid methyl ester.
10-octadecenoic acid methyl ester.
Phytol.
Methyl-18-methylnonadecenoate.
4,8,12,16-tetramethylheptadecan-4-olide.
Campesterol.
All trans farnesyl acetate.
Stigmasterol.
Gamma tocopherol.
Octadecane.
Heneicosane.
Docosanoic acid methyl ester.
Vitamin E.
Eicosane 1-docosanol.
2-bromotetradecane.
Methyl lignocerate.

Further studies identified specific flavonoids in *B. tripartita,* such as flavanomarein, cynaroside, and luteolin, which possess antioxidant properties (Wolniak et al. 2007). Moreover, Uysal et al. (2018) emphasized the presence of chlorogenic acid, luteolin 7-glucoside, and epicatechin as major compounds in *B. tripartita* extracts, indicating its potential health benefits.

Kotov et al. (2020) focused on analyzing the content of key biologically active substances (BASs) in *B. tripartita* extracts, including extractable matter, polyphenols, flavonoids, and polysaccharides. This investigation provides valuable insights into the medicinal potential of *B. tripartita*, highlighting its richness in bioactive compounds with diverse therapeutic effects.

In summary, the phytochemical profile of *B. tripartita* encompasses a wide array of compounds with significant pharmacological importance. Further research into its bioactivity and therapeutic applications is warranted to fully harness the potential benefits of this botanical resource.

20.3.2 Traditional Uses

Bidens tripartita (three-lobe beggar-ticks) boasts a plethora of traditional applications in diverse cultural contexts.

20.3.2.1 Medicinal Purposes

Historically, traditional medicine has long valued the possible anti-inflammatory qualities of *B. tripartita.* Its leaves are often prepared into remedies for external application on wounds, skin irritations, and insect bites. Internally, it has been utilized to alleviate minor gastrointestinal discomfort.

20.3.2.2 Diuretic Properties

Traditional practices involve utilizing *B. tripartita* as a diuretic agent. Infusions or decoctions derived from the plant are consumed to stimulate urination, aiding in the elimination of excess bodily fluids. This usage addresses mild edema and urinary tract concerns.

20.3.2.3 Culinary Use

While less widespread, certain parts of *B. tripartita*, such as its tender leaves and shoots, have culinary value in select cultures. They are cooked and incorporated into dishes like soups and stews. Caution is advised when foraging wild plants for consumption to avoid inadvertent ingestion of toxic species.

20.3.2.4 Folklore and Rituals

Beyond its practical applications, *B. tripartita* holds cultural significance in folklore, ceremonies, and rituals across various regions. Symbolizing concepts like healing, purification, or protection, it plays a role in traditional customs and beliefs.

20.3.2.5 Dyeing

The flowers of *B. tripartita* serve as a natural dye source, yielding yellow or orange pigments. These dyes are utilized to color fabrics or fibers, showcasing the plant's versatility in artisanal practices.

20.3.3 Pharmacological Activities

Bidens tripartita, like many other plants, has been studied for its potential pharmacological properties. Various BASs in the plant result in numerous pharmacological and biological effects. The extracts exhibit antimycotic, antiradical, anti-inflammatory, antithrombotic, antidiabetic, antioxidant, antibacterial, antiulcerogenic, and antitumor properties. Additionally, they have analgesic, antiallergic, antipyretic, immunostimulating, hepatoprotective, and hypotensive effects (Wolniak et al. 2007) (Pozharitskaya et al. 2010) (Orhan et al. 2016). These extracts may also inhibit key enzymes and protect DNA (Uysal et al. 2018). The medicinal raw material of *B. tripartita* is highly valuable for the development and creation of new drugs.

While research on its medicinal uses is ongoing and may vary in different cultural and scientific contexts, here are some reported pharmacological uses and properties associated with *B. tripartita.*

20.3.3.1 Anti-inflammatory Properties

Some studies suggest that extracts from *Bidens tripartita* possess anti-inflammatory properties. These properties may be beneficial for reducing inflammation associated with conditions such as arthritis or skin inflammation. The presence of flavonoids, tannins, polysaccharides, phenols, amino acids, ascorbic acid, organic acids, and polyacetylenes was discovered through a phytochemical investigation and qualitative–quantitative tests. When compared to indomethacin, infusion doses of 20 ml/kg body weight significantly reduced inflammation in rats. Furthermore, the infusion demonstrated antipyretic effects in carrageenan-induced localized hyperthermia and analgesic effects in a hot-plate test, both in rats. Dosage-dependent effects were also seen (Pozharitskaya et al. 2010). The early phase inflammatory mediators (histamine, serotonin, and kinins) are inhibited by the tincture of *Bidens* plants. Due to the complex interaction between the tincture of researched *Bidens* plants and its biologically active ingredients, the tincture has antiexudative activity. The tincture of *B. tripartita* herb contains hydroxycinnamic acids, organic acids, and polyphenolic chemicals, which are responsible for its membrane-stimulating and anti-inflammatory properties (Oproshanska et al. 2021).

20.3.3.2 Antioxidant Activity

Compounds in *Bidens tripartita* have antioxidant qualities that can assist the body in scavenging dangerous free radicals. Antioxidants may be beneficial for general health and problems linked to aging and have a function in shielding cells from oxidative stress. Effective radical scavengers were the pure flavonoids: flavanomarein (isookanin 7-O-glucoside), cynaroside (luteolin 7-O-glucoside), and luteolin that were isolated from this plant (Wolniak et al. 2007).

20.3.3.3 Antitumor and Antithrombotic Action

The methylene chloride extract of *Bidens tripartita* exhibits significant antitumor activity against L1210 cancer cells (mouse leukemia) and possesses antithrombotic properties. Samples of *B. tripartita* collected from Krasnodar, Khabarovsk, and Perm in Russia have demonstrated this significant tumor inhibition activity against L1210 cancer cells and notable antithrombotic effects (Goun et al. 2002).

20.3.3.4 Antimicrobial Activity

The essential oil extracted from the roots of *Bidens tripartita* has moderate effectiveness against bacteria, especially Gram-positive bacteria. It is most effective against two specific Gram-negative bacteria strains: *N. gonorrhoeae* and *M. catarrhalis.* However, it had a limited effect on some other Gram-negative bacteria, like *P. aeruginosa.* It also had a stopping effect on the growth of fungi, with the best result seen against *C. albicans.* Overall, it showed promise in fighting both bacteria and fungi, but its effectiveness varied depending on the type of microorganism (Tomczykowa et al. 2011). This suggests potential applications in the development of antimicrobial agents for treating infections. Excellent efficacy against several bacteria, including Gram-positive (*S. aureus*) and Gram-negative (*E. coli, P. aeruginosa*), as well as fungus (yeast: *C. albicans*; molds: *A. niger*; dermatophytes: *T. rubrum*), has been exhibited by gold nanoparticles generated from *B. tripartita* (Dobrucka and Dlugaszewska 2020). The antibacterial properties of gold nanoparticles on microorganisms could be maintained by the electrostatic interaction between the negatively charged microbe cell surface and the positively charged gold (Bindhu and Umadevi 2015).

20.3.3.5 Anti-angiogenetic Activity

In comparison to the control group, *Bidens tripartita* herb dramatically decreases cell migration and proliferation while increasing cell invasiveness. *B. tripartita* herb extract is antiangiogenic due to its antimigrative, antiproliferative, and comparatively modest influence on endothelial invasiveness properties (Antoniak et al. 2023).

20.3.3.6 Antidiabetic Activity

When compared to a healthy control group, the ethanolic extract of *Bidens tripartita* markedly (11.5%–25.8%) lowered blood glucose levels in normoglycemic and glucose-loaded mice. Furthermore, on streptozotocin-induced diabetic rats, the ethanol extract of *B. tripartita* showed encouraging (14.0–32.7%) antidiabetic action. The extract's considerable antidiabetic efficacy at 500 mg/kg (24.3–32.7%) outperformed the reference medication tolbutamide's (0.0–20.0%) effect across all measurements (Orhan et al. 2016). Comparing the ethanol extract of *B. tripartita* to the reference medication Acarbose (71.11–47.99 at 1–0.25 µg/ml), the α-glucosidase inhibitory activity was found to be moderate (49.87–25.02%) at the tested dosages. At 2 mg/ml, all subextracts exhibited inhibitory action against the α-glucosidase enzyme. The subextract of ethyl acetate had the highest activity (65.56–27.38%) out of all of them. The α-amylase enzyme did not react with *B. tripartita* ethanol extract, and its subextracts (chloroform, ethyl acetate, and n-butanol) only showed a modest inhibitory effect at a dose of 2 mg/ml (Orhan et al. 2016). The methanol extract had the greatest α-glucosidase inhibitory effect, but the ethyl acetate extract had the most α-amylase inhibitory activity (Uysal et al. 2018).

20.3.3.7 Antimycotic Action

The essential oils exhibit strong fungistatic effects, particularly against *Candida albicans* and *Candida parapsilosis*, with the lowest effectiveness against *Aspergillus fumigatus* (Tomczykowa et al. 2008). The significant antimycotic activity of these essential oils is noteworthy, as they could serve as potential alternatives to synthetically produced fungicides for combating pathogenic fungi inhabiting in plants. Thiophenes, a major component of these oils, likely contribute to their antimycotic properties. The essential oil of *B. tripartita* demonstrates fungistatic effects, with the most pronounced inhibitory action against *C. albicans* and the least against *C. krusei*. Essential oil extracted from *B. tripartita* roots shows significant fungistatic activity, indicating its potential use as a treatment for plant pathogenic fungi (Tomczykowa et al. 2011).

20.3.3.8 Intestinal Contractility

The evidences showed that *Bidens tripartita* extracts, and their flavonoids had a strong prokinetic impact on the jejunum's smooth muscle. Both acetylcholine-induced and spontaneous contractions showed this effect. There were minimal variations in the strength of this effect among methanolic, methanolic 50%, aqueous, ethyl acetate, diethyl ether, and butanol extracts, except for the butanol extract, which exhibited a slightly stronger prokinetic effect. The observed myocontractile effect of individual flavonoids supports the hypothesis that these compounds are responsible for the prokinetic activity of the extracts. Among the flavonoids tested, cynaroside appeared to be the most potent, with the response to acetylcholine increasing by over 250% in its presence compared to the control. Based on these findings, *B. tripartita* may be beneficial in treating functional gastrointestinal disorders characterized by constipation, such as irritable bowel syndrome, and dyspeptic complaints that typically involve the use of gastric prokinetics in treatment protocols (Mendel et al. 2020).

20.3.3.9 Psychomotor Abilities and Cognitive Function

The psychomotor abilities and cognitive functions were assessed using the T-maze test to investigate the effects of the aqueous and alcoholic plant extracts on short-term memory in the laboratory animals. The *B. tripartita* aqueous extract significantly increased the spontaneous alternation percentage, though less markedly than magnesium in the T-maze test. The *B. tripartita* alcoholic extract also significantly increased the spontaneous alternation rate, but to a lesser extent than the aqueous extract in this behavioral model. Both aqueous and alcoholic extracts of *B. tripartita* facilitated spatial learning, improved short-term memory, and exhibited a mild anxiolytic effect in rats (Lupuşoru et al. 2017).

20.3.3.10 Cytotoxic Action

The viability of three different cancer cell lines—cervical carcinoma HeLa, breast cancer MDA-MB-231, and liver cancer Huh7—was analyzed using the MTT (3-(4,5-dimethylthiazol-2-yl)-2,5-diphenyltetrazolium bromide) assay to study the cytotoxic effects of *B. tripartita* (Icoz et al. 2017). The ethyl acetate and methanol extracts contain high concentrations of luteolin-7-glucoside, which causes reactive oxygen species (ROS) to build up and decreases β-catenin expression in COLO 320 DM cells, hence inducing programmed cell death. Rectal cancer treatment with luteolin-7-glucoside exhibits promise as a strong anticancer medication. A dose-dependent antiproliferative effect of luteolin-7-glucoside was observed in human hepatocellular carcinoma HepG2 cells, corresponding to the effects seen in Huh7 cells treated with ethyl acetate and methanol extracts (Uysal et al. 2018) (Hwang et al. 2013).

20.3.3.11 Enzyme Inhibiting Activity and Enzyme Inhibiting Effect

There are published studies on the inhibitory effects of *Bidens tripartita* on several enzymes, including acetylcholinesterase, α-amylase, butyrylcholinesterase, and α-glucosidase (Uysal et al. 2018). The ethyl acetate extract shows the highest α-amylase inhibitory activity, whereas the methanol extract exhibits the strongest inhibition of α-glucosidase (Uysal et al. 2018). 20.3.3.12. DNA Protection Effects All *Bidens tripartita* extracts protect the supercoiled form of plasmid pUC19 (> 70%). The aqueous and ethyl acetate extracts (both at 10 mg/ml) are the most effective, providing 80% protection. HPLC analysis identified chlorogenic acid and luteolin-7-glucoside as the predominant phenolic compounds found in the *B. tripartita* extracts. Chlorogenic acid has in vitro antioxidant and antimutagenic activity, while luteolin-7-glucoside has demonstrated radical scavenging activity against the DPPH radical (Wolniak et al. 2007; Rice-Evans et al. 1996).This study demonstrates that the DNA protection properties of *B. tripartita* extracts are due to their high concentrations of luteolin-7-glucoside and chlorogenic acid (Uysal et al. 2018).

20.4 Conclusion

Bidens tripartita emerges as a botanical treasure trove with a rich pharmacological profile and diverse traditional uses. Its exploration unveils a promising source of bioactive substances with significant health and medical implications. Further research into its therapeutic potential and mechanisms of action is imperative to fully harness the benefits of this remarkable plant. *Bidens frondosa* demonstrates potential in areas such as anti-inflammatory, antimicrobial, and antioxidant actions. Nevertheless, further investigation is necessary to comprehensively assess its safety and effectiveness for targeted health conditions. It is imperative to seek clinical trials before incorporating *B. frondosa* into medicinal practices. They can offer insights into possible risks and interactions tailored to individual health circumstances. While this review outlines the traditional uses and potential health advantages of *B. frondosa*, it underscores the necessity for additional research before its widespread adoption as a therapeutic remedy.

References

Abbas Q, Saleem M, Phull AR, Rafiq M, Hassan M, Lee KH, Seo SY, IJPR. (2017) Green synthesis of silver nanoparticles using Bidens frondosa extract and their tyrosinase activity. Iran J Pharm Res 16(2):763–770

Ahn D, Kim DK (2016) Antioxidant components of the aerial parts of Bidens frondosa L. Kor J Pharmacogn 47(2):110–116

Antoniak K, Studzińska-Sroka E, Szymański M, Dudek-Makuch M, Cielecka-Piontek J, Korybalska K (2023) Antiangiogenic, anti-inflammatory and antioxidant properties of *Bidens tripartite* herb, *Galium verum* herb and *Rumex hydrolapathum* root. Molecules 28(13):4966. https://doi.org/10.3390/molecules28134966

Bindhu MR, Umadevi M (2015) Antibacterial and catalytic activities of green synthesized silver nanoparticles. Spectrochim Acta A Mol Biomol Spectrosc 135:373–378. https://doi.org/10.1016/j.saa.2014.07.045

Brandão MGL, Krettli AU, Soares LSR, Nery CGC, Marinuzzi HC (1997) Antimalarial activity of extracts and fractions from *Bidens pilosa* and other *Bidens* species (Asteraceae) correlated with the presence of acetylene and flavonoid compounds. J Ethnopharmacol 57(2):131–138. https://doi.org/10.1016/s0378-8741(97)00060-3

Dobrucka R, Dlugaszewska J (2020) Biomimetic synthesis of nanogold using *Bidens tripartitus* extract and evalution of their antimicrobial and catalytic activities. Inorg Nano-Met Chem 50(6):459–468. https://doi.org/10.1080/24701556.2020.1720722

Goun EA, Petrichenko VM, Solodnikov SU, Suhinina TV, Kline M, Cunningham GN, Nguyen CL, Miles H, JoE. (2002) Anticancer and antithrombin activity of Russian plants. J Ethnopharmacol 81(3):337–342. https://doi.org/10.1016/s0378-8741(02)00116-2

Hassan BAR (2012) Medicinal plants (importance and uses). Pharmaceut Anal Acta 3(10):2153–2435. https://doi.org/10.4172/2153-2435.1000e139

Hwang YJ, Lee E, Kim HR, Hwang K (2013) Molecular mechanisms of luteolin-7-O-glucoside-induced growth inhibition on human liver cancer cells: G2/M cell cycle arrest and caspase-independent apoptotic signaling pathways. BMB Rep 46(12):611–616. https://doi.org/10.5483/bmbrep.2013.46.12.133

Icoz UG, Orhan N, Altun L, Aslan M (2017) In vitro and in vivo antioxidant and antidiabetic activity studies on standardized extracts of two Bidens species. J Food Biochem 41(6):e12429. https://doi.org/10.1111/jfbc.12429

Jebaraj NB, Chakraborty P, Manhas SV (2024) Investigating the relationship between soil moisture levels and the population distribution of *Leersia virginica*. https://doi.org/10.13140/RG.2.2.32666.04801

Karikome H, Ogawa K, Sashida Y (1992) New acylated gluscosides of chalcone from the leaves of Bidens frondosa. Chem Pharm Bull 40(3):689–691. https://doi.org/10.1248/cpb.40.689

Kotov S, Kotova E, Bezruk I, Gontova T, Kotov A, JH. (2020) The study of the extraction dynamics of biologically active substances from the Bidens tripartita L. Herb and antioxidant activity of the obtained extracts. In: Antioxidant activity of the obtained extracts. EUREKA: Health Sciences, vol 6, pp 95–101. https://doi.org/10.21303/2504-5679.2020.001408

Le J, Lu W, Xiong X, Wu Z, Chen W (2015) Anti-inflammatory constituents from Bidens frondosa. Molecules 20(10):18496–18510. https://doi.org/10.3390/molecules201018496

Li HY, Chen XB, Liu QZ, Liu ZL (2017) Chemical composition and insecticidal properties of the essential oil of *Bidens frondosa* L (Asteraceae) against booklice (*Liposcelis bostrychophila*). Trop J Pharm Res 16(1):171–171. https://doi.org/10.4314/tjpr.v16i1.23

Lupuşoru CE, Popa EG, Sandu RB, Buca BR, Mititelu-Tarţău L, Lupuşoru RV (2017) The influence of Bidens tripartita extracts on psychomotor abilities and cognitive functions in rats. Farmacia 65(2):284–288

Mendel M, Chłopecka M, Latek U, Karlik W, Tomczykowa M, Strawa J, Tomczyk M, JoE. (2020) Evaluation of the effects of Bidens tripartita extracts and their main constituents on intestinal motility–An ex vivo study. J Ethnopharmacol 259:112982. https://doi.org/10.1016/j.jep.2020.112982

Oproshanska T, Mishchenko O, Shapoval O, Khvorost O, Koiro O, Bondariev Y (2021) Study of membrane-stabilizing and anti-inflammatory activity of tincture from herba *Bidens tripartita L.*

Organization WH (2004) Rheumatic fever and rheumatic heart disease: report of a WHO expert consultation. World Health Organization, Geneva, 29 October–1 November, 2001

Orhan N, İçöz ÜG, Altun L, Aslan M (2016) Anti-hyperglycaemic and antioxidant effects of *Bidens tripartita* and quantitative analysis on its active principles. Iran J Basic Med Sci 19(10):1114–1124

Pozharitskaya ON, Shikov AN, Makarova MN, Kosman VM, Faustova NM, Tesakova SV (2010) Anti-inflammatory activity of a HPLC-fingerprinted aqueous infusion of aerial part of Bidens tripartita L. Phytomedicine 17(6):463–468. https://doi.org/10.1016/j.phymed.2009.08.001

Rahman A, Bajpai VK, Dung NT, Kang SC (2011) Antibacterial and antioxidant activities of the essential oil and methanol extracts of Bidens frondosa Linn. Int J Food Sci Technol 46(6):1238–1244. https://doi.org/10.1111/j.1365-2621.2011.02615.x

Rice-Evans CA, Miller NJ, Paganga G (1996) Structure-antioxidant activity relationships of flavonoids and phenolic acids. Free Radic Biol Med 20(7):933–956. https://doi.org/10.1016/0891-5849(95)02227-9

Sisein EA, Owaba AD, Faith RO (2022) GC-MS profiling and phytochemical quantification in methanolic extract of Bidens tripartita. Glob J Res Chem Pharm 01:1–6. https://doi.org/10.58175/gjrcp.2022.1.1.0021

Tomczykowa M, Tomczyk M, Jakoniuk P, Tryniszewska E, FHC. (2008) Antimicrobial and antifungal activities of the extracts and essential oils of Bidens tripartita. Folia Histochem Cytobiol 46(3):389–393. https://doi.org/10.2478/v10042-008-0082-8

Tomczykowa M, Leszczyńska K, Tomczyk M, Tryniszewska E, Kalemba D, JoMF. (2011) Composition of the essential oil of Bidens tripartita L roots and its antibacterial and antifungal activities. J Med Food 14(4):428–433. https://doi.org/10.1089/jmf.2010.0066

Uysal S, Ugurlu A, Zengin G, Baloglu MC, Altunoglu YC, Mollica A, Custodio L, Neng NR, Nogueira JMF, Mahomoodally MF, FCT. (2018) Novel in vitro and in silico insights of the multi-biological activities and chemical composition of Bidens tripartita L. Food Chem Toxicol 111:525–536. https://doi.org/10.1016/j.fct.2017.11.058

Wolniak M, Tomczykowa M, Tomczyk M, Gudej J, Wawer I, APP. (2007) Antioxidant activity of extracts and flavonoids from Bidens tripartita. Acta Pol Pharm 64(5):441–447

Xuan TD, Khanh TD (2016) Chemistry and pharmacology of *Bidens pilosa*: an overview. J Pharm Investig 46:91–132. https://doi.org/10.1007/s40005-016-0231-6

The Phytochemistry and Pharmacology of *Bidens alba*

21

Great Iruoghene Edo, Agatha Ngukuran Jikah,
Emad Yousif, Khalid Zainulabdeen,
Athraa Abdulameer Mohammed, Winifred Ndudi,
Susan Chinedu Nwachukwu, Rapheal Ajiri Opiti,
Irene Ebosereme Ainyanbhor, Patrick Othuke Akpoghelie,
Joseph Oghenewogaga Owheruo, Priscillia Nkem Onyibe,
Ufuoma Ugbune, Gracious Okeoghene Ezekiel,
Benedicta Ediri Thomas, Helen Avuokerie Ekokotu,
Ephraim Evi Alex Oghroro,
Lauretta Dohwodakpo Ekpekpo, Endurance Fegor Isoje,
Ufuoma Augustina Igbuku, Joel Okpoghono, Arthur Efeog
hene Athan Essaghah, and Joy Johnson Agbo

Abstract

Bidens alba, a dicotyledonous plant of the *Compositae* family, commonly goes by other names such as butterfly needles, beggar's-ticks and Shepherd's needles. The plant not only originates from Florida but is also distributed across other subtropical and tropical regions of America. *B. alba* grows ubiquitously but may

G. I. Edo (✉)
Department of Chemistry, Faculty of Science, Delta State University of Science and Technology, Ozoro, Delta State, Nigeria

Department of Chemistry, College of Sciences, Al-Nahrain University, Baghdad, Iraq

Department of Petroleum Chemistry, Faculty of Science, Delta State University of Science and Technology, Ozoro, Nigeria

A. N. Jikah
Faculty of Pharmacy, Department of Pharmacy, Near East University, Nicosia, Cyprus

E. Yousif · K. Zainulabdeen · A. A. Mohammed
Department of Chemistry, College of Sciences, Al-Nahrain University, Baghdad, Iraq

W. Ndudi · U. Ugbune · L. D. Ekpekpo · U. A. Igbuku
Department of Chemistry, Faculty of Science, Delta State University of Science and Technology, Ozoro, Delta State, Nigeria

Y. Bibi et al. (eds.), *The Genus Bidens*,
https://doi.org/10.1007/978-981-96-4257-1_21

grow more robustly in temperate weather, under abundant rainfall and on moist and fertile lands. This perennial plant distributes comfortably in wetlands, forests, plantations, roadsides streamlines, agricultural fields, nurseries and coastal lines. The Bidens species are generally reported to exhibit high growth ability and high fruit (achene) production capacity to be self-pollinating etc. *B. alba* possesses immense therapeutic properties and is thus sometimes utilized as a compliment and at other times as an alternative to pharmacological medicines. The plant has been conventionally utilized against a variety of ailments such as swellings, injuries, cuts, jaundice, fevers, colds and flu, intestinal problems, earaches, eye infections, urinary tract problems and dermatological issues. Moreover results from botanical investigations regarding the phytochemical, biological and toxicological properties of *B. alba* in over 40 diseases authenticate the reports of its medicinal uses. *B. alba* has a rich phytochemistry comprising polyacetylene glycosides, flavones, polyacetylenes, flavonoids, alkaloids, aurones, phenolic acids, okanin glycosides, phytosterols, pheophytins, terpenes, fatty acids etc. which have been obtained from diverse parts of the plant. Most of these isolated compounds are believed to be bioactive and therefore responsible for the pharmacological activities of anti-oxidation, anti-inflammation, antimicrobial, anti-carcinogenic, anti-malaria, anti-allergy, anti-diabetic, smooth muscle relaxation and antihypertension which the plant elicits. Besides its conventional use against aliments, the leaves and vegetative parts of *B. alba* have also been used as food

S. C. Nwachukwu · P. O. Akpoghelie · J. O. Owheruo
Faculty of Science, Department of Food Science and Technology, Delta State University of Science and Technology, Ozoro, Delta State, Nigeria

R. A. Opiti · H. A. Ekokotu · E. E. A. Oghroro
Department of Petroleum Chemistry, Faculty of Science, Delta State University of Science and Technology, Ozoro, Nigeria

I. E. Ainyanbhor · J. Okpoghono
Faculty of Science, Department of Biochemistry, Delta State University of Science and Technology, Ozoro, Delta State, Nigeria

P. N. Onyibe
Faculty of Art, Design & Information Technology, Department of Health Informatics, George Brown College, Toronto, Canada

G. O. Ezekiel · B. E. Thomas
Faculty of Science, Department of Biotechnology, Delta State University, Abraka, Nigeria

E. F. Isoje
Faculty of Science, Department of Science Laboratory Technology (Biochemistry Option), Delta State University of Science and Technology, Ozoro, Nigeria

A. E. A. Essaghah
Faculty of Environmental Sciences, Department of Urban and Regional Planning, Delta State University, Abraka, Nigeria

J. J. Agbo
Faculty of Health Sciences, Department of Nursing, Cyprus International University, Nicosia, Turkey

without any obvious record of adverse effects. This chapter intends to present a comprehensive overview of the *B. alba* plant, its bioactive compounds, their biological activities and the traditional and modern pharmacological investigations of the plant. Comparative studies of the plant with other medicinal plants have also been provided.

Keywords

Bidens alba · Flavonoids · Terpenes · *Asteraceae* · Phytosterols

21.1 Introduction

Bidens alba first originated from South America but was later transported to Taiwan where it is currently a common roadside plant. *B. alba* appears very similar in biology and general appearance with *Bidens pilosa* such that the plants can even be inter-bread. The one distinction however between them is that while *B. alba* has petals on its flowers, *B. pilosa* does not (Kato-Noguchi and Kurniadie 2024). *B. alba* produces abundant numbers of flowers all yearlong thus generating enormous amounts of pollen for honey-bees to collect and produce nectar (Khalifa et al. 2021). The plant often colonizes an area via seed distribution and grows into an upright (36 inches high) annual or short-lived perennial plant robustly spreading during warm weather and under abundant rainfall conditions. *B. alba* has strap-shaped cotyledons of about 0.5–1 inch long with green to purple hypocotyls, and its stem is angled, smooth or lightly hairy (Soudzilovskaia et al. 2013). It has dark green leaves that are arranged in opposite style on the stem, and its root is the deep taproot system which may be extensive on plant age and size. Its zygomorphic or actinomorphic flowers are clustered in terminal heads, and each flower has 4–5 elongated white petals adjoining several yellow disc florets (Kato-Noguchi and Kurniadie 2024). The fruits of *B. alba* are called achenes and consist of a single dark brown or black, sparsely haired and linear seed. The species of this genus are represented by other common names as farmer's friends, cobbler's pegs, beggar-ticks and black-jack (Zhang et al. 2019). Moreover, this plant species may occur in several habitats varying from moist rich soils to desiccated, infertile and sandy soils. Thus they may be found in pastures, grasslands, roadsides, wetlands, landscapes, nurseries, ditch banks and potting media of container grown ornamental being distributed by animal and human activities. *B. alba* is nutritionally composed of calories 43, fat 0.5 g, water 80 g, protein 3.8 g, carbohydrate 8.4 g, beta carotene 1800 μg, moisture 85%, fibre 3.9, calcium 0.34, ash 2.2, magnesium 135, zinc 0.80, ascorbic acid 23 and energy 43 kcal and as such has been utilized as a medicinal plant for ages (Khalifa et al. 2021). Its extracts are used against conjunctivitis, burns, ulcers, diuretics etc., and its leaves are also used against abdominal distress, sore eyes, toothaches and swollen glands. Investigations conducted on the extracts of *B. alba* have concluded that the plant is also effective as an anti-ulcerogenic, anti-hyperglycemic, hepatoprotective, antihypertensive, anti-leukemic, antipyretic, antibacterial, anti-tumour, antioxidant and immunosuppressive properties (Abu-Odeh and Talib 2021). These confirmed activities of the plant have made certain countries (e.g. Brazil) to place it in the official list of therapeutic herbs that can be used by the health system. This chapter intends to discuss the *B. alba* plant, its bioactive compounds, the biological activities these compounds elicit and the traditional and modern pharmacological investigations of the plant. Comparative studies of the plant with other medicinal plants have also been provided.

21.2 The Origin and Nature of *B. alba*

B. alba is an upright annual or transitory perennial herb native to South America, West Indies Florida and all islands found within the tropical, subtropical and Lucayan Archipelago regions of the world. The plant was introduced to South Australia, Queensland, East Aegean Island, Marshall Isles, Hawaii, Caroline Island, Ogasawara-shoto, Samoa, Madeira and west Australia. It is regarded as a chief weed on the Marshall Isles (Deshmukh and Gaikwad 2014). *B. alba* does not grow well understory of forests or in shades but copiously grows in open and sandy areas like atolls. *B. alba* spreads by the help of its barbed seeds which stick to clothing, fur and feathers. It may also be conveyed in soil on machinery (Kumar et al. 2018).

21.3 Phytochemistry of *B. alba*

B. alba is an extreme source of phytochemicals which have proven useful in several therapeutic applications. The phytochemical compounds of *B. alba* comprise several compounds including flavonoids, terpenes, porphyrins, aromatics, fatty acids, phenolic acids, phytosterols, phenylpropanoids and pheophytins (Chen et al. 2021). The main active compounds thought to be responsible for its biological activities however are flavonoids, essential oils, polyacetylenes and triterpenes (Kumar et al. 2023). The following section briefly considers some of these compounds. Table 21.1 gives a summary of the phytocompounds found in *B. alba*.

21.3.1 Polyacetylenes

These are chemically reactive compounds isolated from the *B. alba* plant. Above 1400 different polyacetylenes and their derivatives have been identified and extracted from the plant. Several of the polyacetylenes are aliphatic acetylenes and contain double or triple bonds with their glucoside rings, cyclic and aromatic groups (Lai et al. 2023). Complex polyacetylenes, however, contain a single triple bond that has heterocyclic moieties (Onyibe et al. 2021). The polyacetylene components in the roots, flowers and leaves of *B. alba* are 1-phenyl-hepta-5-ene-1,3-dyne and

Table 21.1 An overview of some compounds in *B. alba*

Compound name	Compound class
Benzoic acid	Phenolic acid
Phenylheptatriyne	Polyacetylene
Chlorogenic acid	Phenolic acid
Bornyl acetate	Essential oil
Luteoside	Flavonoid
Sulfuretin	Flavonoid
Ethyl caffeate	Phenolic compound
Eugenol	Phenolic compound
6-Phynyl-1,3-dyne-5-en-7-ol-acetate	Polyacetylene
Isophorone	Essential oil
Rutin	Flavonoid
Caffeic acid	Phenolic acid
Astragalin	Flavonoid
Jacein	Flavonoid
Dimethoxyphenol	Phenolic compound
Megastigmatrienone	Essential oil
Hexadecanol	Essential oil
Limonene	Monoterpene
Hexadecyl acetate	Essential oil
6-Pheylhex-1-ene-3,5-diyn-1-ol	Polyacetylene
Sabinene	Monoterpene
Neochlorogenic acid	Phenolic acid
Quercetin 3-O-ribinobioside	Flavonoid
Esculetin	Coumarin
Myrcene	Monoterpene
Pentacosane	Aliphatics
Hexacosane	Aliphatics
Camphene	Monoterpene
Perillene	Monoterpene
Tridec-1-ene-3,5,7,9,11-pentayne	Polyacetylene
Quercetin 3-O-glucoside	Flavonoid
Terpinolene	Monoterpene
4-O-Caffeoylquinic acid	Phenolic acid
m-Cymol	Monoterpenes
Diphenylmethane	Essential oil

phenylheptatriyne and are contained in the aromatic oils extracted from these parts of the plant. Furthermore, recent studies have reported that several derivatives of polyacetylenes and their glycosides are also located in large quantities in the whole *B. alba* (Myo et al. 2023). Some examples include trideca-2,12-diene-4,6,8,10-tetrayn-1-ol, phenylheptatriyne, 6-Phenylhexa-1,3,5-triyn-1-ol, 6- Phenylhex-1-ene-3,5-diyn-1ol, 6-Phenylhexa-1,3,5-triyn-1-yl acetate, tridec-1-ene-3,5,7,9,11-pentayne etc. (Nwosu et al. 2022).

21.3.2 Flavonoids

Flavonoids number among the largest groups of phytochemicals that are present in diverse species of plants (Edo 2022). The compounds show potent health benefiting antioxidant properties (Hassan et al. 2021). Flavonoids and their derivatives like agycosides, aglycones, okanin and aurone glycosides are located in several parts of *B. alba*. A total of 20 flavonoid glycosides have been extracted from the leaves of *B. alba*. Of these 12 flavonoids are aurone glycosides, 16 are okanin glycosides located in the flowers, leaves and aerial parts of the plant and 10 flavonoids having the quercetin skeleton have also been found all over the plant. Rutin, jacein, axillaroside, astragalin, centaurein and querciturone are some examples of *B. alba*-contained flavonoids (Ullah et al. 2020).

21.3.3 Glycosides

Glycosides are chemical compounds in which a sugar moiety is linked to a non-sugar moiety (Akpoghelie et al. 2022). Glycosides are found inactive within the body and only get activated by the work of enzymes (Erhonyota et al. 2023). The classification of these compounds depends on the aglycone and glycone moiety, and their examples include α-glycosides, fructoside, β-glycosides and glucoside (Kytidou et al. 2020). Glycosides are largely implicated in stimulating the cardiac system and the immune system. The compounds also exhibit antimicrobial properties (Škubník et al. 2021).

21.3.4 Terpenoids

Terpenoid compounds are composed of a simple structure of C5 isoprene units and are classified by the isoprene units implicated in their synthesis (Edo et al. 2023f). In general, 99 terpene compounds including monoterpenes, diterpenes, sesquiterpenes and triterpenes have been reported in *B. alba*. Specifically, 28 monoterpenes (cyclic and acyclic), 6 diterpenes, 6 triterpenes and 58 sesquiterpenes (mono-, bi- and tricyclic and linear forms) with varying types of terpenoid derivatives (Masyita et al. 2022). *B. alba*-contained terpenoids contain both oxygenated compounds and hydrocarbons. Tricyclic diterpenes are components of essential oils from this plant. One acyclic polyterpene and pentacyclic triterpenes have been found to be present in the whole plant. For the leaves, only tetraterpene has been found contained in them (Stéphane et al. 2022).

21.3.5 Tannins

Tannins are water soluble dense weight molecular compounds (Edo et al. 2023d). They are often found as a composite with polysaccharides, alkaloids and proteins

(Jikah and Edo 2023). Tannins are classified into phlorotannins, proanthocyanidins and hydrolysable tannins depending on their hydrolysis products and solubility (Fraga-Corral et al. 2020). Hydrolysable tannins are produced from gallic acid esters. The condensed tannins that are made up of a joint structure of polyhydroxy flavan-3-ol monomers and phlorotannins are produced from phloroglucinol which can be extricated from brown algae. Grapes, apples, peach and berries are some main sources of tannins (Ali et al. 2023).

21.3.6 Fatty Acids

A sum of 12 long-chain fatty acids is reported to be present in *B. alba.* Some are particularly present in the essential oils obtained from the leaves: behenic acid, elaidic acid, myristic acid, capric acid, methyl linolenate, lauric, linoleic acid, linolenic acid etc. (Moore et al. 2020).

21.4 Pharmacological Activities of *B. alba*

Medicinal plants are vital sources of phytochemical compounds that elicit biological activities, and for this reason they are widely utilized against ailments (Edo et al. 2023g). Pre-clinical and clinical studies carried out on *B. alba* report that the plant has potential for the following biological activities (El Haouari et al. 2021).

21.4.1 Anticancer Activities

In vivo and in vitro studies that were carried out on *B. alba* between 1970 and 1980 reported its anticancer properties. For instance, Craddock and Driver (1987) carried out an investigation on rats which had oesophageal tumours. Dried leaves of *B. alba* were fed the rats in the oesophageal epithelium using the semi-quantitative assay called [3H]-thymidine incorporation. A 2.3-fold decrease in cell proliferation was observed from the results of the thymidine incorporation. Thus the administration of dried *B. alba* powder elicited carcinogenic activity against oesophageal tumours (Chowdhury et al. 2006). Another investigation conducted by Posadino et al. (2023) reported that *B. alba* has potent anticancer properties, especially as an anti-angiogenic drug. According to them, the ethyl acetate extracts of the plant elicited strong suppressive actions against cell growth in human umbilical vein of endothelium cells. The water extracts of the plant also inhibited five leukemic cell lines (K562, P3HR1, Raji, L1210 and U937) with a dose-dependent half-maximal inhibitory concentration of 145–586 μg/mL. Chloroform extracts obtained from the plant also showed strong cytotoxicity via Neutral Red Uptake and tetrazolium salt assays with a half-maximal inhibitory concentration of 97.0 and 83.0, respectively (Zakaria et al. 2023). The combination of the hydroalcoholic and ethyl acetate extracts of *B. alba* was also seen to exhibit cytotoxicity with a half-maximal inhibitory concentration less than 200 μg/mL (Rajayan et al. 2024). Additionally, the ethanol extract and essential oil-derived polyacetylenic compounds of *B. alba* showed strong actions against the proliferation of human umbilical vein endothelia cells and upregulated p21 or p27 genes by 2.2–3.0 times augment in western blot analysis. The polyacetylenic compound *β*-D-glucopyranosyloxy-3-hydroxy-6(E)-tetradecen-8,10,12-triyne identified in the methanolic extracts of *B. alba* was observed to cause both transformed and normal human cells to overgrow in culture. Phenylheptatriyne isolated from hexane extracts also demonstrates cytotoxicity in several tumour cell lines, especially cancer lines like Caco-2 and HepG2 (Thusyanthan et al. 2022). Moreover, Arafat et al. (2023) showed that in 80–85% of cases, lung cancer patients who were treated with either a part or the pure essential oil of *B. alba* which contained phenylheptatriyne showed improvement. These results prove the potential antitumour effects of *B. alba*-contained polyacetylenes. Furthermore, the flavonoid contents of *B. alba* also exhibit some actions against tumours, for instance, the compounds reduced both tissue urokinase plasminogen and plasminogen activator in addition to inhibiting tumours. The compounds centaurein and apigenin are known to induce tumour cell death through inhibition of tubulin polymerization (Pasaoglulari Aydinlik et al. 2021). The compounds also augment the activities of the interferon-gamma prompter by fourfold and regulate the transcription of the promoter via the nuclear factor-kB in T cells and nuclear factor of activated T cells (Zhai et al. 2022). Furthermore, apigenin was seen in one experiment to notably inhibit tumour cells obtained from colon cancer, human lymphoma and melanoma (Rahmani et al. 2022). Okanin 4′-glucoside and butein compounds also induce cell death in several tumours; they possibly work by blocking the growth of human cancer cells in vivo and in vitro. Growth in human hepatoma cancer cells was also suppressed by okanin 4′-glucoside through induction of apoptosis and G/M phase arrest, thus minimizing the activity of Cdc22, generating reactive oxygen species and promoting inactivated phosphorylated Cdc2 (Arfin et al. 2021). Ensuing experiments also confirm that butein inhibits the activity of inducible NF-kB which when downregulated represses the angiogenesis of prostate cancer. The monoterpenoids 1,8-cineole and alpha-terpineol also brought about morphological alternations of DNA fragmentation when applied against human leukaemia HL-60 cells and also induced apoptosis that is dependent on caspase enzyme in M14 WT melanoma cells (Piechulla et al. 2016). The aforementioned benefits notwithstanding, incomplete inhibition in human cancer cells maybe experienced with the use of these compounds because they are only present in low amounts in the leaves of *B. alba.*

21.4.2 Anti-inflammation Activities

Inflammation is a form of biological reaction of cells to injurious stimuli like irritants and pathogens (Edo et al. 2023b). Recent studies have shown the regulation of numerous inflammatory cytokines that induce both inducible nitric oxide synthase (iNOS) and cyclooxygenase-2 (COX-2) by *B. alba* extracts. The methanolic extracts of the plant showed strong immunosuppressive actions on murine and human lymphocytes on which inflammation was triggered by the administration of 5 μg/mL of phytohemagglutinin (L. Chen et al. 2018). Additionally, administering the same extracts to mice at 10 mg reduced the magnitude of the popliteal lymph node following a zymosan-induced inflammation. The ethyl acetate extracts of the plants were also seen to significantly inhibit the production of nitric oxide in RAW 264.7 cells to which inflammation was induced by lipopolysaccharide (Yoo et al. 2022). The *B. alba*-contained compound, 2-*O*-*β*-*D*-glucosyltrideca-11E-en-3,5,7,9-tetrayn-1,2-diol, also shows immunomodulatory and anti-inflammatory effects in arthritis mouse models to which arthritis was induced by zymosan. These same effects are observed with the compound in lymphocyte proliferation assays (Sheik et al. 2023). Treating mice for 5 days with *B. alba* extracts stopped the increase in size of the popliteal lymph nodes. This action was thought to be through the repression of lymphocyte proliferation. These reports imply that *B. alba* has anti-inflammatory properties that set it up as a relevant raw material for anti-inflammatory drugs (Kenney et al. 2022). The dried powder of *B. alba* also significantly repressed hydrochloric acid-induced progressive gastric mucosal lesions. The plant also shows cytoprotective actions on the gastric mucosa for the repression of cyclooxygenase-2 exhibiting anti-ulcerogenic actions (Jabbar et al. 2023). Ethyl caffeate a phenolic compound found in *B. alba* has anti-inflammatory activities against lipopolysaccharide-induced macrophages (Suriyaprom et al. 2023). Dimethoxyphenol also significantly inhibits the production of lipopolysaccharide-stimulated nitric oxide with a notable suppression of cyclooxygenase-2 expression. Moreover, 1 μg/mL dose of caffeate also considerably reduced the production of iNOS mRNA in macrophages treated with lipopolysaccharide (Min et al. 2009). The growth hormone PGE2, which promotes development in some carcinoma cell lines and mediates inflammation, is remarkably repressed by dimethoxyphenol. The flavonoid compounds butein, luteoside, apigenin 7-O-glucoside and okanin 3′-glucoside all have noteworthy anti-inflammatory properties; butein particularly inhibits inflammatory responses like nitric oxide expression and the production of lipopolysaccharide-stimulated pro-inflammatory cytokines. The same compound is also reported to have potential against inflammatory diseases of the intestines as it prevents the expression of matrix metalloproteinase-7 and also prevents TNF-alpha-induced interleukin 8 in HT-29 cells effects (Finetti et al. 2020). The compound quercetin is another flavonoid found in the plant that prevents bronchial hyperactivity and bronchial obstruction induced by platelet-activating factor in guinea pig asthma model. Moreover, antipyretic actions are also recorded of *B. alba* in in vivo investigations that are similar to paracetamol (Botting 2004). These reports that *B. alba* has potent anti-inflammatory activities that may be used to produce anti-inflammatory drugs.

21.4.3 Anti-allergic Activity

The aqueous extracts of dried cellulosine enzyme powder displayed strong anti-allergic actions by inhibiting the release of histamine from mast cells (Edo et al.

2023c). These three compounds luteolin 7-*O*-*β*-D-glucopyranoside, 5-*O*-methylhoslundin and quercetin 3-*O*-glucoside that are present in *B. alba* were found contained in large amounts in cellulosine enzyme, showing that *B. alba* in combination with an enzyme has potent anti-allergic actions (Shaik et al. 2018). Furthermore, the effects of cellulosine enzyme on histamine release in rat peritoneal mast cells and on histamine-triggered contraction in the ileum of Guinea pigs were examined. The results showed that the cellulosine enzyme exhibits anti-allergic activity by repressive actions on the H1 receptor leading to a reduction in the release of histamine (Bani-Sacchi et al. 1986). The same action on H1 receptor can be carried out by rutin, luteolin 7-*O*-*β*-D-glucopyranoside and quercetin 3-*O*-glucoside flavonoids from *B. alba*. Currently, a market product, ClearGuard, used for the treatment of nasal allergies is extracted from *B. alba* among other plants (Rehfeldt et al. 2022).

21.4.4 Anti-diabetic Activities

The *B. alba* plant has potential anti-diabetic properties among the *Bidens* species and is globally known for diabetic therapy. The dried plant was boiled in 15% water and ethanol for 5 min, and the solution was administered to normoglycaemic and alloxan-induced mildly diabetic mice. Hypoglycaemic activities were recorded in the mildly diabetic mice although there were no observed effects against severely diabetic mice. This shows that the plant requires the mediator effects of insulin for its hypoglycaemic effects (Zanzabil et al. 2023). The butanol extracts of *B. alba* inhibited the differentiation of human naive helper T cells into type 1 helper T cells by up to 50%. Cell differentiation was completely halted by 500 μg/mL extracts. Administration of the same extracts to non-obese diabetic NOD mice at a 3 m/kg showed a less diabetes occurrence than in the control (Vandamme and Kinnunen 2020). In other investigations, the butanol extricate of *B. alba* was used against type 1 and type 2 diabetes (Lee et al. 2021). In addition, the extracts of this plant may also enhance autoimmune diabetes mediated by Th1 in NOD mice. Other additional investigations where the water extracts of *B. alba* were used reported that the plant mitigated type 2 diabetes in diabetic mice by regulating the secretion of insulin and protection of the islet cells (Persia et al. 2023). The anti-diabetic actions of the methanolic extracts of *B. alba* were also investigated using a mice model of type 2 diabetes, and it was found out that one variant of *B. alba* exerted higher insulin-releasing activity and glucose lowering-properties in a long-term experiment as compared to the other two variants. *B. alba* extracts thus exceptionally decrease the levels of glycosylated haemoglobin A1c in diabetic mice (Zeleke et al. 2023). Polyacetylenic glucosides extracted from the plant also show significant hypoglycaemic effects by mitigating the harmful effects of type 2 diabetes in mice. The compounds are also known to show potent preventive effects at the commencement of diabetes and to regulate levels of blood sugar in NOD mice (Yi et al. 2023).

Cytophiloye gotten from *B. alba* also hinder type 1 diabetes in mice via the regulation of T cells. It also suppresses the growth of CD4 + T cells within the pancreatic lymph nodes and spleen of mice without touching CD8 + T cells. The anti-diabetic activities of this compound (cytophiloyne) are elicited mainly through regulation of T cell, a different mechanism of action compared to other type 1 anti-diabetic drugs. The compound also has lower toxicity and lower inhibitory actions on the immune system. Cytophiloyne is also found to inhibit the expression of IFN- γ in a dose-dependent manner (Utami et al. 2023). Results may vary depending on the solvent used for extraction. Polyacetylenic components of *B. alba* are thought to be the most potent phytochemicals in the management of diabetes.

21.4.5 Antioxidant Activities

Some common *B. alba*-contained phenolic acids such as procatechin, ethyl caffeate, eugenol, 5-*O*-methylhoslundin and p-hydroxybenzoic acid exhibit significant antioxidant properties in lipid peroxidation, β-carotene and reducing powder assays. The warm ethanol and aqueous extracts obtained from the plant also display more antioxidant actions than trolox C, a tocopheroxyl vitamin E homologue that is soluble in water (Liu et al. 2020). The ethanol and ethyl acetate extracts of *B. alba* also defend human erythrocytes from oxidative harm in vitro. These extracts additionally reduce the depletion of cytosolic ATP and glutathione and decrease the decline of superoxide dismutase activity and in red blood cells (Hernández-Cruz et al. 2023). Investigations on the hepatoprotective properties of the total flavonoids of *B. alba* showed that the plant significantly restores glutathione peroxidase and reduces carbon tetrachloride-induced liver injury in rats. This could be as a result of their antioxidant properties and free radical scavenging activities which inhibit NF-KB activation (Kim et al. 2024). The butanolic and ethyl acetate extracts of *B. alba* also showed significant free radical scavenging activity against the production of nitric oxide in lipopolysaccharide-induced RAW264.7 cells (García Díaz et al. 2022). Okanin 4′-glucoside is another very potent antioxidant found in different parts of *B. alba* that is more effective against DPPH than vitamin E (Charlton et al. 2023). Butein is another very effective antioxidant compound; however it shows less pro-oxidant potential than quercetin. Chromane is also very potent in protecting cell membranes against free radical damage in in vivo experiments (Edo et al. 2023e). Similarly, the *B. alba*-contained essential oils are also potent antioxidants and show suppressive effects on DPPH leading to the production of diphenylpicylhydrazine. These effects are a pointer that *B. alba* has strong antioxidant properties and thus may be a main source of flavonoid glycosides, caffeoylquinic acid derivatives and polyphenolics which can be utilized for the production of antioxidant drugs. It is worthy of note, however, that the antioxidant capacity of *B. alba* is less than that of synthetic antioxidants (Baliyan et al. 2022).

21.4.6 Wound Healing Properties

The ethanol extracts of *B. alba* upon investigation were found to have wound healing properties. When compared to the control group that received neomycin sulphate, the extracts of *B. alba* exhibited faster wound closure (Zishan et al. 2024). The histological examinations also showed better angiogenesis, organization and collagenation of the wound tissue in 7 days following application. The total healing time and epithelialization in the rats treated with *B. alba* extracts could be compared to those of neomycin sulphate. These studies indicate that *B. alba* extracts may be possible alternatives to neomycin lotion in wound treatment (Balderas-Cordero et al. 2023). In another investigation, the cyclohexane, methyl chloride and methanol extracts of the plant were used on the gastric ulcers of Wistar rats. Prior to this experiment, the rats were fed with ethanol/HCl gastric necrotizing solution. The results showed that the extracts by methyl chloride had the highest manifestation of lesion inhibition followed by the methanol extracts, while the cyclohexane extracts showed least inhibition of lesion inhibition (Shareef et al. 2022).

21.4.7 Anti-microbial Activities

The potent antimicrobial properties of *B. alba* have been reported by several studies including its activities against viruses. Aqueous extracts (100 μg/mL) of *B. alba* were seen to inhibit the replication of herpes simplex virus (Garber et al. 2021). Further experiments in which the medicinal drug acyclovir was used alongside *B. alba* extracts (500 μg/mL) against herpes simplex virus showed a 39.02% inhibition with the extracts of the plant and a 45.07% inhibition by acyclovir (Xu et al. 2022b). Additionally, the methanolic extracts of *B. alba* extremely repressed all gram-positive and gram-negative microbes at 5–10 mg/mL. The growth of *Penicillium notatum* was also completely retarded by the same extracts. The methanolic, aqueous and petroleum ether extracts of *B. alba* showed powerful antibacterial actions, particularly acting against gram-positive bacteria. Other bacteria strains against which the aqueous and ethanol extracts of the plant are effective are *Staphylococcus aureus* and *Escherichia coli*. The activity of these extracts on these strains was seen to be more potent than those of gentamycin sulphate (Aqel and Farah 2024). Furthermore, 100 μg/mL of the same extracts demonstrated repressive effects on *Mycobacterium intracellulare,* the essential oils and water extracts of the flowers also considerably reduce the growth of fungal and bacterial strains. The essential oil of the flowers is more potent against gram-negative than gram-positive bacteria. Additionally, the antimicrobial activities of phenylheptatriyne against several microbes were also examined; the studies indicated that the compound inhibits the growth of several microbes including yeast, moulds and bacteria (Angane et al. 2022). Another study reported that phenylheptatriyne could repress the germination and development of *myclia* and *macroconidia* and photosensitize the cereal pathogen *F. culmorum* (Błaszczyk et al. 2023). The same compound elicits protozoacidal and anthelminthic activities on infected mice in vitro. It should be noted that phenylhepatriyne has no effects against microbes in the dark, a fluorescent lamp may be necessary to induce its actions and the compound is also unable to or only partially decreases phototoxicity of microbes when an oxygen atom is removed from it (Zupin et al. 2022). Early work on pheylheptatriyne also revealed antifeedant activity of the plant as the compound exhibited potent ovicidal actions against *Drosophila melanogaster*. Light dependant toxicity was also exhibited by the compound against the larvae of some mosquitoes. The polyacetylenic compounds (11E), trideca-1,11-diene-3,5,7,9-tetrayne and tridec-1-ene-3,5,7,9,11-pentayne, also exhibit phytotoxic effects against fungi and bacteria particularly in the dark. Phototoxic effects against bacteria and yeast were also observed as the compound strongly suppressed the growth of *F. culmorum* (Xuan and Khanh 2016). Additionally, the fatty acids, linolenic and linoleic acids, extracted from the plant with petroleum ether solvents are found to be bacteriostatic at 5–50 ppm. Fresh or aqueous extracts of *B. alba* may also contain adequate amounts of unsaturated fatty acids that inhibit the growth of bacteria and other microbes. It is however unknown if these compounds are present in the plant as free fatty acids are generated from the enzymatic breakdown of fats or oils upon drying of the plant material (Shah et al. 2022). Phytosterols like linoleic acid, campesterol and palmitoleic acid and mixtures of n-alkanes extracted using petroleum ether solvent show antibacterial activities. Other key monoterpene and sesquiterpene compounds like alpha-pinene, germacrene-D and 14-Oxy-a-muurolene also display potent inhibitory actions against numerous bacterial and fungal strains (Ferreira et al. 2021). Moreover, anti-human immune deficiency virus activity has also been reported of *B. alba,* and these activities are thought to be linked to butein and okanin 4-glucoside. These compounds hinder the integrase action of the HIV-1 protease and the human immunodeficiency virus (Rotich et al. 2022). The centaurein compound present in the whole plant is also effective against herpes simplex viruses suppressing infections caused by the virus (Xu et al. 2022b).

21.4.8 Anti-malaria Activity

B. alba is a very promising anti-malaria botanical as it strongly inhibited parasitemia in in vitro experiments although against *Plasmodium berghei*, only little inhibition was reported of the plant (Camara et al. 2019). Above all, the ethanol, chloroform and butanol extracts of the dried plant material of *B. alba* inhibited the fatal malarial strain *Plasmodium falciparum* by an up to 90% inhibition (Bizuneh et al. 2023). Another study conducted on mice confirmed the inhibitory actions of *B. alba* against *P. berghei* as the extracts inhibited the organism by up to 60% (Adigo Shibeshi et al. 2021). Even mefloquine- or chloroquine-resistant species of *P.*

falciparum are susceptible to *B. alba* in vitro. In human blood, *P. falciparum* can be appreciably inhibited using the hexane leave extracts of *B. alba* at IC50 of 32.8 µg/mL. The potent anti-malaria properties of *B. alba* are likely to be due to its abundant contents of flavonoids and polyacetylenes. Phenylheptatriyne, for instance, is a key example of a polyacetylenic compound that is present in high concentrations within the plant; the compound is bioactive against several strains of malaria including *P. falciparum*. The toxicity of polyacetylenes against microbes is however light dependent and requires ultraviolent light for their actions (Ali et al. 2021). Campesterol is another compound which is present in all parts of the plant and was found to be active against *P. berghei* in mice models. Up to 58% inhibition activity is observed with this compound usually after 8 days following parasite inoculation (Sirak et al. 2021). Okanin 4′-glucoside present in the fresh parts of *B. alba* also shows inhibitory actions against *Leishmania*. The compound may thus be considered for the production of novel anti-Leishmania and anti-malaria drugs in time to come (Edo et al. 2023a).

21.4.9 Anti-arterial Hypertensive Activity

B. alba is traditional utilized against arterial hypertension the world over, the water and methanolic extracts from the plant's leaves particularly exhibited antihypertensive actions in un-anaesthetized rats. No effect was however observed on the pulse (Ndjenda II et al. 2021). These extracts were further seen to be effective against both salt-loaded and spontaneously hypertensive rats as blood pressure was drastically reduced. Subsequent in vitro research that evaluated the vasorelaxant activities of the neutral extracts of *B. alba* on the isolated aorta of rats that was contracted with KCL showed that there was a decrease in the resting tone of the aorta, validation of vasodilatory effects on the tissue and suppressions of KCL contractions (Xu et al. 2022a). The neutral extracts of *B. alba* thus have an endothelium-independent relaxant effect that possibly results from its ability to block calcium channels. The methylene chloride and aqueous extracts of the plant in another experiment were seen to reverse hypertriglyceridemia and high blood pressure without affecting plasma glucose or insulin levels in fructose fed rat models (Shin et al. 2024). Additionally, the methanolic extracts of the plant not only lowered blood pressure and hindered the establishment of hypertension but also mitigated high fructose diet-triggered high plasma insulin levels in Wistar rats. High plasma triglycerides were however not decreased (Dimo et al. 2002). Some reports provide that *B. alba* if taken in sufficient amounts may reduce and or prevent fructose diet-induced hyperinsulinemia and arterial hypertension. Some flavonoids contained in the plant are also potent at lowering high blood pressure, causing functional vascular modifications in spontaneously hypertensive rats and decreasing cardiac and renal hypertrophy without affecting normal blood pressure (Ciumărnean et al. 2020).

21.4.10 Anti-GIT Problems

The extracts by ethanol of *B. alba*, especially those from its aerial parts, attenuate the volume of gastric juice and pepsin and acid secretion. Antiulcer properties were also observed with the plant which actions are traceable to its quercetin content (Shams and Eissa 2022). An investigation proved that *B. alba* has anti-ulcer and anti-secretary effects, the water and methanolic extracts of the *B. alba* were administered to rats (intragastric administration) in which ulcers were induced by aspirin and the ulcers were seen to heal 10 days following administration (Chowdhury and Kumar De 2023). Flavonoids extracted from the aerial parts of *B. alba* were also seen to considerably induce choleretic activity. The compounds provoked an increase of cholesterol and cholic acids in bile. The methylene chloride, cyclohexane and methanolic extracts of *B. alba* leaves have also been reported by experimentation to have ulcer lesion inhibition effects of up to 47%, 46% and 41% in that order.

21.5 Traditional Uses and Modern Pharmacological Investigations of *B. alba*

As earlier stated, *B. alba* is used in traditional medicine as an anti-inflammatory agent, a sudorific, to stimulate the immunological system and as a diuretic. The *B. alba* herb is also used as a remedy against bleeding and ruptured blood vessel at any level. Against skin ailments, gravel, fevers, kidney and bladder stones, also the plant extracts have been utilized. Other uses in traditional medicine include the use of the plant in treating of liver, bronchioles, spleen diseases etc. Additionally, the water extracts of the leaves of *B. alba* are also harnessed by the Zulu people of South Africa against diarrheal, colic and dysentery (Nur et al. 2023). The plant is also used popularly in China as tea and in traditional remedies against several ailments like inflammation, bacillary dysentery, diabetes, pharyngitis and enteritis (Wahab et al. 2021). In Brazil, the plant is utilized in folk medicine to cure a range of sicknesses counting fever, oedema, angina, inflammation, pain inflammation, infection and diabetes. Additionally, the people of the Amazon region in Southern Brazil use the hydroalcoholic extracts from the roots of *B. alba* in treating malaria and tumours (Ribeiro et al. 2018). In India, it is utilized as traditional treatment for urinary tract infections, glandular sclerosis, wounds, colds, hepatitis and flu (Mammari et al. 2023).

21.6 Comparative Studies with Other Medicinal Plants

D. Zhang et al. (2023) studied the chloroplast genomes of *B. alba*, *B. pilosa* and *Bidens bipinnata*. The team intended to distinguish their phylogenetic and molecular characterization relationships. Comparable numbers of simple sequence repeats were found in all three, but forward, palindromic and mononucleotide repeats were the most abundant. This study may offer relevant information for analysis in genetic diversity and for the molecular classification of *Bidens* species (Duan et al. 2023). Other studies were conducted by Wang et al. (2020). *Bidens alba* and *Bidens tripartite* from different tropical climates; Tartarstan and Burundi. The plants both being possible sources of phytochemicals that exhibit anti-diabetic, anti-inflammatory, anticancer, antimicrobial and other effects were investigated for their phenolic contents. A qualitative analysis of the phenolic compounds showed that the phenolic contents were slightly varied by climate differences in the different locations of growth. Rainfall, temperature and soil composition were also found to affect the phenolic contents of these plants (Brahmi et al. 2022).

21.7 Conclusion

B. alba is distributed widely across countries and continents and is known to be a great source of both medicine and food. Diverse compounds have been obtained from the plant via extraction by different solvents. These extracted compounds elicit various biological and pharmacological activities; among them, polyacetylenes stand out as the most biologically active compounds that are responsible for most of the pharmacological actions of anti-inflammation, anti-oxidation, antimicrobial, antihypertension, anti-diabetes, immune regulation, allergy and gastrointestinal problems which the plant elicits. However, other phytochemicals such as phenolic acids, phenylpropanoids, flavonoids, phytosterols, fatty acids and terpenes are also implicated in its pharmacological actions. This chapter gave an overview of the biological compounds and the biological activities associated with them. Nevertheless, the number of studies conducted on *B. alba* is inadequate, and more research is definitely needed to establish these acclaimed benefits of the plant in therapy. Toxicological studies on the plant will also be important for establishing its efficacy dosage.

Ethics Approval and Consent to Participate Not applicable.

Consent for Publication Not applicable.

Availability of Data and Material All data will be made available upon reasonable request.

Competing Interests The authors have no competing interests.

Funding This research did not receive any specific grant from funding agencies in the public, commercial or not-for-profit sectors.

References

Abu-Odeh AM, Talib WH (2021) Middle East medicinal plants in the treatment of diabetes: a review. Molecules 26(3):742. https://doi.org/10.3390/molecules26030742

Adigo Shibeshi M, Fentahun Enyew E, Mequanint Adinew G, Jemere Aragaw T (2021) Antimalarial activity of Methanolic extracts and solvent fractions of Combretum molle leaves in plasmodium berghei infected mice. J Exp Pharmacol 13:69–89. https://doi.org/10.2147/JEP.S285117

Akpoghelie PO, Edo GI, Akhayere E (2022) Proximate and nutritional composition of beer produced from malted sorghum blended with yellow cassava. Biocatal Agric Biotechnol 45:102535. https://doi.org/10.1016/j.bcab.2022.102535

Ali AH, Sudi S, Shi-Jing N, Hassan WRM, Basir R, Agustar HK et al (2021) Dual anti-malarial and GSK3β-mediated cytokine-modulating activities of quercetin are requisite of its potential as a plant-derived therapeutic in malaria. Pharmaceuticals 14(3):248. https://doi.org/10.3390/ph14030248

Ali S, Ekbbal R, Salar S, Yasheshwar, Ali SA, Jaiswal AK et al (2023) Quality standards and pharmacological interventions of natural oils: current scenario and future perspectives. ACS Omega 8(43):39945–39963. https://doi.org/10.1021/acsomega.3c05241

Angane M, Swift S, Huang K, Butts CA, Quek SY (2022) Essential oils and their major components: an updated review on antimicrobial activities, mechanism of action and their potential application in the food industry. Food Secur 11(3):464. https://doi.org/10.3390/foods11030464

Aqel H, Farah H (2024) Seriphidium herba-alba (Asso): A comprehensive study of essential oils, extracts, and their antimicrobial properties. PLoS One 19(4):e0302329. https://doi.org/10.1371/journal.pone.0302329

Arafat K, Al-Azawi AM, Sulaiman S, Attoub S (2023) Exploring the anticancer potential of Origanum majorana essential oil monoterpenes alone and in combination against non-small cell lung cancer. Nutrients 15(23):5010. https://doi.org/10.3390/nu15235010

Arfin S, Jha NK, Jha SK, Kesari KK, Ruokolainen J, Roychoudhury S et al (2021) Oxidative stress in cancer cell metabolism. Antioxidants 10(5):642. https://doi.org/10.3390/antiox10050642

Balderas-Cordero D, Canales-Alvarez O, Sánchez-Sánchez R, Cabrera-Wrooman A, Canales-Martinez MM, Rodriguez-Monroy MA (2023) Anti-inflammatory and histological analysis of skin wound healing through topical application of Mexican Propolis. Int J Mol Sci 24(14):11831. https://doi.org/10.3390/ijms241411831

Baliyan S, Mukherjee R, Priyadarshini A, Vibhuti A, Gupta A, Pandey RP, Chang C-M (2022) Determination of antioxidants by DPPH radical scavenging activity and quantitative phytochemical analysis of Ficus religiosa. Molecules 27(4):1326. https://doi.org/10.3390/molecules27041326

Bani-Sacchi T, Barattini M, Bianchi S, Blandina P, Brunelleschi S, Fantozzi R et al (1986) The release of histamine by parasympathetic stimulation in Guinea-pig auricle and rat ileum. J Physiol 371(1):29–43. https://doi.org/10.1113/jphysiol.1986.sp015960

Bizuneh GK, Tadege G, Sirak B, Gurmu AE, Adamu BA, Tefera AM, Anagaw YK (2023) Antimalarial activity of the 80%methanol extract and solvent fractions of Cucumis ficifolius A. Rich roots against plasmodium berghei in mice. Heliyon 9(2):e13690. https://doi.org/10.1016/j.heliyon.2023.e13690

Błaszczyk L, Ćwiek-Kupczyńska H, Hoppe Gromadzka K, Basińska-Barczak A, Stępień Ł, Kaczmarek J, Lenc L (2023) Containment of Fusarium culmorum and its mycotoxins in various biological systems by antagonistic Trichoderma and Clonostachys strains. J Fungi 9(3):289. https://doi.org/10.3390/jof9030289

Botting R (2004) Antipyretic therapy. Front Biosci 9(1–3):956. https://doi.org/10.2741/1303

Brahmi F, Lounis N, Mebarakou S, Guendouze N, Yalaoui-Guellal D, Madani K et al (2022) Impact of growth sites on the phenolic contents and antioxidant activities of three Algerian Mentha species (M. pulegium L., M. rotundifolia (L.) Huds., and M. spicata L.). Front Pharmacol 13. https://doi.org/10.3389/fphar.2022.886337

Camara A, Haddad M, Reybier K, Traoré MS, Baldé MA, Royo J et al (2019) Terminalia albida treatment improves survival in experimental cerebral malaria through reactive oxygen species scavenging and anti-inflammatory properties. Malar J 18(1):431. https://doi.org/10.1186/s12936-019-3071-9

Charlton NC, Mastyugin M, Török B, Török M (2023) Structural features of small molecule antioxidants and strategic modifications to improve potential bioactivity. Molecules 28(3):1057. https://doi.org/10.3390/molecules28031057

Chen L, Deng H, Cui H, Fang J, Zuo Z, Deng J et al (2018) Inflammatory responses and inflammation-associated diseases in organs. Oncotarget 9(6):7204–7218. https://doi.org/10.18632/oncotarget.23208

Chen C, Mohamad Razali UH, Saikim FH, Mahyudin A, Mohd Noor NQI (2021) Morus alba L. plant: bioactive compounds and potential as a functional food ingredient. Food Secur 10(3):689. https://doi.org/10.3390/foods10030689

Chowdhury M, Kumar De P (2023) Novel gastroprotective activity of basella alba mucilage-based hydrogel beads on diclofenac sodium induced gastric ulcer. J Med Pharm All Sci 12(5):6082–6090. https://doi.org/10.55522/jmpas.V12I5.5648

Chowdhury U, Zakharyan R, Hernandez A, Avram M, Kopplin M, Aposhian H (2006) Glutathione-S-transferase-omega [MMA(V) reductase] knockout mice: enzyme and arsenic species concentrations in tissues after arsenate administration☆. Toxicol Appl Pharmacol 216(3):446–457. https://doi.org/10.1016/j.taap.2006.06.014

Ciumărnean L, Milaciu MV, Runcan O, Vesa ŞC, Răchişan AL, Negrean V et al (2020) The effects of flavonoids in cardiovascular diseases. Molecules 25(18):4320. https://doi.org/10.3390/molecules25184320

Craddock VM, Driver HE (1987) Sequential histological studies of rat oesophagus during the rapid initiation of cancer by repeated injection of N -methyl- N -benzylnitrosamine. Carcinogenesis 8(8):1129–1132. https://doi.org/10.1093/carcin/8.8.1129

Deshmukh SA, Gaikwad DK (2014) A review of the taxonomy, ethnobotany, phytochemistry and pharmacology of Basella alba (Basellaceae). J Appl Pharm Sci. https://doi.org/10.7324/JAPS.2014.40125

Dimo T, Rakotonirina SV, Tan PV, Azay J, Dongo E, Cros G (2002) Leaf methanol extract of Bidens pilosa prevents and attenuates the hypertension induced by high-fructose diet in Wistar rats. J Ethnopharmacol 83(3):183–191. https://doi.org/10.1016/S0378-8741(02)00162-9

Duan S-D, Liu Y, Hao L-H, Xiang D-Y, Yu W-B, Liang J et al (2023) Complete chloroplast genome sequence of Dahlia imperialis (Asteraceae): comparative analysis and phylogenetic relationships. Horticulturae 10(1):7. https://doi.org/10.3390/horticulturae10010007

Edo GI (2022) Antibacterial, phytochemical and GC-MS analysis of Thevetia peruviana extracts: an approach in drug formulation. Nat Resour Human Health 2(4):418–426. https://doi.org/10.53365/nrfhh/146543

Edo GI, Onoharigho FO, Jikah AN, Oloni GO, Samuel PO, Rapheal OA et al (2023a) Cyperus esculentus (tiger nut): an insight into its bioactive compounds, biological activities, nutritional and health benefits. Food Chem Adv 3:100511. https://doi.org/10.1016/j.focha.2023.100511

Edo GI, Samuel PO, Jikah AN, Oloni GO, Ifejika MN, Oghenegueke O et al (2023b) Proximate composition and health benefit of Roselle leaf (Hibiscus sabdariffa). Insight on food and health benefits. Food Chem Adv 3:100437. https://doi.org/10.1016/j.focha.2023.100437

Edo GI, Samuel PO, Jikah AN, Onoharigho FO, Idu LI, Obasohan P et al (2023c) Biological and bioactive components of bitter leaf (Vernonia amygdalina leaf): insight on health and nutritional benefits. A review. Food Chem Adv 3:100488. https://doi.org/10.1016/j.focha.2023.100488

Edo GI, Samuel PO, Ossai S, Nwachukwu SC, Okolie MC, Oghenegueke O et al (2023d) Phytochemistry and pharmacological compounds present in scent leaf. A review. Food Chem Adv 100300. https://doi.org/10.1016/j.focha.2023.100300

Edo GI, Ugbune U, Ezekiel GO, Onoharigho FO, Agbo JJ (2023e) Cyperus esculentus (tiger nut): its application in agriculture, food, health and nutrition. A review. Vegetos. https://doi.org/10.1007/s42535-023-00672-8

Edo GI, Ugbune U, Onoharigho FO, Ezekiel GO, Agbo JJ (2023f) Antioxidant activities of reissantia indica willd. (mopane paddle-pod) and nephroprotective effect on paracetamol-induced nephrotoxicity in male wistar rats. Forum Nutr 48(1):26. https://doi.org/10.1186/s41110-023-00214-x

Edo GI, Ugbune U, Onoharigho FO, Ezekiel GO, Ugbuwe E, Agbo JJ (2023g) Investigation of the metal complexes and bioactive compound formed by coordination of bioactive phytochemical from ginger (Zingiber officinale) extracts to metal ions. Food Chem Adv 3:100337. https://doi.org/10.1016/j.focha.2023.100337

El Haouari M, Assem N, Changan S, Kumar M, Daştan SD, Rajkovic J et al (2021) An insight into phytochemical, pharmacological, and nutritional properties of Arbutus unedo L. from Morocco. Evid Based Complement Alternat Med 2021:1–19. https://doi.org/10.1155/2021/1794621

Erhonyota C, Edo GI, Onoharigho FO (2023) Comparison of poison plate and agar well diffusion method determining the antifungal activity of protein fractions. Acta Ecol Sin 43(4):684–689. https://doi.org/10.1016/j.chnaes.2022.08.006

Ferreira OO, de Jesus Pereira Franco C, Varela ELP, Silva SG, Cascaes MM, Percário S et al (2021) Chemical composition and antioxidant activity of essential oils from leaves of two specimens of Eugenia florida DC. Molecules 26(19):5848. https://doi.org/10.3390/molecules26195848

Finetti F, Travelli C, Ercoli J, Colombo G, Buoso E, Trabalzini L (2020) Prostaglandin E2 and cancer: insight into tumor progression and immunity. Biology 9(12):434. https://doi.org/10.3390/biology9120434

Fraga-Corral M, García-Oliveira P, Pereira AG, Lourenço-Lopes C, Jimenez-Lopez C, Prieto MA, Simal-Gandara J (2020) Technological application of tannin-based extracts. Molecules 25(3):614. https://doi.org/10.3390/molecules25030614

Garber A, Barnard L, Pickrell C (2021) Review of whole plant extracts with activity against herpes simplex viruses in vitro and in vivo. J Evid Based Integr Med 26:2515690X2097839. https://doi.org/10.1177/2515690X20978394

García Díaz J, González Fernández R, Escalona Arranz JC, Llauradó Maury G, Méndez Rodríguez D, De Vooght L et al (2022) Inhibitory effect on nitric oxide release in LPS-stimulated macrophages and free radical scavenging activity of Croton linearis Jacq. leaves. Antioxidants 11(10):1915. https://doi.org/10.3390/antiox11101915

Hassan F, Edo GI, Nwosu LC, Jalloh AA, Onyibe PN, Itoje-akpokiniovo LO, Irogbo PU (2021) An inventory of medicinal plants used as sedative, analgesic and blood tonic in Abeokuta, Ogun State, Nigeria. Acta Ecol Sin. https://doi.org/10.1016/j.chnaes.2021.11.003

Hernández-Cruz EY, Eugenio-Pérez D, Ramírez-Magaña KJ, Pedraza-Chaverri J (2023) Effects of vegetal extracts and metabolites against oxidative stress and associated diseases: studies in Caenorhabditis elegans. ACS Omega 8(10):8936–8959. https://doi.org/10.1021/acsomega.2c07025

Jabbar AAJ, Mothana RA, Ameen Abdulla M, Othman Abdullah F, Abdul-Aziz Ahmed K, Rizgar Hussen R et al (2023) Mechanisms of anti-ulcer actions of Prangos pabularia (L.) in ethanol-induced gastric ulcer in rats. Saudi Pharm J 31(12):101850. https://doi.org/10.1016/j.jsps.2023.101850

Jikah AN, Edo GI (2023) Moringa oleifera : a valuable insight into recent advances in medicinal uses and pharmacological activities. J Sci Food Agric. https://doi.org/10.1002/jsfa.12892

Kato-Noguchi H, Kurniadie D (2024) The invasive mechanisms of the noxious alien plant species Bidens pilosa. Plan Theory 13(3):356. https://doi.org/10.3390/plants13030356

Kenney HM, Peng Y, Bell RD, Wood RW, Xing L, Ritchlin CT, Schwarz EM (2022) Persistent popliteal lymphatic muscle cell coverage defects despite amelioration of arthritis and recovery of popliteal lymphatic vessel function in TNF-Tg mice following anti-TNF therapy. Sci Rep 12(1):12751. https://doi.org/10.1038/s41598-022-16884-y

Khalifa SAM, Elshafiey EH, Shetaia AA, El-Wahed AAA, Algethami AF, Musharraf SG et al (2021) Overview of bee pollination and its economic value for crop production. Insects 12(8):688. https://doi.org/10.3390/insects12080688

Kim M, Jee S-C, Sung J-S (2024) Hepatoprotective effects of flavonoids against benzo[a]Pyrene-induced oxidative liver damage along its metabolic pathways. Antioxidants 13(2):180. https://doi.org/10.3390/antiox13020180

Kumar BR, Anupam A, Manchikanti P, Rameshbabu AP, Dasgupta S, Dhara S (2018) Identification and characterization of bioactive phenolic constituents, anti-proliferative, and anti-angiogenic activity of stem extracts of Basella alba and rubra. J Food Sci Technol 55(5):1675–1684. https://doi.org/10.1007/s13197-018-3079-0

Kumar A, Nirmal P, Kumar M, Jose A, Tomer V, Oz E et al (2023) Major phytochemicals: recent advances in health benefits and extraction method. Molecules 28(2):887. https://doi.org/10.3390/molecules28020887

Kytidou K, Artola M, Overkleeft HS, Aerts JMFG (2020) Plant glycosides and Glycosidases: a treasure-trove for therapeutics. Front Plant Sci 11. https://doi.org/10.3389/fpls.2020.00357

Lai J-X, Dai S-F, Xue B-X, Zhang L-H, Chang Y, Yang W, Wu H-H (2023) Plant polyacetylenoids: phytochemical, analytical and pharmacological updates. Arab J Chem 16(10):105137. https://doi.org/10.1016/j.arabjc.2023.105137

Lee J, Noh S, Lim S, Kim B (2021) Plant extracts for type 2 diabetes: from traditional medicine to modern drug discovery. Antioxidants 10(1):81. https://doi.org/10.3390/antiox10010081

Liu Y, Ragaee S, Marcone MF, Abdel-Aal E-SM (2020) Composition of phenolic acids and antioxidant properties of selected pulses cooked with different heating conditions. Food Secur 9(7):908. https://doi.org/10.3390/foods9070908

Mammari N, Albert Q, Devocelle M, Kenda M, Kočevar Glavač N, Sollner Dolenc M et al (2023) Natural products for the prevention and treatment of common cold and viral respiratory infections. Pharmaceuticals 16(5):662. https://doi.org/10.3390/ph16050662

Masyita A, Mustika Sari R, Dwi Astuti A, Yasir B, Rahma Rumata N, Emran TB et al (2022) Terpenes and terpenoids as main bioactive compounds of essential oils, their roles in human health and potential application as natural food preservatives. Food Chem X 13:100217. https://doi.org/10.1016/j.fochx.2022.100217

Min H-Y, Kim MS, Jang DS, Park E-J, Seo E-K, Lee SK (2009) Suppression of lipopolysaccharide-stimulated inducible nitric oxide synthase (iNOS) expression by a novel humulene derivative in macrophage cells. Int Immunopharmacol 9(7–8):844–849. https://doi.org/10.1016/j.intimp.2009.03.005

Moore EM, Wagner C, Komarnytsky S (2020) The enigma of bioactivity and toxicity of botanical oils for skin care. Front Pharmacol 11. https://doi.org/10.3389/fphar.2020.00785

Myo H, Liana D, Phanumartwiwath A (2023) Unlocking therapeutic potential: comprehensive extraction, profiling, and pharmacological evaluation of bioactive compounds from Eclipta alba (L.) Hassk for dermatological applications. Plants 13(1):33. https://doi.org/10.3390/plants13010033

Ndjenda MK II, Nguelefack-Mbuyo EP, Atsamo AD, Fofie CK, Fodem C, Nguemo F, Nguelefack TB (2021) Antihypertensive effects of the methanol extract and the ethyl acetate fraction from Crinum zeylanicum (Amaryllidaceae) leaves in L-NAME-treated rat. Evid Based Complement Alternat Med 2021:1–12. https://doi.org/10.1155/2021/2656249

Nur MA, Islam M, Biswas S, Hasan MN, Rahman MM, Uddin MJ et al (2023) Determination of biological activities of Malabar spinach (Basellaalba) fruit extracts and molecular docking against COX-II enzyme. Heliyon 9(11):e21568. https://doi.org/10.1016/j.heliyon.2023.e21568

Nwosu LC, Edo GI, Özgör E (2022) The phytochemical, proximate, pharmacological, GC-MS analysis of Cyperus esculentus (Tiger nut): A fully validated approach in health, food and nutrition. Food Biosci 46:101551. https://doi.org/10.1016/j.fbio.2022.101551

Onyibe PN, Edo GI, Nwosu LC, Ozgor E (2021) Effects of Vernonia amygdalina fractionate on glutathione reductase and glutathione-S-transferase on alloxan induced diabetes wistar rat. Biocatal Agric Biotechnol 36:102118. https://doi.org/10.1016/j.bcab.2021.102118

Pasaoglulari Aydinlik N, Abubakar J, Edo G (2021) Phytochemical and GCMS analysis on the ethanol extract of Foeniculum Vulgare and Petroselinum crispum leaves. Int J Chem Technol. https://doi.org/10.32571/ijct.911711

Persia FA, Abba R, Pascual LI, Hapon MB, Mackern-Oberti JP, Gamarra-Luques C (2023) Prosopis strombulifera aqueous extract reduces T cell response and ameliorates type I diabetes in NOD mice. J Tradit Complement Med 13(1):20–29. https://doi.org/10.1016/j.jtcme.2022.10.001

Piechulla B, Bartelt R, Brosemann A, Effmert U, Bouwmeester H, Hippauf F, Brandt W (2016) The α-Terpineol to 1,8-cineole cyclization reaction of tobacco terpene synthases. Plant Physiol 172(4):2120–2131. https://doi.org/10.1104/pp.16.01378

Posadino AM, Giordo R, Pintus G, Mohammed SA, Orhan IE, Fokou PVT et al (2023) Medicinal and mechanistic overview of artemisinin in the treatment of human diseases. Biomed Pharmacother 163:114866. https://doi.org/10.1016/j.biopha.2023.114866

Rahmani AH, Alsahli MA, Almatroudi A, Almogbel MA, Khan AA, Anwar S, Almatroodi SA (2022) The potential role of Apigenin in cancer prevention and treatment. Molecules 27(18):6051. https://doi.org/10.3390/molecules27186051

Rajayan JS, Chandrasekar V, Duraipandian C, Rajendran K (2024) In vitro evaluation of extracts from Ixora species for a potential Phytosomal formulation. Cureus. https://doi.org/10.7759/cureus.55396

Rehfeldt SCH, Silva J, Alves C, Pinteus S, Pedrosa R, Laufer S, Goettert MI (2022) Neuroprotective effect of Luteolin-7-O-glucoside against 6-OHDA-induced damage in undifferentiated and RA-differentiated SH-SY5Y cells. Int J Mol Sci 23(6):2914. https://doi.org/10.3390/ijms23062914

Ribeiro VP, Arruda C, Abd El-Salam M, Bastos JK (2018) Brazilian medicinal plants with corroborated anti-inflammatory activities: a review. Pharm Biol 56(1):253–268. https://doi.org/10.1080/13880209.2018.1454480

Rotich W, Mas-Claret E, Sadgrove N, Guantai A, Padilla-González GF, Langat MK (2022) HIV-1 integrase inhibitory effects of major compounds present in CareVid™: an anti-HIV multi-herbal remedy. Life 12(3):417. https://doi.org/10.3390/life12030417

Shah AM, Yang W, Mohamed H, Zhang Y, Song Y (2022) Microbes: a hidden treasure of polyunsaturated fatty acids. Front Nutr 9. https://doi.org/10.3389/fnut.2022.827837

Shaik Y, Caraffa A, Ronconi G, Lessiani G, Conti P (2018) Impact of polyphenols on mast cells with special emphasis on the effect of quercetin and luteolin. Central Eur J Immunol 43(4):476–481. https://doi.org/10.5114/ceji.2018.81347

Shams SGE, Eissa RG (2022) Amelioration of ethanol-induced gastric ulcer in rats by quercetin: implication of Nrf2/HO1 and HMGB1/TLR4/NF-κB pathways. Heliyon 8(10):e11159. https://doi.org/10.1016/j.heliyon.2022.e11159

Shareef SH, Al-Medhtiy MH, Ibrahim IAA, Alzahrani AR, Jabbar AA, Galali Y et al (2022) Gastroprophylactic effects of p-cymene in ethanol-induced gastric ulcer in rats. PRO 10(7):1314. https://doi.org/10.3390/pr10071314

Sheik A, Kim E, Adepelly U, Alhammadi M, Huh YS (2023) Antioxidant and antiproliferative activity of Basella alba against colorectal cancer. Saudi J Biol Sci 30(4):103609. https://doi.org/10.1016/j.sjbs.2023.103609

Shin S, Park J, Choi H-Y, Bu Y, Lee K (2024) Blood-pressure-lowering and endothelium-dependent Vasorelaxant effects of nutgall tree in rats. Food Secur 13(7):1041. https://doi.org/10.3390/foods13071041

Sirak B, Mann L, Richter A, Asres K, Imming P (2021) In vivo antimalarial activity of leaf extracts and a major compound isolated from Ranunculus multifidus Forsk. Molecules 26(20):6179. https://doi.org/10.3390/molecules26206179

Škubník J, Pavlíčková V, Rimpelová S (2021) Cardiac glycosides as immune system modulators. Biomol Ther 11(5):659. https://doi.org/10.3390/biom11050659

Soudzilovskaia NA, Elumeeva TG, Onipchenko VG, Shidakov II, Salpagarova FS, Khubiev AB et al (2013) Functional traits predict relationship between plant abundance dynamic and long-term climate warming. Proc Natl Acad Sci 110(45):18180–18184. https://doi.org/10.1073/pnas.1310700110

Stéphane FFY, Kezetas Jean Jules B, El-Saber Batiha G, Ali I, Ndjakou Bruno L (2022) Extraction of bioactive compounds from medicinal plants and herbs. In: Natural medicinal plants. IntechOpen. https://doi.org/10.5772/intechopen.98602

Suriyaprom S, Srisai P, Intachaisri V, Kaewkod T, Pekkoh J, Desvaux M, Tragoolpua Y (2023) Antioxidant and anti-inflammatory activity on LPS-stimulated RAW 264.7 macrophage cells of white mulberry (Morus alba L.) leaf extracts. Molecules 28(11):4395. https://doi.org/10.3390/molecules28114395

Thusyanthan J, Wickramaratne NS, Senathilake KS, Rajagopalan U, Tennekoon KH, Thabrew I, Samarakoon SR (2022) Cytotoxicity against human hepatocellular carcinoma (HepG2) cells and anti-oxidant activity of selected endemic or medicinal plants in Sri Lanka. Adv Pharmacol Pharm Sci 2022:1–9. https://doi.org/10.1155/2022/6407688

Ullah A, Munir S, Badshah SL, Khan N, Ghani L, Poulson BG et al (2020) Important flavonoids and their role as a therapeutic agent. Molecules 25(22):5243. https://doi.org/10.3390/molecules25225243

Utami AR, Maksum IP, Deawati Y (2023) Berberine and its study as an antidiabetic compound. Biology 12(7):973. https://doi.org/10.3390/biology12070973

Vandamme C, Kinnunen T (2020) B cell helper T cells and type 1 diabetes. Scand J Immunol 92(4). https://doi.org/10.1111/sji.12943

Wahab S, Annadurai S, Abullais SS, Das G, Ahmad W, Ahmad MF et al (2021) Glycyrrhiza glabra (licorice): A comprehensive review on its Phytochemistry, biological activities, clinical evidence and toxicology. Plants 10(12):2751. https://doi.org/10.3390/plants10122751

Wang Y, Lian J, Shen H, Ni Y, Zhang R, Guo Y, Ye W (2020) The effects of Bidens alba invasion on soil bacterial communities across different coastal ecosystem land-use types in southern China. PLoS One 15(10):e0238478. https://doi.org/10.1371/journal.pone.0238478

Xu J-F, Xia J, Wan Y, Yang Y, Wu J-J, Peng C, Ao H (2022a) Vasorelaxant activities and its underlying mechanisms of Magnolia volatile oil on rat thoracic aorta based on network pharmacology. Front Pharmacol 13. https://doi.org/10.3389/fphar.2022.812716

Xu L, Zhong X-L, Xi Z-C, Li Y, Xu H-X (2022b) Medicinal plants and natural compounds against acyclovir-resistant HSV infections. Front Microbiol 13. https://doi.org/10.3389/fmicb.2022.1025605

Xuan TD, Khanh TD (2016) Chemistry and pharmacology of Bidens pilosa: an overview. J Pharm Investig 46(2):91–132. https://doi.org/10.1007/s40005-016-0231-6

Yi X, Dong M, Guo N, Tian J, Lei P, Wang S et al (2023) Flavonoids improve type 2 diabetes mellitus and its complications: a review. Front Nutr 10. https://doi.org/10.3389/fnut.2023.1192131

Yoo J, Kim D, Park J, Kim Y-K, Park Choo H-Y, Woo HA (2022) Novel small molecule inhibitors targeting the IL-6/STAT3 pathway or IL-1β. Molecules 27(9):2696. https://doi.org/10.3390/molecules27092696

Zakaria NH, Saad N, Che Abdullah CA, Mohd. Esa, N. (2023) The antiproliferative effect of chloroform fraction of Eleutherine bulbosa (mill.) Urb. On 2D- and 3D-human lung cancer cells (A549) model. Pharmaceuticals 16(7):936. https://doi.org/10.3390/ph16070936

Zanzabil KZ, Hossain MS, Hasan MK (2023) Diabetes mellitus management: an extensive review of 37 medicinal plants. Diabetology 4(2):186–234. https://doi.org/10.3390/diabetology4020019

Zeleke YG, Atnafie SA, Aragaw TJ (2023) Anti-diabetic activities of hydro-Methanolic crude extract and solvent fractions of Heteromorpha arborescens (Apiaceae) leaves in mice. J Exp Pharmacol 15:107–121. https://doi.org/10.2147/JEP.S392742

Zhai B-T, Tian H, Sun J, Zou J-B, Zhang X-F, Cheng J-X et al (2022) Urokinase-type plasminogen activator receptor (uPAR) as a therapeutic target in cancer. J Transl Med 20(1):135. https://doi.org/10.1186/s12967-022-03329-3

Zhang K, Yao L, Zhang Y, Tao J (2019) Achene heteromorphism in Bidens pilosa (Asteraceae): differences in germination and possible adaptive significance. AoB Plants 11(3). https://doi.org/10.1093/aobpla/plz026

Zhang D, Tu J, Ding X, Guan W, Gong L, Qiu X et al (2023) Analysis of the chloroplast genome and phylogenetic evolution of Bidens pilosa. BMC Genomics 24(1):113. https://doi.org/10.1186/s12864-023-09195-7

Zishan SA, Uddin MM, Mohammad M, Asadul Karim Azad SM, Naima J, Ibban SS, Saiful Islam Arman M (2024) Costus speciosus leaf and seed extracts for wound healing: a comparative evaluation using mice excision wound models. Clin Phytoscience 10(1):5. https://doi.org/10.1186/s40816-024-00368-9

Zupin L, dos Santos-Silva CA, Al Mughrbi ARH, Vilela LMB, Benko-Iseppon AM, Crovella S (2022) Bioactive antimicrobial peptides: A new weapon to counteract zoonosis. Microorganisms 10(8):1591. https://doi.org/10.3390/microorganisms10081591

Part VI

Miscellany

Ecological Significance of *Bidens* Species 22

Huma Qureshi and Tauseef Anwar

Abstract

Bidens, a diverse genus of flowering plants within the Asteraceae family, holds significant ecological importance across various habitats worldwide. This chapter aims to provide a comprehensive exploration of the ecological significance of *Bidens* species, shedding light on their multifaceted roles in ecosystem functioning, biodiversity conservation, and provision of ecosystem services. *Bidens* plants interact with a wide array of organisms, including pollinators, seed dispersers, and soil microbes, thereby playing crucial roles in maintaining ecosystem processes and supporting biodiversity. Their attractive flowers and nectar resources make them important for pollinator attraction, contributing to the pollination services crucial for the reproduction of numerous plant species. Furthermore, *Bidens* species act as providers of food and habitat for various wildlife, enhancing biodiversity in their respective ecosystems. Moreover, *Bidens* plants contribute to soil stabilization and nutrient cycling through their root systems and litter decomposition processes. Their ability to colonize disturbed habitats and thrive under diverse environmental conditions underscores their ecological resilience and importance in ecological succession. This chapter discusses existing literature and presents case studies to illustrate *Bidens'* contributions to ecosystem dynamics and resilience. Additionally, this chapter discusses *Bidens'* adaptive strategies to environmental stressors, including climate change and habitat disturbance, highlighting their capacity to persist and thrive in changing environments. By elucidating the ecological roles of *Bidens,* this chapter

H. Qureshi (✉)
Department of Botany, University of Chakwal, Chakwal, Pakistan
e-mail: huma.qureshi@uoc.edu.pk

T. Anwar
Department of Botany, The Islamia University of Bahawalpur, Bahawalpur, Pakistan

Y. Bibi et al. (eds.), *The Genus Bidens*,
https://doi.org/10.1007/978-981-96-4257-1_22

provides valuable insights into the conservation and management of ecosystems where these plants occur. Understanding and conserving *Bidens* habitats are essential for sustaining ecological processes, promoting biodiversity resilience, and mitigating the impacts of global environmental challenges.

Keywords

Ecosystem functioning · Pollinator attraction · Wildlife habitat · Soil stabilization · Ecological resilience

22.1 Introduction

The genus *Bidens*, commonly known as beggarticks or bur-marigolds, comprises approximately 230 species of flowering plants in the family Asteraceae. Widely distributed across temperate and tropical regions, *Bidens* species thrive in various habitats, including wetlands, grasslands, forests, and disturbed areas. This chapter explores the ecological significance of *Bidens*, focusing on its role in ecosystem functioning, species interactions, and contributions to biodiversity.

22.2 Taxonomy and Distribution

Bidens species are characterized by their composite flower heads, opposite leaves, and distinctive achene fruits equipped with barbs that facilitate seed dispersal. The genus is divided into several sections based on morphological and genetic characteristics. Notable species include *Bidens pilosa*, *Bidens tripartita*, and *Bidens alba*, each exhibiting diverse ecological adaptations (see Table 22.1).

22.3 Ecological Roles and Adaptations

22.3.1 Pollination Services

Bidens species, with their attractive flowers and abundant nectar resources, serve as vital pollinator attractants. Bees, butterflies, moths, and other pollinators visit *Bidens* flowers, facilitating the reproduction of numerous plant species (Budumajji and Solomon Raju 2018). By providing pollination services, *Bidens* species enhance the genetic diversity and reproductive success of co-occurring flora, thereby contributing to ecosystem stability and resilience (Table 22.2).

Table 22.1 Notable *Bidens* species and their habitats

Species	Common name	Habitat	Geographic distribution
Bidens pilosa	Blackjack	Disturbed areas, fields	Tropical and subtropical regions
Bidens tripartita	Trifid bur-marigold	Wetlands, marshes	North America, Europe
Bidens alba	White beggarticks	Grasslands, roadsides	Southeastern United States
Bidens cernua	Nodding beggarticks	Wetlands, shorelines	North America
Bidens frondosa	Devil's beggarticks	Wetlands, riparian zone	North America, Europe
Bidens bipinnata	Spanish needles	Disturbed areas, fields	Worldwide
Bidens aristosa	Bearded beggarticks	Wet meadows, ditches	Eastern North America
Bidens connata	Purple-stem beggarticks	Wetlands, ditches	North America
Bidens laevis	Smooth beggarticks	Wetlands, shores	North America, South America
Bidens vulgata	Tall beggarticks	Wetlands, open woods	North America

Table 22.2 Pollinators associated with *Bidens* species

Pollinator type	Species	Role in ecosystem
Bees	*Apis mellifera*	Pollination, honey production
Butterflies	*Danaus plexippus*	Pollination, indicator species
Beetles	*Meligethes aeneus*	Pollination, pest control
Wasps	*Vespula vulgaris*	Pollination, natural pest control
Flies	*Syrphidae*	Pollination, biocontrol
Moths	*Helicoverpa zea*	Pollination, nocturnal pollinators
Ants	*Formica rufa*	Pollination, soil aeration
Birds	*Trochilidae*	Pollination, seed dispersal
Bats	*Glossophaga soricina*	Pollination, seed dispersal
Beetles	*Coccinellidae*	Pollination, aphid control

22.3.2 Wildlife Habitat and Food Source

Bidens plants serve as important providers of food and habitat for various wildlife species. Birds, mammals, and insects rely on *Bidens* seeds, foliage, and shelter for sustenance and shelter. The presence of *Bidens* in ecosystems enhances biodiversity by supporting diverse assemblages of wildlife, contributing to the ecological richness and resilience of natural communities (Table 22.3).

Table 22.3 Wildlife associated with *Bidens* species

Wildlife type	Species	Role in ecosystem
Birds	*Cardinalis cardinalis*	Seed dispersal, habitat provision
Mammals	*Odocoileus virginianus*	Seed dispersal, herbivory
Insects	*Danaus plexippus*	Pollination, habitat provision
Amphibians	*Lithobates catesbeianus*	Habitat provision, insect control
Reptiles	*Thamnophis sirtalis*	Habitat provision, prey species
Mammals	*Sylvilagus floridanus*	Herbivory, habitat provision
Birds	*Sturnus vulgaris*	Insect control, seed dispersal
Insects	*Harmonia axyridis*	Aphid control, pollination
Birds	*Zenaida macroura*	Seed dispersal, habitat provision
Mammals	*Sciurus carolinensis*	Seed dispersal, herbivory

Table 22.4 Soil stabilization and nutrient cycling functions of *Bidens*

Function	Ecological benefit	Example *Bidens* species
Soil stabilization	Prevents erosion, improves soil structure	*Bidens pilosa*
Organic matter addition	Enhances soil fertility, supports plant growth	*Bidens alba*
Nutrient release	Recycles essential nutrients, supports microbial activity	*Bidens frondosa*
Root penetration	Enhances soil aeration, supports root health	*Bidens cernua*
Soil microbe support	Promotes beneficial soil microbes	*Bidens bipinnata*
Carbon sequestration	Stores carbon, mitigates climate change	*Bidens laevis*
Moisture retention	Improves water-holding capacity of the soil	*Bidens connata*
Soil pH regulation	Balances soil pH, supports diverse plant communities	*Bidens vulgata*
Heavy metal remediation	Phytoremediation, detoxifies soil	*Bidens aristosa*
Soil texture improvement	Enhances soil structure, supports diverse plant roots	*Bidens tripartita*

22.3.3 Soil Stabilization and Nutrient Cycling

The root systems of *Bidens* plants play a crucial role in soil stabilization, by preventing erosion and promoting soil structure. Additionally, these plants contribute to nutrient cycling through its litter decomposition processes, releasing organic matter and essential nutrients back into the soil (Wei et al. 2017; Chauhan et al. 2019; Wang et al. 2023). By enhancing soil fertility and stability, *Bidens* facilitates the persistence of plant communities and supports ecosystem productivity (Table 22.4).

22.3.4 Ecological Resilience and Succession

Bidens species demonstrate ecological resilience by colonizing disturbed habitats and thriving under diverse environmental conditions. Their ability to rapidly

Table 22.5 Roles of *Bidens* in ecological resilience and succession

Habitat type	Disturbance type	*Bidens* species	Role in ecological succession
Wetlands	Flooding	*Bidens cernua*	Early colonizer, stabilizes soil
Grasslands	Fire	*Bidens alba*	Pioneer species, supports recovery
Riparian zones	Erosion	*Bidens frondosa*	Soil stabilizer, supports plant diversity
Disturbed areas	Urban development	*Bidens pilosa*	Soil restoration, initiates succession
Forest edges	Logging	*Bidens bipinnata*	Understory colonizer, supports biodiversity
Agricultural fields	Soil degradation	*Bidens alba*	Bioindicator, supports soil health
Coastal areas	Storm surges	*Bidens laevis*	Erosion control, habitat recovery
Meadows	Grazing	*Bidens aristosa*	Pollinator support, soil fertility
Urban areas	Construction	*Bidens vulgata*	Green space, urban biodiversity
High-altitude zones	Climate change	*Bidens connata*	Adaptation to extreme conditions

establish and reproduce contributes to ecological succession processes, facilitating the recovery of ecosystems following disturbances. *Bidens* plays a key role in maintaining ecosystem integrity and function, particularly in habitats subject to human-induced or natural disturbances (Table 22.5).

22.3.5 Invasiveness

Bidens species are known for their invasive potential in certain regions, impacting local biodiversity and ecosystem dynamics. They exhibit rapid colonization and high reproductive rates, often outcompeting the native vegetation and altering the ecosystem structure and function. Understanding the invasiveness of *Bidens* species is crucial for effective management and conservation strategies in invaded habitats (Ronzhina et al. 2021; Sharma et al. 2023).

22.4 Adaptation to Environmental Stressors

Bidens exhibits adaptive strategies to environmental stressors, including climate change and habitat disturbance. Some species display phenotypic plasticity, adjusting their morphology and physiology in response to changing environmental conditions. Others exhibit genetic diversity, enabling populations to adapt to local environmental gradients. Understanding *Bidens*' adaptive capacity is essential for predicting its responses to future environmental changes and informing conservation strategies (Table 22.6).

Table 22.6 Adaptive strategies of *Bidens* species

Environmental stressor	Adaptive strategy	Example *Bidens* species
Temperature increase	Phenotypic plasticity, heat tolerance	*Bidens pilosa*
Drought	Deep rooting, drought-resistant varieties	*Bidens alba*
Salinity	Salt tolerance, selective ion uptake	*Bidens bipinnata*
Habitat disturbance	Rapid colonization, high reproductive rate	*Bidens cernua*
Soil nutrient deficiency	Efficient nutrient uptake, symbiosis with soil microbes	*Bidens frondosa*
Flooding	Aerenchyma formation, waterlogging tolerance	*Bidens laevis*
Heavy metal pollution	Phytoremediation, heavy metal sequestration	*Bidens aristosa*
Herbivory	Chemical defenses, rapid regrowth	*Bidens tripartita*
Pathogen attack	Resistance genes, secondary metabolite production	*Bidens vulgata*
UV radiation	UV-absorbing compounds, protective leaf structures	*Bidens connata*

Table 22.7 Potential impacts of climate change on *Bidens* species

Climate change factor	Potential impact	Adaptive strategy
Temperature increase	Altered flowering times	Breeding for heat tolerance
Changes in precipitation	Reduced water availability	Drought-resistant varieties
Extreme weather events	Habitat destruction	Habitat restoration and protection
Sea-level rise	Loss of coastal habitats	Coastal zone management
Increased CO_2 levels	Enhanced growth, altered competition	Monitoring and adaptive management

22.5 Challenges and Future Directions

Climate changes pose significant challenges to the survival and distribution of *Bidens* species. The shifts in temperature and precipitation patterns can affect their phenology, reproductive success, and interactions with other species. Research on the ecological responses of *Bidens* to climate change is essential for developing adaptive conservation strategies (Khapugin et al. 2022) (Table 22.7).

22.6 Integrating Ecological Knowledge into Management

Integrating ecological knowledge of *Bidens* into land management practices can enhance ecosystem resilience and biodiversity conservation. Collaborative efforts between scientists, conservationists, and local communities are needed to develop effective management plans that balance the ecological benefits and potential risks associated with *Bidens* species.

22.7 Conclusion

Bidens species occupy a central position in ecosystems worldwide, contributing to biodiversity conservation, ecosystem functioning, and the provision of ecosystem services. By elucidating *Bidens'* ecological significance, this chapter underscores the importance of conserving and managing habitats where these plants occur. Sustaining *Bidens* populations is essential for promoting biodiversity resilience, maintaining ecosystem stability, and mitigating the impacts of global environmental challenges. Through interdisciplinary research and conservation efforts, we can ensure the continued ecological relevance and resilience of *Bidens* species in a rapidly changing world.

References

Budumajji U, Solomon Raju A (2018) Pollination ecology of *Bidens pilosa L.*(Asteraceae). Taiwania 63(2):89–100

Chauhan BS, Ali HH, Florentine S (2019) Seed germination ecology of *Bidens pilosa* and its implications for weed management. Sci Rep 9(1):16004

Khapugin AA, Kuzmin IV, Ivanov LA, Ronzhina DA, Ivanova LA (2022) Environmental preferences of an invasive plant species, Bidens frondosa (Asteraceae), in European Russia and Western Siberia. Diversity 14(8):598

Ronzhina DA, Ivanova LA, Ivanov LA, Khapugin AA (2021) *Bidens frondosa* (Asteraceae), a new alien invasive plant species in the Kurgan region (Russia). Contrib Bot 56:53–58

Sharma A, Kaur A, Kohli RK, Singh HP, Batish DR (2023) *Bidens pilosa* (Asteraceae) invasion reshapes the pattern of plant communities and edaphic properties across the north-western Himalayan landscape. Ecol Inform 77:102281

Wang S, Dai H, Ji D, Cui S, Jiang C, Skuza L et al (2023) Influencing factors of *Bidens pilosa L* hyperaccumulating cadmium explored by the real-time uptake of Cd^{2+} influx around root apexes under different exogenous nutrient ion levels. Toxics 11(3):227

Wei C, Tang S, Pan Y, Li X (2017) Plastic responses of invasive *Bidens frondosa* to water and nitrogen addition. Nord J Bot 35(2):232–239

Exploring *Bidens* Genomics: Evolutionary Insights and Applications

23

Wajiha Zaka Ansari, Hannan Mukhtar, Haris Ahmed Khan, Hamna Khan, Anam Fatima, Salman Majeed, and Nida Zahra

Abstract

The genus *Bidens,* comprising numerous species distributed across diverse habitats globally, has been a subject of botanical interest due to its ecological significance and medicinal properties. This chapter offers an in-depth exploration of recent advances in *Bidens* genomics, integrating findings from genome sequencing, comparative genomics, and functional analyses. Genomic studies have offered insights into *Bidens* evolutionary history, genetic diversity, and adaptive mechanisms. Variations in genome size and structure, coupled with the genetic basis of adaptive traits such as herbivore defense and stress tolerance, highlight the role of gene duplication and horizontal gene transfer in driving evolutionary innovation. Comparative genomics sheds light on phylogenetic relationships and speciation events within *Bidens*, while population genomics reveals patterns of genetic diversity and demographic history. The integration of genomic data with traditional botanical knowledge holds great promise for various applications. These include improving agricultural productivity, advancing drug discovery, and enhancing strategies for managing ecosystems. By utilizing genomic resources such as high-quality reference genomes and detailed gene expression

W. Z. Ansari · H. Mukhtar (✉)
Department of Botany, Lahore College for Women University, Lahore, Pakistan
e-mail: hannan.mukhtar@lcwu.edu.pk

H. A. Khan · H. Khan
Department of Biotechnology, University of Mianwali, Mianwali, Pakistan

A. Fatima · N. Zahra
Department of Botany, University of Mianwali, Mianwali, Pakistan

S. Majeed
Department of Botany, University of Mianwali, Mianwali, Pakistan

Department of Plant Sciences, Quaid-i-Azam University Islamabad, Islamabad, Pakistan

Y. Bibi et al. (eds.), *The Genus Bidens*,
https://doi.org/10.1007/978-981-96-4257-1_23

profiles, researchers can lead focused breeding programs to make crops more resilient and increase yields, even in the face of worsening climate conditions and growing pest challenges. Moreover, genomic-guided approaches facilitate the sustainable utilization of *Bidens* medicinal compounds for drug discovery and healthcare applications.

Keywords

Beggar-ticks · Genomic · Chromosome · Ecosystems · *Bidens*

23.1 Introduction

The genus *Bidens*, commonly referred to as Beggar-ticks or Spanish needles due to their seed morphology, belongs to the Asteraceae family. This genus includes approximately 230–240 species found worldwide, ranging from the tropics to the subtropics across all continents and throughout Polynesia. *Bidens* plants have garnered attention from researchers in ecology, pharmacology, and biology (Knope et al. 2020a). The genus *Bidens* is widespread in both cultivated and uncultivated areas and is considered one of the most harmful weeds in agriculture, causing crop losses in over 40 countries (Holm et al. 1978). Various species of this genus exhibit rapid growth and allelopathic effects on different crops. Additionally, *Bidens* is utilized as a medicinal plant, cover crop, and source of nectar for bees. Its roots, leaves, and seeds possess antibacterial, antidysenteric, anti-inflammatory, antimalarial, antiseptic, anticancer, antipyretic, hepatoprotective, hypotensive, hypoglycemic, diuretic, and antidiabetic properties (Towers et al. 1984; Tsao and Eto 1996; Hsu and Kao 2009). A literature survey revealed that the primary compounds in this genus (301 in total) include polyacetylenes, polyacetylene glycosides, flavonoids, flavone glycosides, aurones, chalcones, okanine glycosides, phenolic acids, terpenes, pheophytins, fatty acids, and phytosterols. These compounds have been identified and isolated from various parts of the plant and are considered bioactive, potentially responsible for the genus's pharmacological, biological, and allelopathic properties (Lopes et al. 2022). In addition, studying the genomics of medicinal plants is crucial for understanding their molecular mechanisms for disease prevention. This involves using genetic information, species regulatory networks, and omics technologies. The rapid advancements and decreasing costs of sequencing technology have recently transformed whole-genome sequencing and the identification of secondary metabolite production in medicinal plants. Progress in functional genomics, including transcriptomics, proteomics, and metabolomics, is facilitating the discovery of secondary metabolites and key related genes. Multi-omics approaches provide valuable insights into the diversity, distribution, and evolution of biosynthetic gene clusters (Cheng et al. 2021; Alami et al. 2022). This chapter aims to present an overview of the genomic strategies used to study the genus *Bidens*. It will cover how genome sequencing, comparative genomics, and functional analysis are utilized to explore the evolutionary history, genetic diversity, and adaptive mechanisms of these plants. Additionally, the chapter will discuss the application of these genomic approaches in fields such as agriculture, drug discovery, and ecosystem management.

23.2 Overview of Genomic Studies

Genome sequencing is widely recognized as a cornerstone in both genomic studies and legal analyses of biological phenomena, offering crucial insights for genetic research and legal examinations of biological events. Significant advancements in this field have been driven by rapid scientific and technological developments (Pareek et al. 2011). These improvements have enhanced sequencing speed and quality, offering unprecedented opportunities to delve into the intricate complexities of genomes, especially for rare species. Genome sequencing not only reveals the structural organization and functional intricacies of plants but also serves as a valuable resource for uncovering the molecular mechanisms underlying adaptive evolution, stress resilience, and agricultural traits (Duque et al. 2013; Ashraf et al. 2022). Genomic studies offer valuable insights into the potential applications of genomic tools and technologies for crop improvement and sustainable agriculture. They also address fundamental questions regarding the evolution and function of plant genes and genomes (Zhang et al. 2017).

23.3 Genome Sequencing Efforts in Genus *Bidens*

Genome sequencing in the genus *Bidens* is crucial for exploring phylogenetic relationships at the genomic level and applying these evolutionary relationships to study genomic mechanisms such as DNA repair and the functional annotation of unknown genes (Wu et al. 2022). Although *Bidens* species have limited genomic resources, some have been sequenced using organelle sequencing methods. However, issues exist with the local standards of these medicinal plants, such as inconsistencies between their scientific names and their Chinese names and records in the Flora of China. Additionally, *Bidens* species often have homonyms and synonyms and are challenging to identify using traditional methods due to their similar morphologies and chemical components (Bartolome et al. 2013; Yang et al. 2020). Reports on the molecular identification of *Bidens* species are scarce. Some researchers have utilized noncoding regions of the chloroplast genome (trnL intron and trnL-trnF) and nuclear ribosomal DNA (ITS1, 5.8S, and ITS2) to identify *Bidens* species, finding that ITS1, 5.8S, ITS2, and the trnL intron could only distinguish *Bidens biternata* from *Bidens pilosa* var. *pilosa* (Tsai et al. 2008). Other experiments have shown that universal DNA barcodes ITS, ITS2, and psbA-trnH are ineffective in identifying *Bidens* species. These identification challenges significantly hinder the development and medicinal quality of *Bidens* plants (Wu et al. 2022). There is an urgent need for an accurate and effective method to identify medicinal *Bidens* plants. Various *Bidens* species have been sequenced using different approaches, and some of these species are discussed below.

23.4 Genome Profiling of *Bidens hawaiensis*

The 19 hexaploid species native to the Hawaiian Islands are a prime example of adaptive radiation and are of significant conservation concern, with many facing threats to their survival (Ganders and Nagata 1984). This study has provided a model system for examining the evolutionary genomics of rapid plant diversification. A high-quality reference genome for the endemic species *B. hawaiensis A. Gray* was reconstructed using long-read, high-fidelity sequences from the Pacific Biosciences Sequel II System (Wenger et al. 2019). The haplotype-aware draft genome assembly covered approximately 6.67 gigabases (Gb), which is close to the estimated holoploid genome size of 7.56 Gb obtained through flow cytometry (Fig. 23.1). After filtering out alternate haplotigs and contaminants, the consensus haploid reference genome comprised of 15,904 contigs totaling about 3.48 Gb, with a contig N50 value of 422,594. The genome's high interspersed repeat content, around 74%, coupled with its hexaploid nature, contributed to assembly fragmentation. Despite these challenges, both the haplotype-aware and consensus haploid

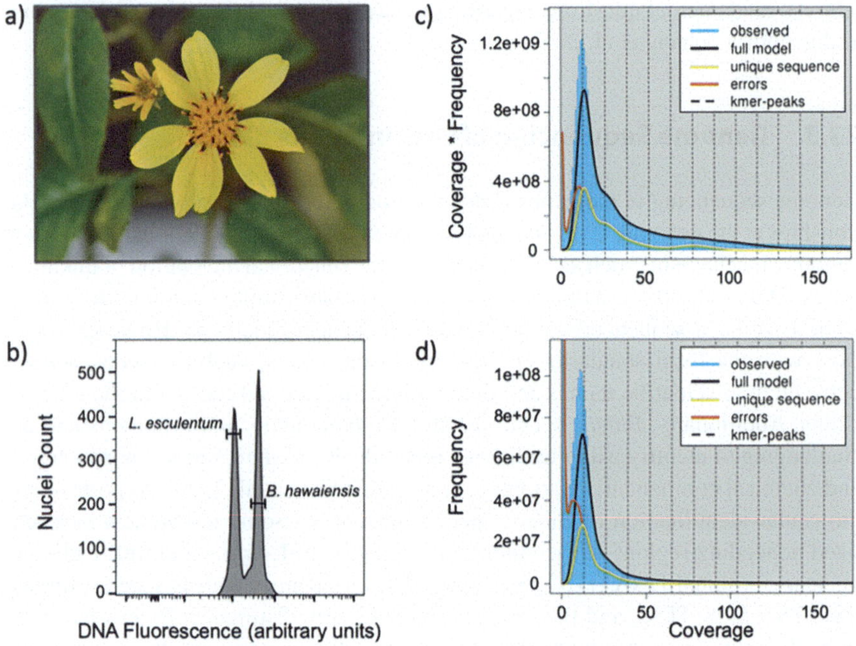

Fig. 23.1 (**a**) A *Bidens hawaiensis* flower from the plant used for flow cytometry and genome sequencing. (**b**) An example flow cytometry data plot from a replicate sample containing both *Lycopersicum esculentum* and *B. hawaiensis*. (**c, d**) K-mer spectrum analysis of HiFi data. (**c**) The three monoploid genomes of *B. hawaiensis* are visible as peaks in the coverage versus frequency plot. (**d**) The plot demonstrates a dominant k-mer frequency peak at approximately 12× coverage, with many k-mers with modest coverage miscategorized as sequencing errors. (Adapted from Bellinger et al. 2022)

assemblies successfully recovered over 96% of Benchmarking Universal Single-Copy Orthologs. However, the removal of alternate haplotigs did not significantly decrease the proportion of duplicated benchmarking genes (~79% vs. ~68%) (Bellinger et al. 2022). This reference genome will be instrumental for future studies on speciation during adaptive radiation, aiding in the resolution of evolutionary relationships, the determination of the genomic basis for trait evolution, and ongoing conservation efforts.

23.5 Chromosome-Scale Haplotype-Resolved Genomes of *Bidens alba*

Wang et al. (2024) investigated the chromosome-scale haplotype-resolved genomes of *Bidens alba* to understand the evolution of polyploidization. *B. alba* is a globally invasive weed, classified among the most severe invasive species in China, and is also widely used in traditional medicine for its anti-inflammatory, antioxidative, and antibacterial properties (Bartolome et al. 2013). Their study found that *B. alba* exhibits A similar morphology to many other *Bidens* species and includes both tetraploid and hexaploid populations, with a karyotype of 2n = 4× = 48. Sequencing was conducted using Pacific Biosciences, High-Fidelity (HiFi), and long-range Illumina Hi-C technologies. By integrating HiFi and Hi-C data with the HiFiASM tool, they achieved a contig assembly of 3.75 Gb with a contig N50 size of 15.05 Mb for *B. alba* (Table 23.1). The higher assembly size of *B. alba* compared to other plants is attributed to its high heterozygosity and tetraploid karyotype. They utilized YaHS (Zhou et al. 2023) and Juicebox (Durand et al. 2016) to anchor contigs onto chromosome-scale scaffolds, resulting in a 95.2% anchoring rate and scaffold N50 size of 75.1 Mb for *B. alba* (Fig. 23.2).

Table 23.1 Statistics of genome assembly for diploid and tetraploid genomes of *B. alba* (Wang et al. 2024)

	B. alba (2×)	*B. alba* (4×)
Estimate genome size (Gb)	1.93	3.86
Total assembly size (bp)	1,874,562,522	3,749,125,044
Number of chromosomes	24	48
Guanine Cytosine (GC percentage)	36.6	36.6
Contig N50 (bp)	15,049,174	15,049,174
Scaffold N50 (bp)	75,126,418	75,126,418
% of sequences anchored to chromosomes	95.2	95.2
% of telomeres assembled	57.3	57.3
BUSCO completeness assembly (%)	98.4	99.6

Fig. 23.2 Circos plots of *B. alba*. Circos plot has four tracks: (**a**) gene density, (**b**) tandem repeat density, (**c**) transposable element density, and (**d**) GC percentage. Feature density and GC percentage were calculated with a 1-Mb window size. (Adapted from Wang et al. 2024).

23.6 Role of Phylogenomic in Evolutionary Studies

Phylogenomic methods are increasingly employed to explore the evolutionary histories of recent plant radiations (Welch et al. 2016). Phylogenetics, which reconstructs the evolutionary history of life on Earth, originally relied solely on morphological data. However, the advent of Sanger sequencing and PCR in the late 1970s allowed for the integration of genetic information into phylogenetic studies. Early multilocus analyses significantly enhanced our understanding of phylogenetic relationships and challenged established views of plant and animal connections (Young and Gillung 2020). Since these foundational studies, advancements in sequencing and analytical methods have greatly improved the robustness of molecular phylogenies. The introduction of next-generation sequencing techniques in the mid-2000s revolutionized DNA sequencing, drastically reducing costs per nucleotide and accelerating data generation. This progress has made it possible to generate vast amounts of sequence data for both model and non-model organisms affordably, leading to the evolution of molecular phylogenetics into phylogenomic, where genome-scale data can be obtained from multiple samples at a significantly lower cost (Mardis 2011).

23.7 Evolutionary History and Adaptive Radiation in Genus *Bidens* from Hawaiian Island

The Hawaiian Islands have become a premier location for studying speciation and adaptive radiation, thanks to their well-documented and relatively straightforward geological history, isolation from mainland sources of immigrant taxa, and a high number of endemic species. Plant radiation in Hawaii often exhibits remarkable species richness and extensive ecological and morphological diversification but typically shows very low genetic variation compared to their mainland counterparts. This low genetic diversity complicates the use of traditional DNA markers to trace their evolutionary history (Ziegler 2002). Iconic examples of such Hawaiian plant radiations include the silverswords, lobeliads, mints, Scaevola, Tetramolopium, and *Bidens*. *Bidens*, along with silverswords and lobeliads, is regarded as one of the best examples of adaptive radiation in Hawaiian flora (Ganders et al. 2000; Knope et al. 2012). The 19 endemic *Bidens* species in the Hawaiian archipelago exhibit greater ecological and morphological diversity compared to the approximately 230 *Bidens* species found across five continents (Ganders and Nagata 1984). Knope et al. (2020b) employed a phylogenomic approach to develop a well-resolved phylogenetic hypothesis for the adaptive radiation of Hawaiian *Bidens* (Asteraceae), including refined estimates for initial colonization and divergence times. They sequenced the chloroplast genome (plastome) and nuclear ribosomal complex for 18 of the 19 endemic Hawaiian *Bidens* species and four outgroup species. Phylogenomic analyses of the combined plastome and nuclear datasets produced identical Bayesian and Maximum Likelihood trees with strong statistical support at most nodes. Dating estimates suggested the crown group emerged approximately 1.76–1.82 million years ago (Fig. 23.3). Biogeographic analyses indicated that colonization within the Hawaiian Islands generally followed the progression rule, with 67–80% of colonization events occurring from older to younger islands; however, only 53% of events followed this rule in the plastome analysis. Evidence of nuclear-plastome conflict suggests that hybridization may play a significant role in the group's evolution, though incomplete lineage sorting cannot be excluded due to the limited number of independent loci analyzed. The radiation of Hawaiian *Bidens* is among the fastest recorded for flowering plants on islands and continental areas per unit area. The ongoing development of a Hawaiian *Bidens* reference genome is expected to facilitate a genome-wide analysis of nucleotide variability, helping to clarify ancient or ongoing interspecific gene flow, incomplete lineage sorting, and the potential role of hybridization in speciation and adaptive radiation (Marques et al. 2019).

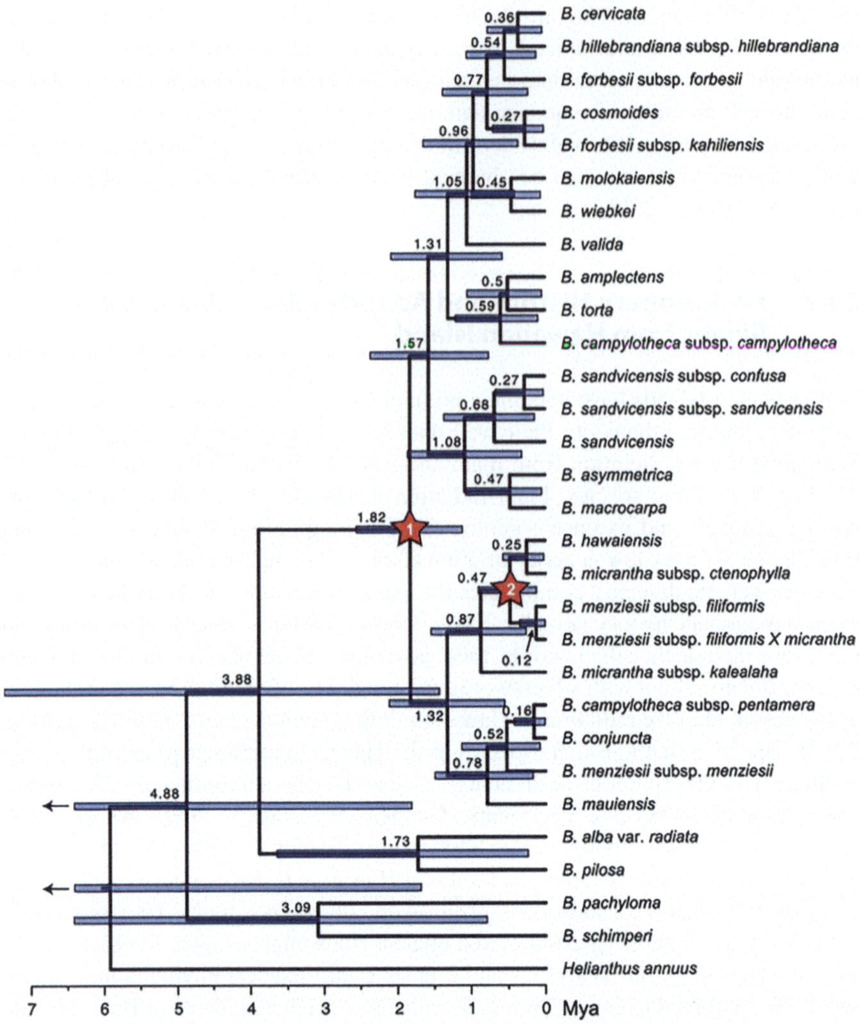

Fig. 23.3 Maximum clade credibility tree based on a BEAST analysis of concatenated whole plastome and nuclear ribosomal complex sequences. (Adapted from Knope et al. 2020b).

23.8 Evolutionary History and Adaptive Radiation in Genus *Bidens* from Polynesia

Polynesian *Bidens* species exhibit a diverse range of achene morphologies compared to their mainland counterparts, including traits that reduce dispersal, such as reduced or absent barbs and shorter awn lengths (Carlquist, 1966a). This creates a paradox: Despite a reduction in dispersal potential in their achenes relative to their continental ancestors, Polynesian *Bidens* have widely dispersed and evolved into one of the largest plant radiations in Polynesia, with 41 recognized species across the Hawaiian, Marquesas, Society, and Austral Islands. Polynesian species also display significant ecomorphological variation (Knope et al. 2020a). However, the genus has a complex taxonomic history, and its phylogenetic and biogeographic relationships remain poorly understood. To clarify the evolutionary history of Polynesian *Bidens*, one study included 152 individuals representing 91 species, including 39 of the 41 described Polynesian species. Four chloroplast and two nuclear DNA markers were used to estimate phylogenetic relationships, divergence times, and biogeographic history. The study confirmed that *Bidens* is polyphyletic within Coreopsis, consistent with previous findings. The Polynesian radiation was identified as monophyletic, with initial dispersal into the Pacific potentially originating from South America to either the Hawaiian or Marquesas Islands (Fig. 23.4). From the Marquesas, *Bidens* spread to the Society Islands and eventually to the Austral Islands. The initial diversification of the crown group in the Pacific is estimated to have occurred approximately 1.63 million years ago (0.74–2.72, 95% HPD), marking Polynesian *Bidens* as one of the youngest and most rapid plant diversification events in the Pacific.

23.9 Population Genomics

Population genomics is a rapidly changing science that employs whole-genome data to investigate population genetic diversity, organization, and evolution. This chapter focuses on the use of population genetics to better understand the genus *Bidens*, a group of plants with important ecological and therapeutic importance. By combining genome sequencing, comparative genomics, and functional analysis, this study was designed to learn about the evolutionary history, genetic diversity, and adaptive processes of *Bidens* species.

23.10 Patterns of Genetic Diversity

Bidens is estimated to comprise approximately to 230–240 species globally (Karis and Ryding 1994; Pozharitskaya et al. 2010). *Biden pilosa* has been found as an erect perennial herb with widespread distribution across temperate and tropical zone. This species exhibits variable pubescence, ranging from glabrous to hairy. The leaves are arranged oppositely and possess a green coloration. Their morphology can be described as serrate, lobed, or dissected. Inflorescences bear flowers with white or yellow petals. The fruits of *B. pilosa* are achenes, characterized by their elongated, narrow shape, black color, and ribbed texture (Fig. 23.5). The typical growth range for this herb is up to 60 cm, with a maximum potential of 150 cm under optimal environmental conditions (Alcaraz and Jimenez 1988).

Recent advancement in molecular techniques, such as restriction site-associated DNA sequencing (RADseq), has illuminated the patterns of genetic diversity within and between populations of *Bidens*. A study focusing on *B. pilosa* revealed substantial genetic variation, with 12% of this variation attributable to interpopulation

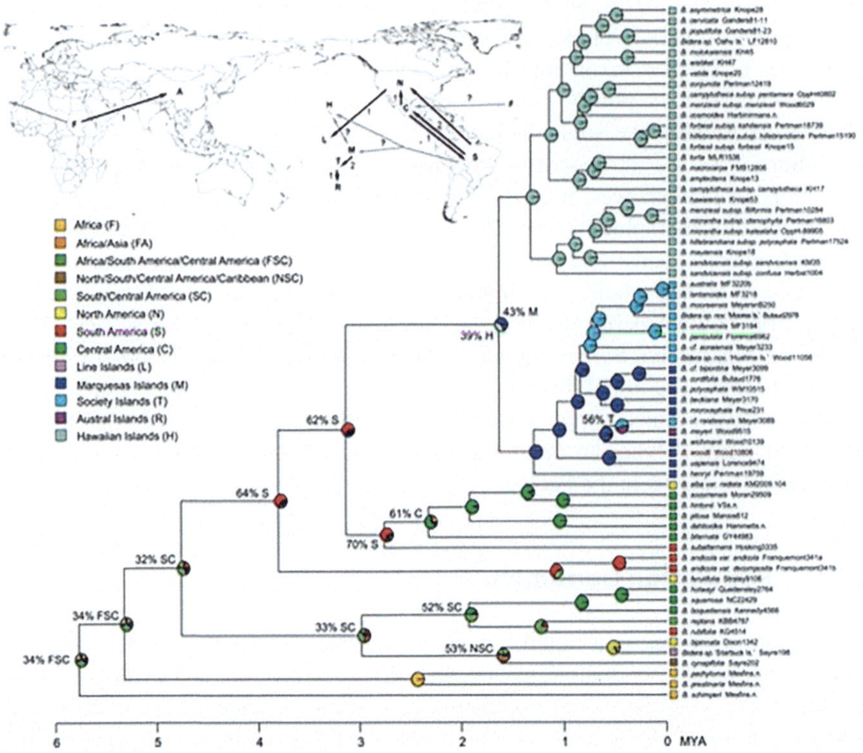

Fig. 23.4 Ancestral range estimation for *Bidens* species under the best model determined in BioGeoBEARS (DEC + j + x), based on the divergence times estimated in BEAST and the distribution of extant species. (Adapted from Knope et al. 2020a)

differences and the remaining 88% residing within individual plants. This observation signifies a high degree of intrapopulation genetic diversity, a characteristic commonly associated with species exhibiting extensive geographic distribution and adaptability, $F_{ST} = 0.115$ ($p < 0.001$) and $F_{IT} = 0.013$ ($p > 0.05$) (Reyes-Ardila et al. 2024; Tables 23.2 and 23.3).

The examination of the chloroplast genomes of *B. pilosa*, *B. bipinnata*, and *B. alba* uncovered a consistent quadripartite structure present in these *Bidens* species. Within this structure, there are both unchanging areas and differences in GC content. These findings indicate that chloroplast genome analysis can be valuable in uncovering the phylogenetic connections and evolutionary past of *Bidens* (Zhang et al. 2023). Knope et al. reported that Hawaiian *Bidens* radiation took place in the last two million years. Due to the relatively short divergence time and significant niche differences among *Bidens* species, there is low genetic variation and high morphological variation, making it difficult to use traditional DNA and morphological markers to identify and understand the evolutionary history of these species. The

(A) (B) (C)

Fig. 23.5 (**a**) *Bidens pilosa* (**b**) Flower (**c**) Achenes

Table 23.2 Analyses of molecular variance for three populations of *B. pilosa*

Source of variation	df.	Sum of squares	Variance components	Percentage of variation	Fixation indices
Among populations	2	1200.527	50.537	12	$F_{ST} = 0.115$ ($p < 0.001$)
Within individuals	7	4363.083	388.927	88	$F_{IT} = 0.013$ ($p > 0.05$)

Table 23.3 Pairwise F_{ST} values for three populations of *B. pilosa* F_{ST} values are shown above diagonal and their *p*-values are below diagonal

	Bogotá	Sibundoy	Palmira
Bogotá	0	0.007	0.056
Sibundoy	0.329 ± 0.011	0	0.087
Palmira	0.067 ± 0.008	0.061 ± 0.008	0

classification of *Bidens* remains a topic of controversy, and more comprehensive studies are needed to bridge the gap between existing knowledge of their evolutionary history and the actual situation (Knope et al. 2020a). This review explores the diversity found both within and among populations, emphasizing the impact of geographic and environmental elements.

Studies indicate that *Bidens* populations frequently contain significant genetic variation within species. For example, investigations into *Bidens pilosa*, an abundant weed, demonstrated the presence of considerable genotypic diversity within populations. This diversity is probably a result of factors such as outcrossing breeding systems, extensive population sizes, and the capability to inhabit diverse environments (Tanahara and Maki 2010; Vyšniauskienė et al. 2018). *Bidens* populations can vary greatly, but they can also differ significantly genetically from one another. This divergence shows that populations that are geographically isolated or that are subject to diverse environmental pressures have limited gene exchange

(Vyšniauskienė et al. 2018). In general, adaptive radiation is the process of divergence of organisms from a common ancestor to exploit different ecological niches. *Bidens* in the Hawaiian Islands exhibit extreme environmental and morphological differentiation, although hybridization is relatively infrequent because it tends to grow in various environments. Genetic studies indicate that a few genes could play a role in trait differences (Helenurm 1983; Wen et al. 2022). There are minimal reports on the molecular identification of *Bidens* species. To determine *Bidens* species, the noncoding portions of the genome of the chloroplast (trnL intron and trnL-trnF) and nuclear ribosomal DNA (ITS1, 5.8S, and ITS2) were employed, and only *B. biternata* and *B. pilosa var. pilosa* could be distinguished. Our preliminary investigations revealed that the universal DNA barcodes ITS, ITS2, and psbA-trnH are all inefficient for distinguishing *Bidens* species. The challenges in identifying *Bidens* species impede the development of therapeutic *Bidens* plants and limit their medicinal value. A reliable and efficient approach for identifying therapeutic *Bidens* plants is urgently required (Tsai et al. 2008).

23.11 Demographic History of *Bidens* Populations

The genus *Bidens*, consisting of 230 species across five continents, has a complex taxonomic history. A study involving 152 individuals from Polynesia revealed that *Bidens* is polyphyletic within Coreopsis and monophyletic in the Polynesian radiation. The initial diversification occurred around 1.63 million years ago, making it one of the youngest and most rapid plant diversification events in the Pacific. The study suggests that rare long-distance dispersal and founder event speciation can explain the rapid adaptive radiation of *Bidens* throughout Polynesia (Tsai et al. 2008). Nuclear ribosomal DNA from 24 plants, comprising 20 species and 2 hybrids from 6 sections of *Bidens* and 1 species of Cosmos, was sequenced in its entirety, yielding a phylogenetic tree. Even the most morphologically diverse species in the genus, *Bidens* cosmoides, shares the same sequence with all other Hawaiian species. Additionally, the study discovered a basal dichotomy in *Bidens* that separates subtropical and north temperate species, indicating their lack of close kinship (Ganders et al. 2000). Migration increases genetic diversity, reduces genetic differentiation and the presence of nonnative genetic material, and affects population size and growth (Magdoń 1999). A population bottleneck, also known as a genetic bottleneck, occurs when human activity or environmental factors cause a population to shrink, which lowers genetic diversity. Bottlenecks in *Bidens* populations can result in decreased fitness and loss of alleles, increased inbreeding and genetic homogeneity, accumulation of harmful mutations, diminished toughness and adaptability, and increased genetic drift and random distribution of alleles (Magdoń 1999).

23.12 Factors Influencing the Genetic Diversity of *Bidens* Populations

23.12.1 Geographical Distribution and Environmental Adaptation

B. pilosa is regarded as a problematic invasive species in numerous countries worldwide, including but not limited to Belgium, Botswana, Canada, China, Egypt, France, Greece, India, Indonesia, Israel, Japan, Kenya, Nigeria, Portugal, the UK, and the USA (Sharma et al. 2023). *B. pilosa* exhibits numerous traits that enable it to thrive in diverse environmental conditions, such as abundant seed production, effective spread, a broad germination range, and the capacity to thrive in different soil types (Gurvich et al. 2004; Zhang et al. 2020). Furthermore, there have been accounts of the plant reproducing asexually by generating adventitious roots from the lower nodes on its lateral branches (Hsu HsiaoMei and Kao WenYuan 2014). The weed's fast expansion and use of resources allows it to rival other species for resources (Yu et al. 2021).

23.12.2 Climate Changes

Climate impacts the *Bidens* population, is multifaceted, and depends on the location and environmental changes. The biodiversity and loss of habitats are greatly impacted by climate change, which is essential for the survival of many species, including those in the *Bidens* genus. Alterations in temperature and precipitation can change the environments in which these plants flourish, possibly resulting in a decline in their numbers. For instance, severe weather occurrences and changes in climate zones can disturb the delicate equilibrium necessary for their development (Habibullah et al. 2022).

23.12.2.1 Agriculture

Climate change can impact the growing of crops, such as *Bidens* species used in agriculture. Variations in climate can lead to reduced or unsuccessful harvests caused by extreme temperatures, which can affect the plant's population density and distribution. Additionally, changes in climate can impact the occurrence of pests and diseases, creating an additional obstacle in sustaining stable populations (Maja and Ayano 2021).

23.12.2.2 Reduction in Suitable Habitats

Loss of habitat, caused by urban development, expansion of agriculture, and deforestation, results in a decrease in the suitable habitats available for *Bidens* species. This decrease in both the size and quality of habitats can directly lower their population as they have fewer opportunities to thrive and reproduce (Habibullah et al. 2022).

23.13 Applications of *Bidens* Genomics

Genomic insights will greatly improve the productivity and resilience of *Bidens*-based crops, as certain traits will be precisely manipulated with their help. For example, gene editing technologies in plants, such as CRISPR/Cas9, can edit genes related to gibberellin biosynthesis to increase productivity, therefore, obtaining plants that have an increased rate of growth and upregulated biomass production (Rasheed et al. 2021; Castro-Camba et al. 2022). Secondly, improving light energy conversion efficiency through genetic optimization of photosynthesis can increase plant yields. This will increase not only the capacity of crops to absorb soil nutrients, for instance, by increasing nitrogen fixation or improvement in phosphorus solubilization, using less fertilizer, but also their capacities that provide gain in productivity (Croce et al. 2024). One of them is the manipulation of flowering-time regulatory genes for synchronizing crop cycles to improve the efficiency and predictability of harvests (Hill and Li 2016). Genomic insight into fine-tuning drought-responsive genes, such as DREB and NAC transcription factors, can aid in enhancing resilience in crops by offering better drought tolerance without affecting yield stability when there is a limitation of water (Singh et al. 2021). The identification of pathogen recognition genes, characterized by R genes conferring resistance to a particular pathogen, would also limit losses caused by these crop constraints, thus reducing the chemical pesticides applied to crops. The genes of such antimicrobial peptides can be designed in plants to fight against several pathogens (Zhang et al. 2022). Other strategies for making transgenic plants resistant include the use of heat shock proteins to assist adaptation to high temperatures by maintaining protein expression and antifreeze genes such as CBF genes to make crops more tolerant to frost or low temperatures (Ijaz et al. 2020).

Genes that control root development can be engineered to develop deeper and more extensive roots, enhancing water and nutrient uptake (Li et al. 2016). Modifying genes that affect light capture efficiency can reduce unnecessary water loss by adjusting leaf structure and function (Drewry et al. 2014). For the addition of economic value and nutritional benefits, genes associated with seed size, oil content, and nutritional quality can be enhanced (Hannoufa et al. 2014). Additionally, as *Bidens* has various uses in traditional medicine, overexpressing genes responsible for the synthesis of secondary metabolites such as flavonoids and alkaloids will improve the medicinal effects of this plant (Ramabulana 2020; Chung et al. 2021). This will allow researchers and the agricultural community to leverage *Bidens* genomics knowledge for crop improvement, combining desired productivity factors with increased sustainability under our changing climate and genetic pest/disease pressures. This, in turn, will lead to sustainable agricultural practices and food security. Genomics leads to a better knowledge of genetics and facilitates improving breeding programs by selecting and combining favorable traits within *Bidens* species. This allows MAS (marker-assisted selection), which transgresses the selection of plants in the seedling stage by 2 years, for the characterization and diagnosis of biological agricultural properties such as resistance to disease or drought tolerance.

Genomic selection (GS) is a statistical method that uses genome-wide marker information to estimate phenotypic values, enabling the selection of the best candidates early in the breeding process, even when target traits are not yet observed. This pathway equally enhances breeding programs in getting more precise predictions for gene combinations conditioning continuous (polygenic) traits, such that focused populations detected for large-effect genes with base effect maps lead to the design of molecular QTL, providing improved prediction on allele interactions underlying the architecture of complex morphological and physiological characteristics like yield or resilience (Alemu et al. 2024). When performing GWAS (Genome-Wide Association Study) with genome sequencing of various populations of *Bidens* species, it provides the identification of genes/alleles that control traits. It has the potential to identify any genes that regulate key agricultural traits and facilitate the discovery of novel traits by finding underlying genetics that aid in the identification of candidate loci in crop improvement programs (Qi et al. 2024). In a way, CRISPR/Cas9 is a next-generation gene editing technology that can be a very strong tool in the improvement of plants directly targeting a package of elite genes into the *Bidens* genome or eliminating undesirable ones, for example, making crops more acceptable to consumers and regulators, for instance, through the introduction of disease resistance or drought tolerance by editing the corresponding genes without incorporating any foreign DNA. MAS has been widely applied to develop cucumber varieties resistant to powdery mildew with markers closely related to resistance genes selected by breeders (Li et al. 2023). This process also used selection for GS to increase the genetic makeup of grain yield related to water-deficit conditions in maize for planting tolerance to drought (Campos et al. 2004). The gene editing technology to fight vitamin A deficiency was used to produce Golden Rice with its enhanced content levels of beta-carotene (Dubock et al. 2021). GWAS, applied to soybeans, developed disease-resistant and high-yield genotypes. In breeding programs, genomic tools and associated technologies must be applied in the interest of fast-tracking the development of *Bidens* species with enhanced traits, such as incorporating genes that confer resistance to diseases, tolerance to drought, and increasing yield. Such improvements will make the cultivation of *Bidens* crops sustainable and productive, therefore answering the continuous demand for foodstuffs sourced from medicinal plants (Ravelombola et al. 2021). Based on the genomic tools developed, scientists will have a better understanding of bioactive compounds in *Bidens* spp. and hence make it an ideal platform for the detection and enhancement of natural lead molecules. Investigation on this genomics must identify the basic genes and metabolic pathways that facilitate the synthesis of putative therapeutic agents and guide future variety development programs through breeding or genetic engineering strategies.

Anti-inflammatory, antimicrobial, or antifungal metabolites, including flavonoids, terpenoids, and polyacetylenes, may be potential research fronts in secondary metabolites. For example, flavonoids such as quercetin and kaempferol present in *Bidens* species are shown to have antioxidant and anti-inflammatory abilities. The study has contributed to the knowledge on the biosynthetic pathway for such molecules and thus opens possibilities for their production using a genetic engineering strategy (Chauhan et al. 2020). The most recent research is also dedicated to the possibility of using *Bidens* species to synthesize absolutely new compounds for several therapeutic applications. An example includes some antitumor polyacetylenes found in *B. pilosa*. Genomic approaches more quickly identify genes involved in the biosynthesis, acting like tools of genetic engineering with increased final yield in production (Reyes-Ardila et al. 2024). Terpenoids from *Bidens* species have also been effective targets against the antimicrobial and anti-inflammatory activity, against which genomics can help optimize their biosynthesis through the unmasking probable genes encoding for the enzyme steps involved (Xuan and Khanh 2016). With further advanced tools in transcriptomics and metabolomics, more information is gained on the networks of genes interacting with compounds that confer resistance to weedy *Bidens* populations. This information could also be used for the identification of the key regulatory genes and transcription factors involved in the expression of metabolic pathways related to medicinally active compounds. These genomic resources will help conventional breeding strategies and biotechnological approaches for developing improved varieties of *Bidens*, considering specific medicinal properties toward their antispecific therapeutic applications (Tyagi et al. 2022).

23.14 Integrating Genomics with Traditional Botanical Knowledge

The *Bidens* species have long been utilized by indigenous and local communities for their medicinal properties, and to date, there is quite an inventory of ethnobotanical knowledge. This means identification, therapeutic applications, methods of preparation, and dosages are very important and will help guide scientific studies today. This information thus represents the integration of ethnobotanical knowledge with genomic data concerning facilitation and guidance to let a researcher know which *Bidens* species and compounds to study for greater efficiency and relevance of this genomics work. For example, knowing about the traditional use of some *Bidens* species or plant parts to cure diseases can guide genomic studies into these specific areas in the genome. This targeted approach can simplify both bioactive compound discovery and gene identification. Moreover, as indigenous knowledge of traditional preparation methods and uses can guide the extraction and testing of compounds (Theodoridis et al. 2023), it is useful for scientists to be attentive to this when designing research that aims at increasing scientific understanding while also being aligned with some aspects of tribal practices. The ethnobotanical tool would provide researchers with insights into environmental conditions and ecological interactions, enabling the survival of *Bidens* species, which could serve as a useful lead to understanding gene expression and regulation concerning stress responses and secondary metabolites production. This holistic knowledge will help develop improvement strategies through breeding and genetic engineering to produce *Bidens* varieties containing their metabolic molecules with stronger antimicrobial properties (Haq et al. 2023). In *Bidens* research, the success of putting together genomics with traditional knowledge has brought unprecedented breakthroughs in the elucidation of these medicinal properties. Some examples include:

23.15 *Bidens pilosa* and Anti-inflammatory Properties

Traditional knowledge of the anti-inflammatory activity of *Bidens* properties can be found in many cultures. Traditional knowledge of the anti-inflammatory properties of *Bidens* can be found in many cultures. However, more significant are the tropical societies that have utilized this medicinal plant for generations This ethnobotanical knowledge informed subsequent genomic studies seeking to identify individual bioactive compounds underlying these effects. Other classes of anti-inflammatory phytocompounds extracted from *Bidens pilosa*, including flavonoids and polyacetylenes, have been characterized at the molecular level concerning their requirements. Such knowledge of biosynthetic pathways, made possible by genomics tools, has allowed production levels to be improved through selective breeding and genetic manipulation (Rodríguez-Mesa et al. 2023).

23.16 *Bidens pilosa* and Antimalarial Activity

Indigenous knowledge from Africa and South America was reported for the use of *B. pilosa* against malaria. Many findings have been confirmed by genomic work identifying key genes for the biosynthesis of antimicrobial polyacetylenes and flavonoids. For example, research has demonstrated the compound phenylheptatriyne in *B. pilosa* to exhibit antimalarial activities. So, using traditional knowledge combined with genomics enabled the isolation of these compounds—phenylheptatriyne—and enhanced their accumulation in the plant through targeted breeding programs (Cortés-Rojas et al. 2021).

23.17 *Bidens alba* and Anticancer Properties

The multiplicity of the folk medicinal application of *B. alba* in treating various diseases, including cancer, has pushed it to the frontline for genomic investigations to ascertain its medicinal properties. Researchers have already identified a number of terpenoids and flavonoids with possible anticancer activities. Genomic studies have evidenced expression mapping for the meta-pathways of those compounds and, at the very same time, unveiled key genes as regulators. This information has allowed breeding varieties of *B. alba* containing higher contents of these bioactive compounds to provide better natural resources for cancer treatment (Bijauliya et al. 2017).

23.18 *Bidens* Species and Antioxidant Properties

Traditional knowledge in many cultures identified the *Bidens* species as plants with antioxidant activity that could be utilized against oxidative stress in various diseases. Genomic studies have identified genes responsible for coding quercetin and luteolin antioxidant compounds. Improved *Bidens* lines with enhanced antioxidant activity have been developed using this genetic lineage, hence improving their therapeutic value (Vera Saltos et al. 2015).

23.19 Future Prospects and Challenges

While incorporating traditional botanical knowledge with genomics in *Bidens* research appears to have a promising future, it also brings some challenges. Future research should further exploit the rich ethnomedicinal data by combining this knowledge with cutting-edge genomic tools to identify novel bioactive molecules and their biosynthetic routes, which may eventually evolve into more efficacious drugs derived from various *Bidens* species (Pirintsos et al. 2022). Further development of this integration could lay the foundation for personalized medicine approaches, where medicinal products are tailored to specific health needs and genetic profiles on a population-specific basis (Hartmanshenn et al. 2016). Moreover, integration of genomic insights with local agronomic and management practices in systematizing traditional production systems for *Bidens* species can contribute to improving crop productivity, resistance against pests or diseases, while minimizing their ecological footprint (Khan et al. 2024). Genomic data can help detect and preserve the genetic diversity present in *Bidens* species, which is crucial for maintaining their adaptive capacity especially in the face of current environmental changes as well as ensuring the conservation of this valuable gene pool. Additionally, integrating traditional knowledge and genomics can confirm ethnopharmacological uses for *Bidens* species and add possible applications to clinical use, ensuring the exploitation of these plant-based resources is widespread in common medical practices.

In sharing data among a multitude of researchers, there is a bigger need to achieve uniformity in data formats and to integrate different forms of information under a single roof. Detailed research into genomics, coupled with documentation about traditional knowledge, is a laborious process; such ventures are normally hard to finance at the best of times. Another way to close the gap between genomic sciences and ethnobotany is through training and capacity building, which provides an avenue for transaction across the boundaries of various disciplines. It would also call for ascertaining that the increased exploitation of *Bidens* species for their medicinal value does not have adverse effects on the natural ecosystems, a need that calls for sustainable practices to balance demand with conservation. Such integration of genomics and traditional botanical knowledge into research on *Bidens* finally opens vast opportunities for advancements in the discovery of compounds, personalized medicine, agriculture, and conservation. The authors caution that careful attention must be given to the different challenges related to concerns about intellectual property, cultural sensitivity, data integration, resource constraints, and environmental impact. If researchers think fully work through these issues collaboratively, they will be able to unlock this integrative approach to its full potential, benefitting not just science and medicine but indigenous communities as well.

23.20 Conclusion

In other words, an integrated genomics approach using conventional knowledge related to *Bidens* has open-ended potential for the discovery of medicinal compounds, personalized medicine, and agricultural sustainability and, very importantly, for conservation efforts. Combining traditional knowledge developed over centuries with state-of-the-art genomic approaches will uncover new bioactive compounds, reveal their genetic basis, and enable enhancements in their productivity through targeted breeding and genetic engineering. This therefore links the validation and increases *Bidens* species usage through their association with modern scientific validation in traditional remedies. Key challenges to be mitigated include those of intellectual property rights, cultural sensitivity, integration of data, and resource constraints while keeping low environmental impacts by ethical and sustainable management. This can be achieved through collaboration, transparency, and fair sharing of benefits with indigenous and local communities. Managing these challenges will require investment in training and capacity building in genomic sciences and ethnobotany to enhance interdisciplinarity and thereby effectively integrate traditional knowledge into genomic research. Further research into the genomics of *Bidens* will continue to unlock secrets of evolution that will further enable sustainable use while conserving genetic diversity. Adopting an integrated approach respecting indigenous knowledge and merging it with scientific breakthroughs is the way forward to unlock the full potential of the *Bidens* species for improving global health, agriculture, and biodiversity conservation.

References

Alami MM et al (2022) The current developments in medicinal plant genomics enabled the diversification of secondary metabolites' biosynthesis. Int J Mol Sci 23(24):15932

Alcaraz MJ, Jimenez MJ (1988) Flavonoids as anti-inflammatory agents. Available at: https://www.cabidigitallibrary.org/doi/full/10.5555/19880351320. Accessed 4 Aug 2024

Alemu A et al. (2024) Genomic selection in plant breeding: Key factors shaping two decades of progress. Molecular Plant [Preprint]. Available at: https://www.cell.com/molecular-plant/fulltext/S1674-2052(24)00080-7. Accessed 5 Aug 2024

Ashraf MF et al (2022) Entailing the next-generation sequencing and metabolome for sustainable agriculture by improving plant tolerance. Int J Mol Sci 23(2):651

Bartolome AP, Villaseñor IM, Yang W-C (2013) *Bidens pilosa* L. (Asteraceae): botanical properties, traditional uses, phytochemistry, and pharmacology. In: Evidence-based complementary and alternative medicine, pp 1–51. https://doi.org/10.1155/2013/340215

Bellinger MR et al (2022) A genome for Bidens hawaiensis: a member of a hexaploid Hawaiian plant adaptive radiation. J Hered 113(2):205–214

Bijauliya RK et al (2017) A comprehensive review on cancer and anticancer herbal drugs. Int J Pharm Sci Res 8(7):2740–2761

Campos H et al (2004) Improving drought tolerance in maize: a view from industry. Field Crop Res 90(1):19–34

Carlquist SJ. 1966a. The biota of long-distance dispersal. I. Principlesof dispersal and evolution. The Quarterly Review of Biology 41:247–270.

Castro-Camba R et al (2022) Plant development and crop yield: the role of gibberellins. Plan Theory 11(19):2650

Chauhan DS et al (2020) Secondary metabolites in the treatment of diabetes mellitus: a paradigm shift. Curr Drug Metab 21(7):493–511

Cheng Q-Q et al (2021) Review on the development and applications of medicinal plant genomes. Front Plant Sci 12:791219

Chung H-H et al (2021) Elucidation of enzymes involved in the biosynthetic pathway of bioactive polyacetylenes in Bidens pilosa using integrated omics approaches. J Exp Bot 72(2):525–541

Cortés-Rojas DF et al (2021) Antimalarial activity of Bidens pilosa root extract co-spray dried in the presence of β-Cyclodextrin or aerosil:microcrystalline cellulose blend. Planta Med Int Open 8(01):e1–e9. https://doi.org/10.1055/a-1348-5707

Croce R et al (2024) Perspectives on improving photosynthesis to increase crop yield. Plant Cell:koae132

Drewry DT, Kumar P, Long SP (2014) Simultaneous improvement in productivity, water use, and albedo through crop structural modification. Glob Chang Biol 20(6):1955–1967. https://doi.org/10.1111/gcb.12567

Dubock AC et al (2021) Golden rice, VAD, Covid and public health: saving lives and money. In: Integrative advances in rice research. IntechOpen, p 307. Available at: https://books.google.com.au/books?hl=en&lr=&id=AaRcEAAAQBAJ&oi=fnd&pg=PA307&dq=+Dubock. Accessed 5 Aug 2024

Duque AS et al (2013) Abiotic stress responses in plants: unraveling the complexity of genes and networks to survive. In: Abiotic stress-plant responses and applications in agriculture, pp 49–101

Durand NC et al (2016) Juicebox provides a visualization system for Hi-C contact maps with unlimited zoom. Cell Syst 3(1):99–101

Ganders FR, Nagata KM (1984) Role of hybridization in the evolution of Bidens on the Hawaiian islands. In: Grant WF (ed) Plant biosystematics. Chapman & Hall, New York, pp 179–194

Ganders FR, Berbee M, Pirseyedi M (2000) ITS base sequence phylogeny in Bidens (Asteraceae): evidence for the continental relatives of Hawaiian and Marquesan Bidens. Syst Bot 25(1):122–133

Gurvich DE et al (2004) Seed mass, seed production, germination and seedling traits in two phenological types of Bidens pilosa (Asteraceae). Aust J Bot 52(5):647–652

Habibullah MS et al (2022) Impact of climate change on biodiversity loss: global evidence. Environ Sci Pollut Res 29(1):1073–1086

Hannoufa A, Pillai BVS, Chellamma S (2014) Genetic enhancement of Brassica napus seed quality. Transgenic Res 23(1):39–52. https://doi.org/10.1007/s11248-013-9742-3

Haq SM et al (2023) Integrating traditional ecological knowledge into habitat restoration: implications for meeting forest restoration challenges. J Ethnobiol Ethnomed 19(1):33. https://doi.org/10.1186/s13002-023-00606-3

Hartmanshenn C, Scherholz M, Androulakis IP (2016) Physiologically-based pharmacokinetic models: approaches for enabling personalized medicine. J Pharmacokinet Pharmacodyn 43(5):481–504. https://doi.org/10.1007/s10928-016-9492-y

Helenurm K (1983) Genetic differentiation of Hawaiian Bidens. PhD thesis. University of British Columbia. Available at: https://open.library.ubc.ca/soa/cIRcle/collections/ubctheses/831/items/1.0095801. Accessed 4 Aug 2024

Hill CB, Li C (2016) Genetic architecture of flowering phenology in cereals and opportunities for crop improvement. Front Plant Sci 7:1906

Holm LG et al (1978) The world's worst weeds. Distribution and biology. Pedobiologia 18(4):296–297

Hsu HsiaoMei HH, Kao WenYuan KW (2014) Vegetative and reproductive growth of an invasive weed Bidens pilosa L. var. radiata and its noninvasive congener Bidens bipinnata in Taiwan. Available at: https://www.cabidigitallibrary.org/doi/full/10.5555/20143265950. Accessed 4 Aug 2024

Hsu H-M, Kao W-Y (2009) Contrasting effects of aqueous tissue extracts from an invasive plant, Bidens pilosa L. var. radiata, on the performance of its sympatric plant species. Taiwania 54(3):255–260

Ijaz S et al (2020) Transgenic approaches in plants: strategic control for disease management. In: Ul Haq I, Ijaz S (eds) Plant disease management strategies for sustainable agriculture through traditional and modern approaches, Sustainability in plant and crop protection. Springer, Cham, pp 187–215. https://doi.org/10.1007/978-3-030-35955-3_9

Karis PO, Ryding O (1994) Asteraceae cladistics and classification. K. Bremer, ed, pp 559–569

Khan AA et al (2024) Advanced molecular approaches for improving crop yield and quality: a review. J Plant Growth Regul 43(7):2091–2103. https://doi.org/10.1007/s00344-024-11253-7

Knope ML et al (2012) Area and the rapid radiation of Hawaiian *Bidens* (Asteraceae). J Biogeogr 39(7):1206–1216. https://doi.org/10.1111/j.1365-2699.2012.02687.x

Knope ML, Funk VA et al (2020a) Dispersal and adaptive radiation of *Bidens* (Compositae) across the remote archipelagoes of Polynesia. J Syst Evol 58(6):805–822. https://doi.org/10.1111/jse.12704

Knope, M.L., Bellinger, M.R., et al. (2020b) 'Insights into the evolutionary history of the Hawaiian Bidens (Asteraceae) adaptive radiation revealed through phylogenomics', J Hered, 111(1), pp. 119–137

Li X, Zeng R, Liao H (2016) Improving crop nutrient efficiency through root architecture modifications. J Integr Plant Biol 58(3):193–202. https://doi.org/10.1111/jipb.12434

Li T et al (2023) CRISPR/Cas9 therapeutics: progress and prospects. Signal Transduct Target Ther 8(1):36

Lopes AD et al (2022) Potential allelopathic effect of species of the Asteraceae family and its use in agriculture. In: Vegetation dynamics, changing ecosystems and human responsibility. IntechOpen. Available at https://www.intechopen.com/chapters/84835. Accessed 1 July 2024

Magdoń MS (1999) Effect of migration on population dynamics. Int J Modern Phys C 10(06):1163–1174. https://doi.org/10.1142/S0129183199000942

Maja MM, Ayano SF (2021) The impact of population growth on natural resources and farmers' capacity to adapt to climate change in low-income countries. Earth Syst Environ 5(2):271–283

Mardis ER (2011) A decade's perspective on DNA sequencing technology. Nature 470(7333):198–203

Marques DA, Meier JI, Seehausen O (2019) A combinatorial view on speciation and adaptive radiation. Trends Ecol Evol 34(6):531–544

Pareek CS, Smoczynski R, Tretyn A (2011) Sequencing technologies and genome sequencing. J Appl Genet 52(4):413–435. https://doi.org/10.1007/s13353-011-0057-x

Pirintsos S et al (2022) From traditional ethnopharmacology to modern natural drug discovery: a methodology discussion and specific examples. Molecules 27(13):4060

Pozharitskaya ON et al (2010) Anti-inflammatory activity of a HPLC-fingerprinted aqueous infusion of aerial part of Bidens tripartita L. Phytomedicine 17(6):463–468

Qi T et al (2024) From genetic associations to genes: methods, applications, and challenges. Trends Genet. Available at: https://www.cell.com/trends/genetics/fulltext/S0168-9525(24)00095-7?dgcid=raven_jbs_aip_email. Accessed 5 Aug 2024

Ramabulana A-T (2020) Phytochemical profiling of secondary metabolites from Bidens pilosa plants and cell cultures. University of Johannesburg (South Africa). Available at: https://search.proquest.com/openview/2674ce19e592099e74b07a09e527ea89/1?pq-origsite=gscholar&cbl=18750&diss=y. Accessed 5 Aug 2024

Rasheed A et al (2021) A critical review: recent advancements in the use of CRISPR/Cas9 technology to enhance crops and alleviate global food crises. Curr Issues Mol Biol 43(3):1950–1976

Ravelombola W et al (2021) Genome-wide association study and genomic selection for yield and related traits in soybean. PLoS One 16(8):e0255761
Reyes-Ardila WL et al (2024) Exploring genomics and microbial ecology: analysis of Bidens pilosa L. genetic structure and soil microbiome diversity by RAD-Seq and metabarcoding. Plan Theory 13(2):221
Rodríguez-Mesa XM et al (2023) Immunomodulatory properties of natural extracts and compounds derived from Bidens pilosa L.: literature review. Pharmaceutics 15(5):1491
Sharma A et al (2023) Bidens pilosa (Asteraceae) invasion reshapes the pattern of plant communities and edaphic properties across the north-western Himalayan landscape. Eco Inform 77:102281
Singh S et al (2021) The biotechnological importance of the plant-specific NAC transcription factor family in crop improvement. J Plant Res 134(3):475–495. https://doi.org/10.1007/s10265-021-01270-y
Tanahara A, Maki M (2010) Genetic diversity and population genetic differentiation in the endangered annual weed, *Bidens cernua* (Compositae), and two common congeners in Japan. Weed Biol Manag 10(2):113–119. https://doi.org/10.1111/j.1445-6664.2010.00374.x
Theodoridis S et al (2023) Evaluating natural medicinal resources and their exposure to global change. Lancet Planet Health 7(2):e155–e163
Towers GHN et al (1984) Controlling pests using a naturally occurring conjugated polyacetylen. Canadian Patent CA 1173743
Tsai L-C et al (2008) Bidens identification using the noncoding regions of chloroplast genome and nuclear ribosomal DNA. Forensic Sci Int Genet 2(1):35–40
Tsao R, Eto M (1996) Light-activated plant growth inhibitory activity of cis-dehydromatricaria ester, rose bengal and fluoren-9-one on lettuce (Lactuca saliva L.). Chemosphere 32(7):1307–1317
Tyagi P et al (2022) Upcoming progress of transcriptomics studies on plants: an overview. Front Plant Sci 13:1030890
Vera Saltos M et al (2015) Antioxidant and free radical scavenging activity of Phenolics from Bidens humilis. Planta Med 81(12/13):1056–1064. https://doi.org/10.1055/s-0035-1545928
Vyšniauskienė R et al (2018) Genetic diversity of populations of Bidens genera invasive and native species in Lithuania. Žemdirbystė 105(2):183–190
Wang H et al (2024) The genomes of Dahlia pinnata, Cosmos bipinnatus, and Bidens alba in tribe Coreopsideae provide insights into polyploid evolution and inulin biosynthesis. GigaScience 13:giae032
Welch AJ et al (2016) The quest to resolve recent radiations: plastid phylogenomics of extinct and endangered Hawaiian endemic mints (Lamiaceae). Mol Phylogenet Evol 99:16–33
Wen B et al (2022) Variation in tolerance to high temperature and water stress in weedy Bidens pilosa seeds from Yunnan, SW China'. Available at: https://scholar.archive.org/work/d4k2hylhivafthevnnmr7xsc4m/access/wayback/https://assets.researchsquare.com/files/rs-1782585/v1/622f74fe-97e6-4a3e-ac39-f05ef27191cd.pdf?c=1657742481. Accessed 4 Aug 2024
Wenger AM et al (2019) Accurate circular consensus long-read sequencing improves variant detection and assembly of a human genome. Nat Biotechnol 37(10):1155–1162
Wu L et al (2022) Identification of medicinal Bidens plants for quality control based on organelle genomes. Front Pharmacol 13:842131
Xuan TD, Khanh TD (2016) Chemistry and pharmacology of Bidens pilosa: an overview. J Pharm Investig 46(2):91–132. https://doi.org/10.1007/s40005-016-0231-6
Yang X et al (2020) Enrichment of flavonoid-rich extract from *B*idens bipinnata L. by macroporous resin using response surface methodology, UHPLC–Q-TOF MS/MS-assisted characterization and comprehensive evaluation of its bioactivities by analytical hierarchy process. Biomed Chromatogr 34(11):e4933. https://doi.org/10.1002/bmc.4933
Young AD, Gillung JP (2020) Phylogenomics—principles, opportunities and pitfalls of big-data phylogenetics. Syst Entomol 45(2):225–247. https://doi.org/10.1111/syen.12406
Yu Y et al (2021) Response of multiple mountain ecosystem services on environmental gradients: how to respond, and where should be priority conservation? J Clean Prod 278:123264

Zhang H et al (2017) Genome editing—principles and applications for functional genomics research and crop improvement. Crit Rev Plant Sci 36(4):291–309. https://doi.org/10.1080/07352689.2017.1402989

Zhang H-Y et al (2020) Invasion by the weed Conyza canadensis alters soil nutrient supply and shifts microbiota structure. Soil Biol Biochem 143:107739

Zhang H, Huang J, Huang Y (2022) Identification and characterization of plant resistance genes (R genes) in sorghum and their involvement in plant defense against aphids. Plant Growth Regul 96(3):443–461. https://doi.org/10.1007/s10725-022-00797-x

Zhang D et al (2023) Analysis of the chloroplast genome and phylogenetic evolution of Bidens pilosa. BMC Genomics 24(1):113. https://doi.org/10.1186/s12864-023-09195-7

Zhou C, McCarthy SA, Durbin R (2023) YaHS: yet another Hi-C scaffolding tool. Bioinformatics 39(1):btac808

Ziegler AC (2002) Hawaiian natural history, ecology, and evolution. University of Hawaii Press. Available at: https://books.google.com.au/books?hl=en&lr=&id=q5wBEAAAQBAJ&oi=fnd&pg=PP1&dq=Ziegler+AC.+2002.+Hawaiian+natural+history,+ecology,+and+evolution.+University+of+Hawai%CA%BBi+Press,+Honolulu,+HI &ots=VW2745PIln&sig=JDNWpA1eRYvCL6zZRHo8UoN9IA8. Accessed 1 Aug 2024

The manufacturer's authorised representative in the EU is Springer Nature Customer Service Centre GmbH, Europaplatz 3, 69115 Heidelberg, Germany. If you have any concerns regarding our products, please contact ProductSafety@springernature.com

Printed and bound by CPI Group (UK) Ltd, Croydon, CR0 4YY
01/10/2026
02225929-0006